Graduate Texts in Mathematics **140**

Springer
Berlin
Heidelberg
New York
Barcelona
Budapest
Hong Kong
London
Milan
Paris
Singapore
Tokyo

Graduate Texts in Mathematics

1 TAKEUTI/ZARING. Introduction to Axiomatic Set Theory. 2nd ed.
2 OXTOBY. Measure and Category. 2nd ed.
3 SCHAEFFER. Topological Vector Spaces.
4 HILTON/STAMMBACH. A Course in Homological Algebra.
5 MAC LANE. Categories for the Working Mathematician.
6 HUGHES/PIPER. Projective Planes.
7 SERRE. A Course in Arithmetic.
8 TAKEUTI/ZARING. Axiomatic Set Theory.
9 HUMPHREYS. Introduction to Lie Algebras and Representation Theory.
10 COHEN. A Course in Simple Homotopy Theory.
11 CONWAY. Functions of One Complex Variable. 2nd ed.
12 BEALS. Advanced Mathematical Analysis.
13 ANDERSON/FULLER. Rings and Categories of Modules. 2nd ed.
14 GOLUBITSKY/GUILLEMIN. Stable Mappings and Their Singularities.
15 BERBERIAN. Lectures in Functional Analysis and Operator Theory.
16 WINTER. The Structure of Fields.
17 ROSENBLATT. Random Processes. 2nd ed.
18 HALMOS. Measure Theory.
19 HALMOS. A Hilbert Space Problem Book. 2nd ed., revised.
20 HUSEMÖLLER. Fibre Bundles. 2nd ed.
21 HUMPHREYS. Linear Algebraic Groups.
22 BARNES/MACK. An Algebraic Introduction to Mathematical Logic.
23 GREUB. Linear Algebra. 4th ed.
24 HOLMES. Geometric Functional Analysis and its Applications.
25 HEWITT/STROMBERG. Real and Abstract Analysis.
26 MANES. Algebraic Theories.
27 KELLEY. General Topology.
28 ZARISKI/SAMUEL. Commutative Algebra. Vol. I.
29 ZARISKI/SAMUEL. Commutative Algebra. Vol. II.
30 JACOBSON. Lectures in Abstract Algebra I. Basic Concepts.
31 JACOBSON. Lectures in Abstract Algebra II. Linear Algebra.
32 JACOBSON. Lectures in Abstract Algebra III. Theory of Fields and Galois Theory.
33 HIRSCH. Differential Topology.
34 SPITZER. Principles of Random Walk. 2nd ed.
35 WERMER. Banach Algebras and Several Complex Variables. 2nd ed.
36 KELLEY/NAMIOKA et al. Linear Topological Spaces.
37 MONK. Mathematical Logic.
38 GRAUERT/FRITZSCHE. Several Complex Variables.
39 ARVESON. An Invitation to C^*-Algebras.
40 KEMENY/SNELL/KNAPP. Denumerable Markov Chains. 2nd ed.
41 APOSTOL. Modular Functions and Dirichlet Series in Number Theory. 2nd ed.
42 SERRE. Linear Representations of Finite Groups.
43 GILLMAN/JERISON. Rings of Continuous Functions.
44 KENDIG. Elementary Algebraic Geometry.
45 LOÈVE. Probability Theory I. 4th ed.
46 LOÈVE. Probability Theory II. 4th ed.
47 MOISE. Geometric Topology in Dimensions 2 and 3.
48 SACHS/WU. General Relativity for Mathematicians.
49 GRUENBERG/WEIR. Linear Geometry. 2nd ed.
50 EDWARDS. Fermat's Last Theorem.
51 KLINGENBERG. A Course in Differential Geometry.
52 HARTSHORNE. Algebraic Geometry.
53 MANIN. A Course in Mathematical Logic.
54 GRAVER/WATKINS. Combinatorics with Emphasis on the Theory of Graphs.
55 BROWN/PEARCY. Introduction to Operator Theory I: Elements of Functional Analysis.
56 MASSEY. Algebraic Topology: An Introduction
57 CROWELL/FOX. Introduction to Knot Theory.
58 KOBLITZ. p-adic Numbers, p-adic Analysis, and Zeta-Functions. 2nd ed.
59 LANG. Cyclotomic Fields.
60 ARNOLD. Mathematical Methods in Classical Mechanics. 2nd ed.
61 WHITEHEAD. Elements of Homotopy Theory.
62 KARGAPOLOV/MERZLJAKOV. Fundamentals of the Theory of Groups.
63 BOLLOBÁS. Graph Theory.
64 EDWARDS. Fourier Series. Vol. I. 2nd ed.

continued after index

Jean-Pierre Aubin

Optima and Equilibria

An Introduction to Nonlinear Analysis

Translated from the French by Stephen Wilson
With 28 Figures

Second Edition 1998

 Springer

Jean-Pierre Aubin
Université Paris-Dauphine
Centre de Recherche Viabilité, Jeux, Contrôle
F-75775 Paris Cedex 16, France

Title of the French original editions:
L'analyse non linéaire et ses motivations économiques © Masson Paris 1984 and
Exercices d'analyse non linéaire © Masson Paris 1987

Mathematics Subject Classification (1991): 90C, 65K; 90A, 90B, 90D; 47H,
47N10, 49J, 49N.

ISSN 0072-5285
ISBN 3-540-64983-2 Springer-Verlag Berlin Heidelberg New York
ISBN 3-540-52121-6 1st Edition Springer-Verlag Berlin Heidelberg New York

Cataloging-in-Publication Data applied for

Die Deutsche Bibliothek – CIP-Einheitsaufnahme

Aubin, Jean-Pierre: Optima and equilibria: an introduction to nonlinear analysis / Jean-Pierre Aubin. Transl. from the French by Stephen Wilson. – 2. ed. – Berlin; Heidelberg; New York; Barcelona; Budapest; Hong Kong; London; Milan; Paris; Singapore; Tokyo: Springer, 1998 (Graduate texts in mathematics; 140) ISBN 3-540-64983-2

© Springer-Verlag Berlin Heidelberg 1993, 1998
Printed in Germany

The use of general descriptive names, registered names, trademarks, etc. in this publication does not imply, even in the absence of a specific statement, that such names are exempt from the relevant protective laws and regulations and therefore free for general use.

Typesetting: Camera-ready by translator using Springer TeX macropackage
SPIN 10692532   41/3143 - 5 4 3 2 1 0 - Printed on acid-free paper

This book is dedicated to Alain Bensoussan, Ivar Ekeland, Pierre-Marie Larnac and Francine Roure, in memory of the adventure which brought us together more than twenty years ago to found the U.E.R. and the Centre de Recherche de Mathématiques de la Décision (CEREMADE).

Jean-Pierre Aubin

Doubtless you have often been asked about the purpose of mathematics and whether the delicate constructions which we conceive as entities are not artificial and generated at whim. Amongst those who ask this question, I would single out the practical minded who only look to us for the means to make money. Such people do not deserve a reply.

Henri Poincaré
La Valeur de la Science
Chapter V

In his use of mathematical techniques to study general economic phenomena relating to countries or individuals Mr. Léon Walras has truly instituted a science.

Charles Péguy
Un économiste socialiste, Mr. Léon Walras
La Revue Socialiste, no. 146, 1897

It may be that the coldness and the objectivity for which we often reproach scientists are more suitable than feverishness and subjectivity as far as certain human problems are concerned. It is passions which use science to support their cause. Science does not lead to racism and hatred. Hatred calls on science to justify its racism. Some scientists may be reproached for the ardour with which they sometimes defend their ideas. But genocide has never been perpetrated in order to ensure the success of a scientific theory. At the end of this the XXth century, it should be clear to everyone that no system can explain the world in all its aspects and detail. Quashing the idea of an intangible and eternal truth is possibly not the least claim to fame of the scientific approach.

François Jacob
Le Jeu des possibles
Fayard (1981) p. 12

I enjoy talking to great minds and this is a taste which I like to instil in my students. I find that students need someone to admire; since they cannot normally admire their teachers because their teachers are examiners or are not admirable, they must admire great minds while, for their part, teachers must interpret great minds for their students.

Raymond Aron
Le Spectateur engagé
Julliard (1981) p. 302

Foreword

By Way of Warning

As in ordinary language, metaphors may be used in mathematics to explain a given phenomenon by associating it with another which is (or is considered to be) more familiar. It is this sense of familiarity, whether individual or collective, innate or acquired by education, which enables one to convince oneself that one has understood the phenomenon in question.

Contrary to popular opinion, mathematics is not simply a *richer* or *more precise* language. Mathematical reasoning is a separate faculty possessed by all human brains, just like the ability to compose or listen to music, to paint or look at paintings, to believe in and follow cultural or moral codes, etc.

But it is impossible (and dangerous) to compare these various faculties within a *hierarchical* framework; in particular, one cannot speak of the *superiority* of the language of mathematics.

Naturally, the construction of mathematical metaphors requires the autonomous development of the discipline to provide theories which may be substituted for or associated with the phenomena to be explained. This is the domain of pure mathematics. The construction of the mathematical corpus obeys its own logic, like that of literature, music or art. In all these domains, a temporary aesthetic satisfaction is at once the objective of the creative activity and a signal which enables one to recognise successful works. (Likewise, in all these domains, fashionable phenomena – reflecting social consensus – are used to develop aesthetic criteria).

That is not all. A mathematical metaphor associates a mathematical theory with another object. There are two ways of viewing this association. The first and best-known way is to search for a theory in the mathematical corpus which corresponds as precisely as possible with a given phenomenon. This is the domain of *applied mathematics*, as it is usually understood. But the association is not always made in this way; the mathematician should not be simply a purveyor of formulae for the user. Other disciplines, notably physics, have guided mathematicians in their selection of problems from amongst the many arising and have prevented them from continually turning around in the same circle by presenting them with new challenges and encouraging them to be daring and question the ideas of their predecessors. These other disciplines may also provide mathematicians with metaphors, in that they may suggest concepts and arguments, hint at solutions and embody new modes of intuition. This is the domain of what one might call *motivated mathematics* from which the examples you will read about in this book are drawn.

You should soon realize that the work of a *motivated* mathematician is *daring*, above all where problems from the *soft* sciences, such as social sciences and, to a lesser degree, biology, are concerned. Many hours of thought may very well only lead to the mathematically obvious or to problems which cannot be solved in the short term, while the same effort expended on a structured problem of pure or applied mathematics would normally lead to visible results.

Motivated mathematicians must possess a sound knowledge of another discipline and have an adequate arsenal of mathematical techniques at their fingertips together with the capacity to create new techniques (often similar to those they already know). In a constant, difficult and frustrating dialogue they must investigate whether the problem in question can be solved using the techniques which they have at hand or, if this is not the case, they must *negotiate* a deformation of the problem (a possible restructuring which often seemingly leads to the original model being forgotten) to produce an *ad hoc* theory which they sense will be useful later. They must convince their colleagues in the other disciplines that they need a very long period for learning and appreciation in order to grasp the language of a given theory, its foundations and main results and that the proof and application of the simplest, the most naive and the most attractive results may require theorems which may be given in a number of papers over several decades; in fact, one's comprehension of a mathematical theory is never complete. In a century when no more cathedrals are being built, but impressive skyscrapers rise up so rapidly, the profession of the motivated mathematician is becoming rare. This explains why users are very often not aware of how mathematics could be used to improve aspects of the questions with which they are concerned. When users are aware of this, the intersection of their central areas of interest with the preoccupations of mathematicians is often small – users are interested in *immediate* impacts on their problems and not in the mathematical techniques that could be used and their relationship with the overall mathematical structure.

It is these constraints which distinguish mathematicians from researchers in other disciplines who use mathematics, with a *different time constant*. It is clear that the slowness and the esoteric aspect of the work of mathematicians may lead to impatience amongst those who expect them to come up with rapid responses to their problems. Thus, it is vain to hope to *pilot* the mathematics *downstream* as those who believe that scientific development may be *programmed* (or worse still, *planned*) may suggest.

In Part I, we shall only cover aspects of pure mathematics (optimisation and nonlinear analysis) and aspects of mathematics motivated by economic theory and game theory. It is still too early to talk about *applying* mathematics to economics. Several fruitful attempts have been made here and there, but mathematicians are a long way from developing the mathematical techniques (the domains of pure mathematics) which are best adapted to the potential applications.

However, there has been much progress in the last century since pioneers such as Quesnais, Boda, Condorcet, Cournot, Auguste and Léon Walras, despite great opposition, dared to use the tools of mathematics in the economic domain. Brouwer, von Neumann, Kakutani, Nash, Arrow, Debreu, Scarf, Shapley, Ky Fan and many others all contributed to the knowledge you are about to share.

You will surely be disappointed by the fact that these difficult theorems have little relevance to the major problems facing mankind. But, please don't be impatient, like others, in your desire for an overall, all-embracing explanation. Professional mathematicians must be very humble and modest.

It is this modesty which distinguishes mathematicians and scientists in general from prophets, ideologists and modern system analysts. The range of scientific explanations is reduced, hypotheses must be contrasted with logic (this is the case in mathematics) or with experience (thus, these explanations must be falsifiable or refutable). Ideologies are free from these two requirements and thus all the more seductive.

But what is the underlying motivation, other than to contribute to an explanation of reality? We are brains which perceive the outside world and which intercommunicate in various ways, using natural language, mathematics, bodily expressions, pictorial and musical techniques, etc.

It is the consensus on the consistency of individual perceptions of the environment, which in some way measures the degree of reality in a given social group.

Since our brains were built on the same model, and since the ability to believe in explanations appears to be innate and universal, there is a very good chance that a social group may have a sufficiently broad consensus that its members share a common concept of reality. But prophets and sages often challenge this consensus, while high priests and guardians of the ideology tend to dogmatise it and impose it on the members of the social group. (Moreover, quite often prophets and sages themselves become the high priests and guardians of the ideology, the other way round being exceptional.) This continual struggle forms the framework for the history of science.

Thus, research must contribute to the evolution of this consensus, teaching must disseminate it, without dogmatism, placing knowledge in its relative setting and making you take part in man's struggle, since the day when *Homo sapiens, sapiens* ... But we do not know what happened, we do not know when, why or how our ancestors sought to agree on their perceptions of the world to create myths and theories, when why or how they transformed their faculty for exploration into an insatiable curiosity, when, why or how mathematical faculties appeared, etc.

It is not only the utilitarian nature (in the short term) which has motivated mathematicians and other scientists in their quest. We all know that without this permanent, free curiosity there would be no technical or technological progress.

Perhaps you will not use the techniques you will soon master and the results you will learn in your professional life. But the hours of thought which you will have devoted to understanding these theories will (subtly and without you being aware) shape your own way of viewing the world, which seems to be the hard kernel around which knowledge organizes itself as it is acquired. At the end of the day, it is at this level that you must judge the relevance of these lessons and seek the reward for your efforts.

Table of Contents

Part I   Nonlinear Analysis: Theory

1 **Minimisation Problems: General Theorems**   9
 1.1   Introduction   9
 1.2   Definitions   9
 1.3   Epigraph 10
 1.4   Lower Sections 11
 1.5   Lower Semi-continuous Functions 11
 1.6   Lower Semi-compact Functions 13
 1.7   Approximate Minimisation of Lower Semi-continuous Func-
       tions on a Complete Space 15
 1.8   Application to Fixed-point Theorems 17

2 **Convex Functions and Proximation, Projection and Separation Theorems** 21
 2.1   Introduction 21
 2.2   Definitions 21
 2.3   Examples of Convex Functions 24
 2.4   Continuous Convex Functions 25
 2.5   The Proximation Theorem 27
 2.6   Separation Theorems 31

3 **Conjugate Functions and Convex Minimisation Problems** . 35
 3.1   Introduction 35
 3.2   Characterisation of Convex Lower Semi-continuous Functions 37
 3.3   Fenchel's Theorem 39
 3.4   Properties of Conjugate Functions 43
 3.5   Support Functions 48
 3.6   The Cramèr Transform 52

4 **Subdifferentials of Convex Functions** 57
 4.1   Introduction 57
 4.2   Definitions 61
 4.3   Subdifferentiability of Convex Continuous Functions 64

4.4   Subdifferentiability of Convex Lower Semi-continuous Functions   66
4.5   Subdifferential Calculus   67
4.6   Tangent and Normal Cones   70

5 Marginal Properties of Solutions of Convex Minimisation Problems   75
5.1   Introduction   75
5.2   Fermat's Rule   76
5.3   Minimisation Problems with Constraints   80
5.4   Principle of Price Decentralisation   82
5.5   Regularisation and Penalisation   84

6 Generalised Gradients of Locally Lipschitz Functions   87
6.1   Introduction   87
6.2   Definitions   87
6.3   Elementary Properties   91
6.4   Generalised Gradients   95
6.5   Normal and Tangent Cones to a Subset   97
6.6   Fermat's Rule for Minimisation Problems with Constraints .   99

7 Two-person Games. Fundamental Concepts and Examples   101
7.1   Introduction   101
7.2   Decision Rules and Consistent Pairs of Strategies   102
7.3   Brouwer's Fixed-point Theorem (1910)   104
7.4   The Need to Convexify: Mixed Strategies   105
7.5   Games in Normal (Strategic) Form   106
7.6   Pareto Optima   108
7.7   Conservative Strategies   110
7.8   Some Finite Games   112
7.9   Cournot's Duopoly   116

**8 Two-person Zero-sum Games:
Theorems of Von Neumann and Ky Fan**   125
8.1   Introduction   125
8.2   Value and Saddle Points of a Game   125
8.3   Existence of Conservative Strategies   130
8.4   Continuous Partitions of Unity   135
8.5   Optimal Decision Rules   137

9 Solution of Nonlinear Equations and Inclusions   143
9.1   Introduction   143
9.2   Upper Hemi-continuous Set-valued Maps   144
9.3   The Debreu–Gale–Nikaïdo Theorem   148
9.4   The Tangential Condition   149

9.5	The Fundamental Theorem for the Existence of Zeros of a Set-valued Map	150
9.6	The Viability Theorem	152
9.7	Fixed-point Theorems	154
9.8	Equilibrium of a Dynamical Economy	155
9.9	Variational Inequalities	157
9.10	The Leray–Schauder Theorem	159
9.11	Quasi-variational Inequalities	160
9.12	Shapley's Generalisation of the Three-Poles Lemma	162

10 Introduction to the Theory of Economic Equilibrium 167
10.1	Introduction	167
10.2	Exchange Economies	168
10.3	The Walrasian Mechanism	169
10.4	Another Mechanism for Price Decentralisation	173
10.5	Collective Budgetary Rule	174

11 The Von Neumann Growth Model 179
11.1	Introduction	179
11.2	The Von Neumann Model	179
11.3	The Perron–Frobenius Theorem	184
11.4	Surjectivity of the M matrices	187

12 n-person Games 189
12.1	Introduction	189
12.2	Non-cooperative Behaviour	189
12.3	n-person Games in Normal (Strategic) Form	190
12.4	Non-cooperative Games with Constraints (Metagames)	192
12.5	Pareto Optima	193
12.6	Behaviour of Players in Coalitions	196
12.7	Cooperative Games Without Side Payments	197
12.8	Evolutionary Games	205

13 Cooperative Games and Fuzzy Games 209
13.1	Introduction	209
13.2	Coalitions, Fuzzy Coalitions and Generalised Coalitions of n Players	209
13.3	Action Games and Equilibrium Coalitions	214
13.4	Games with Side Payments	216
13.5	Core and Shapley Value of Standard Games	224

Part II   Nonlinear Analysis: Examples

14 Exercises 235
  14.1   Exercises for Chapter 1 – Minimisation Problems: General
          Theorems 235
  14.2   Exercises for Chapter 2 – Convex Functions and Proximation,
          Projection and Separation Theorems 240
  14.3   Exercises for Chapter 3 – Conjugate Functions and Convex
          Minimisation Problems 245
  14.4   Exercises for Chapter 4 – Subdifferentials of Convex Functions 254
  14.5   Exercises for Chapter 5 – Marginal Properties of Solutions of
          Convex Minimisation Problems 261
  14.6   Exercises for Chapter 6 – Generalised Gradients of Locally
          Lipschitz Functions 268
  14.7   Exercises for Chapter 8 – Two-person Zero-sum Games: The-
          orems of Von Neumann and Ky Fan 275
  14.8   Exercises for Chapter 9 – Solution of Nonlinear Equations and
          Inclusions 280
  14.9   Exercises for Chapter 10 – Introduction to the Theory of Eco-
          nomic Equilibrium 285
  14.10 Exercises for Chapter 11 – The Von Neumann Growth Model 290
  14.11 Exercises for Chapter 12 – n-person Games 290
  14.12 Exercises for Chapter 13 – Cooperative Games and Fuzzy
          Games 297

15 Statements of Problems 301
  15.1   Problem 1 – Set-valued Maps with a Closed Graph 301
  15.2   Problem 2 – Upper Semi-continuous Set-valued Maps 301
  15.3   Problem 3 – Image of a Set-valued Map 302
  15.4   Problem 4 – Inverse Image of a Set-valued Map 302
  15.5   Problem 5 – Polars of a Set-valued Map 303
  15.6   Problem 6 – Marginal Functions 303
  15.7   Problem 7 – Generic Continuity of a Set-valued Map with a
          Closed Graph 304
  15.8   Problem 8 – Approximate Selection of an Upper Semi-continuous
          Set-valued Map 304
  15.9   Problem 9 – Continuous Selection of a Lower Semi-continuous
          Set-valued Map 305
  15.10 Problem 10 – Interior of the Image of a Convex Closed Cone 305
  15.11 Problem 11 – Discrete Dynamical Systems 308
  15.12 Problem 12 – Fixed Points of Contractive Set-valued Maps . 310
  15.13 Problem 13 – Approximate Variational Principle 311
  15.14 Problem 14 – Open Image Theorem 311
  15.15 Problem 15 – Asymptotic Centres 313
  15.16 Problem 16 – Fixed Points of Non-expansive Mappings . . . 314

15.17 Problem 17 – Orthogonal Projectors onto Convex Closed Cones   315

15.18 Problem 18 – Gamma-convex functions   316

15.19 Problem 19 – Proper Mappings   317

15.20 Problem 20 – Fenchel's Theorem for the Functions $L(x, Ax)$ 319

15.21 Problem 21 – Conjugate Functions of $x \to L(x, Ax)$ 320

15.22 Problem 22 – Hamiltonians and Partial Conjugates   321

15.23 Problem 23 – Lack of Convexity and Fenchel's Theorem for Pareto Optima   322

15.24 Problem 24 – Duality in Linear Programming   323

15.25 Problem 25 – Lagrangian of a Convex Minimisation Problem   324

15.26 Problem 26 – Variational Principles for Convex Lagrangians   325

15.27 Problem 27 – Variational Principles for Convex Hamiltonians   326

15.28 Problem 28 – Approximation to Fermat's Rule   327

15.29 Problem 29 – Transposes of Convex Processes   327

15.30 Problem 30 – Cones with a Compact Base   329

15.31 Problem 31 – Regularity of Tangent Cones   329

15.32 Problem 32 – Tangent Cones to an Intersection   330

15.33 Problem 33 – Derivatives of Set-valued Maps with Convex Graphs   331

15.34 Problem 34 – Epiderivatives of Convex Functions   332

15.35 Problem 35 – Subdifferentials of Marginal Functions   333

15.36 Problem 36 – Values of a Game Associated with a Covering .   333

15.37 Problem 37 – Minimax Theorems with Weak Compactness Assumptions   334

15.38 Problem 38 – Minimax Theorems for Finite Topologies . . .   335

15.39 Problem 39 – Ky Fan's Inequality   336

15.40 Problem 40 – Ky Fan's Inequality for Monotone Functions .   337

15.41 Problem 41 – Generalisation of the Gale–Nikaïdo–Debreu Theorem   338

15.42 Problem 42 – Equilibrium of Coercive Set-valued Maps . . .   339

15.43 Problem 43 – Eigenvectors of Set-valued Maps   339

15.44 Problem 44 – Positive Eigenvectors of Positive Set-valued Maps   340

15.45 Problem 45 – Some Variational Principles   341

15.46 Problem 46 – Generalised Variational Inequalities   341

15.47 Problem 47 – Monotone Set-valued Maps   343

15.48 Problem 48 – Walrasian Equilibrium for Set-valued Demand Maps   344

16 Solutions to Problems   347

16.1   Problem 1 – Solution. Set-valued Maps with a Closed Graph   347

16.2   Problem 2 – Solution. Upper Semi-continuous set-valued Maps   347

16.3   Problem 3 – Solution. Image of a Set-valued Map   348

16.4   Problem 4 – Solution. Inverse Image of a Set-valued Map . .   348

16.5   Problem 5 – Solution. Polars of a Set-valued Map   350

16.6   Problem 6 – Solution. Marginal Functions   350

16.7   Problem 7 – Solution. Generic Continuity of a Set-valued Map
with a Closed Graph   351

16.8   Problem 8 – Solution. Approximate Selection of an Upper
Semi-continuous Set-valued Map   351

16.9   Problem 9 – Solution. Continuous Selection of a Lower Semi-
continuous Set-valued Map   352

16.10 Problem 10 – Solution. Interior of the Image of a Convex
Closed Cone   352

16.11 Problem 11 – Solution. Discrete Dynamical Systems   356

16.12 Problem 12 – Solution. Fixed Points of Contractive Set-valued
Maps   358

16.13 Problem 13 – Solution. Approximate Variational Principle .   359

16.14 Problem 14 – Solution. Open Image Theorem   360

16.15 Problem 15 – Solution. Asymptotic Centres   362

16.16 Problem 16 – Solution. Fixed Points of Non-expansive Mappings   363

16.17 Problem 17 – Solution. Orthogonal Projectors onto Convex
Closed Cones   365

16.18 Problem 18 – Solution. Gamma-convex Functions   366

16.19 Problem 19 – Solution. Proper Mappings   367

16.20 Problem 20 – Solution. Fenchel's Theorem for the Functions
$L(x, Ax)$ 368

16.21 Problem 21 – Solution. Conjugate Functions of $x \to L(x, Ax)$ 369

16.22 Problem 22 – Solution. Hamiltonians and Partial Conjugates   369

16.23 Problem 23 – Solution. Lack of Convexity and Fenchel's Theorem for Pareto Optima   370

16.24 Problem 24 – Solution. Duality in Linear Programming . . .   372

16.25 Problem 25 – Solution. Lagrangian of a Convex Minimisation
Problem   373

16.26 Problem 26 – Solution. Variational Principles for Convex Lagrangians   374

16.27 Problem 27 – Solution. Variational Principles for Convex
Hamiltonians   374

16.28 Problem 28 – Solution. Approximation to Fermat's Rule . .   375

16.29 Problem 29 – Solution. Transposes of Convex Processes . . .   376

16.30 Problem 30 – Solution. Cones with a Compact Base   377

16.31 Problem 31 – Solution. Regularity of Tangent Cones   378

16.32 Problem 32 – Solution. Tangent Cones to an Intersection . .   379

16.33 Problem 33 – Solution. Derivatives of Set-valued Maps with
Convex Graphs   381

16.34 Problem 34 – Solution. Epiderivatives of Convex Functions .   382

16.35 Problem 35 – Solution. Subdifferentials of Marginal Functions   383

16.36 Problem 36 – Solution. Values of a Game Associated with a
Covering   384

16.37 Problem 37 – Solution. Minimax Theorems with Weak Compactness Assumptions 385

16.38 Problem 38 – Solution. Minimax Theorems for Finite Topologies 386

16.39 Problem 39 – Solution. Ky Fan's Inequality 387

16.40 Problem 40 – Solution. Ky Fan's Inequality for Monotone Functions 388

16.41 Problem 41 – Solution. Generalisations of the Gale–Nikaïdo–Debreu Theorem 389

16.42 Problem 42 – Solution. Equilibrium of Coercive Set-valued Maps 390

16.43 Problem 43 – Solution. Eigenvectors of Set-valued Maps . . . 391

16.44 Problem 44 – Solution. Positive Eigenvectors of Positive Set-valued Maps 391

16.45 Problem 45 – Solution. Some Variational Principles 391

16.46 Problem 46 – Solution. Generalised Variational Inequalities . 393

16.47 Problem 47 – Solution. Monotone Set-valued Maps 395

16.48 Problem 48 – Solution. Walrasian Equilibrium for Set-valued Demand Maps 397

Appendix

17 Compendium of Results 401

17.1 Nontrivial, Convex, Lower Semi-continuous Functions 401

17.2 Convex Functions 403

17.3 Conjugate Functions 404

17.4 Separation Theorems and Support Functions 405

17.5 Subdifferentiability 408

17.6 Tangent and Normal Cones 409

17.7 Optimisation 411

17.8 Two-Person Games 413

17.9 Set-valued Maps and the Existence of Zeros and Fixed Points 415

References 421

Index 425

Introduction

This.is a book on nonlinear analysis and its underlying motivations in economic science and game theory. It is entitled *Optima and Equilibria* since, in the final analysis, response to these motivations consists of perfecting mechanisms for selecting an element from a given set. Such *selection mechanisms* may involve either

- *optimisation of a criterion function* defined on this set (or of several functions, in the case of multi-criterion problems in game theory), or

- searching in this set for an *equilibrium of a given undelying dynamical system*, which is a stationary solution of this dynmical system.

The mathematical techniques used have their origins in what is known as nonlinear analysis, and in particular, in convex analysis.

Progress in nonlinear analysis has proceeded hand in hand with that in the theory of economic equilibrium and in game theory; there is interaction between each of these areas, mathematical techniques are applied in economic science which, in turn, motivates new research and provides mathematicians with new challenges.

In the course of the book we shall have occasion to interrupt the logical course of the exposition with several historical recollections. Here, we simply note that it was Léon Walras who, at the end of the last century, suggested using mathematics in economics, when he described certain economic agents as automata seeking to optimise evaluation functions (utility, profit, etc.) and posed the problem of economic equilibrium. However, this area did not blossom until the birth of nonlinear analysis in 1910, with Brouwer's fixed-point theorem, the usefulness of which was recognised by John von Neumann when he developed the foundations of game theory in 1928. In the wake of von Neumann came the works of John Nash, Kakutani, Aumann, Shapley and many others which provided the tools used by Arrow, Debreu, Gale, Nikaïdo et al. to complete Walras's construction, culminating in the 1950s in the proof of the existence of economic equilibria. Under pressure from economists, operational researchers and engineers, there was stunning progress in optimisation theory, in the area of linear programming after the Second World War and following the work of Fenchel, in the 1960s in convex analysis. This involved the courageous step of differentiating nondifferentiable functions by Moreau and Rockafellar at the dawn of the 60's, and set-valued maps ten years later, albeit in a different way and for different reasons than in distribution theory discovered by Laurent Schwartz in the 1950s. (see for instance (Aubin and Frankowska 1990) and (Rockafellar and Wets 1997)). These works provided for use of the rule hinted at by Fermat more than three hundred years ago, namely that the derivative of a function is zero at points at which the function attains its optimum, in increasingly complicated problems of the calculus of variations and optimal control theory. The 1960s also saw a re-awakening of interest in nonlinear analysis for the different problem of solving nonlinear, partial-differential equations. A profusion of new results were used to clarify many questions and simplify proofs, notably using an inequality discovered in 1972 by Ky Fan.

At the time of writing, at the dawn of the 1980s, it is appropriate to take stock and draw all this together into a homogeneous whole, to provide a concise and self-contained appreciation of the fundamental results in the areas of nonlinear analysis, the theory of economic equilibrium and game theory.

Our selection will not be to everyone's taste: it is partial. For example, in our description of the theory of economic equilibrium, we do not describe consumers in terms of their utility functions but only in terms of their demand functions. A minority will certainly hold this against us. However, conscious of the criticisms made of the present-day formalism of the Walrasian model, we propose an alternative which, like Walras, retains the explanation of prices in terms of their decentralising virtues and also admits dynamic processing.

Our succinct introduction to game theory is not orthodox, in that we have included the theory of cooperative games in the framework of the theory of fuzzy games.

In the book we accept the shackles of the static framework that are at the origin of the inadequacies and paradoxes which serve as pretexts for rejection of the use of mathematics in economic science. J. von Neumann and O. Morgenstern were also aware of this when, in 1944, at the end of the first chapter of *Theory of Games and Economic Behaviour*, they wrote:

'*Our theory is thoroughly static. A dynamic theory would unquestionably be more complete and, therefore, preferable. But there is ample evidence from other branches of science that it is futile to try to build one as long as the static side is not thoroughly understood...*'

'*Finally, let us note a point at which the theory of social phenomena will presumably take a very definite turn away from the existing patterns of mathematical physics. This is, of course, only a surmise on a subject where much uncertainty and obscurity prevail...*'

'*Our static theory specifies equilibria... A dynamic theory, when one is found – will probably describe the changes in terms of simpler concepts.*'

Thus, this book describes the static theory and the tool which may be used to develop it, namely nonlinear analysis.

It is only now that we can hope to see the birth of a dynamic theory calling upon all other mathematical techniques (see (Aubin and Cellina 1984), (Aubin 1991) and (Aubin 1997)). But, as in the past, so too now, and in the future, the static theory must be placed in its true perspective, even though this may mean questioning its very foundations, like March and Simon (who suggested replacing optimal choices by choices that are only satisfactory) and many (less fortunate) others. Imperfect yet perfectible, mathematics has been used to put the finishing touches to the monument the foundation of which was laid by Walras. Even if this becomes an historic monument, it will always need to be visited in order to construct others from it and to understand them once constructed.

Of course, the book only claims to present an *introduction* to nonlinear analysis which can be read by those with the basic knowledge acquired in a first-level university mathematics course. It only requires the reader to have mastered the fundamental notions of topology in metric spaces and vector spaces. Only Brouwer's fixed-point theorem is assumed.

This is a book of *motivated mathematics*, i.e. a book of mathematics motivated by economics and game theory, rather than a book of mathematics *applied* to these fields. We have included a *Foreword* to take up this issue which deals with pure, applied and motivated mathematics. In our view, this is important in order to avoid setting too great store by the importance of mathematics in its interplay with social sciences.

The book is divided into two parts. Part I describes the theory, while Part II is devoted to exercises, and problem statements and solutions. The book ends with an Appendix containing a *Compendium of Results*.

In the first three chapters, we discuss the existence of solutions minimising a function, in the general framework (Chapter 1) and in the framework of convex functions (Chapter 3). Between times, we prove the projection theorem (on which so many results in functional analysis are based) together with a number of separation theorems and we study the duality relationship between convex functions and their conjugate functions.

The following three chapters are devoted to Fermat's rule which asserts that the gradient of a function is zero at any point at which the function attains its minimum. Since convex functions are not necessarily differentiable in the customary sense, the notion of the 'differential' had to be extended for Fermat's rule to apply. The simple, but unfamiliar idea consists of replacing the concept of gradient by that of subgradients, forming a *set* called a *subdifferential*. We describe a subdifferential calculus of convex functions in Chapter 4 and in Chapter 5, we exploit Fermat's rule to characterise the solutions of minimisation problems as solutions of a set-valued equation (called an *inclusion*) or as the subdifferential of another function.

In Chapter 6, we define the notion of the generalised gradient of a locally Lipschitz function, as proposed by F. Clarke in 1975. This enables us to apply Fermat's rule to functions other than differentiable functions and convex functions. It will be useful in the study of cooperative games.

Chapters 7 and 8 are devoted to the theory of two-person games; here, we prove two fundamental minimax theorems due to von Neumann (1928) and Ky Fan (1962).

In Chapter 9, we use Ky Fan's inequality to prove the existence theorems for solutions of the inclusion

$$0 \in C(\overline{x})$$

(where C is a set-valued map) together with the fixed-point theorems which we shall use to prove the existence of economic equilibria and non-cooperative equilibria in the theory of n-person games.

In Chapter 10, we provide two explanations of the role of prices in a decentralisation mechanism which provides economic agents with access to sufficient information for them to take their decisions without knowing the global state of the economic system or the decisions of other agents. The first explanation is provided by the Walrasian model, as formalised since the fundamental work of Arrow and Debreu in 1954. The second explanation is compatible with dynamic models which go beyond the scope of this book and for which we refer to (Aubin, 1997).

Chapter 11 is devoted to a study of the von Neumann growth model and provides us with the opportunity to prove the Perron–Frobenius theorem on the eigenvalues of positive matrices.

In Chapter 12 we adapt the concepts introduced in Chapter 7 for 2-person games to study n-person games.

Chapter 13 deals with standard cooperative games (using the behaviour of coalitions of players) and fuzzy cooperative games (involving fuzzy coalitions of players).

The collection of 165 exercises and 48 problems with solutions in Part II has two objectives in view. Firstly, it will provide the reader of Part I with the wherewithal to practise the manipulation of the new concepts and theorems which he has just read about.

Whilst, once assimilated, the mathematics may appear simple (and even self-evident), a great deal of time (and energy) is needed to familiarise oneself with these new cognitive techniques.

If a passive approach is taken, the assimilation will be difficult; for, strange as it may seem, emotional mechanisms (or, in the terminology of psychologists, motivational mechanisms) play a crucial role in the acquisition of these new methods of thinking. This mathematics book should be read (or skimmed through) quickly when the reader is looking for a piece of information which is indispensable to the solution of problem which is occupying his mind day and night!

Thus, it is best to approach this work as dispassionately as possible. You will then realise how easy it is to acquire a certain mastery of the subject. You will also see that old knowledge takes on a new depth, when it is replaced in a new perspective. You will improve (or at least modify) your understanding of aspects you thought you had already understood, since there is no end to understanding, either in the theory of mathematics or in other areas of knowledge. That is why we advise the reader to skim through the book to determine what it is about. You will then begin to understand it in a more active way by proving for yourself the results listed for each chapter of Part I at the beginning of the relevant section of the Exercises (Chapter 14). Both the pleasure of success and the lessons of partial failure will help you to overcome the difficulties you encounter. The pleasure of discovery is not a vain sentiment; the more ambitious is the challenge, the more intense is the pleasure.

These exercises (and above all the solutions) were also designed to provide the reader with additional information which could not be given in an introductory text. The results which the reader will discover will convince him of the richness of nonlinear analysis.

The exercises (Chapter 14) are grouped according to chapters and follow the order of Part I. Except for certain exceptions (which are explicitly mentioned), they only use results that have already been proved. However, some exercises do assume that one or two immediately preceding exercises have been solved.

The problems (Chapter 15) use a priori all the material in Part I and are largely grouped according to topic.

The first nine problems concern various topological properties of set-valued maps. The description of the notion of set-valued maps and their properties given in Part I is a bare minimum and is insufficient for profound applications of nonlinear analysis. The tenth problem generalises Banach's theorem (closed graph or open image) either to the case of continuous linear operators defined on a closed convex cone or to that of set-valued maps (Robinson–Ursescu theorem). It goes together with Problem 14 which extends the inverse function theorem to set-valued maps and which thus plays an important role in applications. Problem 11 returns to the proof of Ekeland's theorem in the very instructive context of discrete dynamical systems. Problems 12, 13, 14 and 28 provide applications of Ekeland's theorem, which turns out to be the most manageable and the most effective theorem in the whole family of results equivalent to the fixed-point theorem for contractions. This is complemented by a fixed-point theorem for non-expansive mappings (Problem 16) which uses an interesting notion (the asymptotic centre of sequences, which is a sort of virtual limit) which is the subject of Problem 15.

The solution of Problem 17 on the properties of orthogonal projectors onto convex closed cones (discovered by Jean-Jacques Moreau, co-founder with R.T. Rockafellar of convex analysis) is indispensable. Problem 18 studies a class of functions with properties analogous to those of convex functions.

A continuous mapping is 'proper' if it transforms closed sets to closed sets and if its inverse has compact images. As one might imagine, such functions play an important role. Their properties are the subject of Problem 19. Problems 20, 21, 23 and 26 are designed to extend the results of Chapters 3 to 5 for the functions $x \to f(x) + g(Ax)$ to the functions $x \to L(x, Ax)$; they will help the reader to assimilate the above chapters properly. Problem 24 is devoted to the application of Chapter 5 to linear programming. Variational principles form the subject of Problems 26, 27, 45 and 46; these last two problems use Ky Fan's inequality.

The graph of a continuous linear operator is a closed vector subspace. The set-valued maps analogous to continuous linear operators are set-valued maps with graphs a convex closed cone. These are known as 'closed convex processes' and inherit numerous properties of continuous linear operators, as Problems 10 (closed graph) and 29 (transposition) show.

Since the derivatives of differentiable mappings are continuous linear operators, we might expect to look for candidates for the role of the derivative of a set-valued map among such closed convex processes. It is sufficient to return to the origins, that is to say to Pierre de Fermat who introduced the notion of the tangent to a curve. This idea is taken up in Problem 33, which provides an introduction to the differential calculus of set-valued maps. Over recent years, this latter has become the subject of intense activity, because of its intrinsic attraction and its numerous potential applications. This 'geometric' view of the differential calculus is taken up again in Problem 34 to complete the study of subdifferentials of convex functions, whilst Problem 35 leads to a very elegant formula for calculating the subdifferential of a marginal function. This differential calculus of set-valued maps is the topic of (Aubin and Frankowska 1990) which contains a thorough investigation of set-valued maps. Problems 36, 37, 38, 39 and 40 describe refinements of the minimax inequalities of von Neumann and Ky Fan which are very useful in infinite-dimensional spaces. Problems 41 and 48 provide variants and applications of the Gale–Nikaïdo–Debreu theorem, whilst Problem 42 shows how to trade the compactness of the domain of a set-valued map for 'coercive' properties. The existence of eigenvectors of set-valued maps forms the subject of Problems 43 (general case) and 44 (positive set-valued maps).

Problem 47 provides an introduction to maximum monotonic set-valued maps and their numerous properties.

We could have included many other problems, but forced ourselves to make a difficult selection. One area of applications of nonlinear analysis, namely the calculus of variations and optimal control, is not touched on by this collection of problems, although it is a most rich and exciting area which remains the subject of active research.

This requires a reasonable mastery of topological vector spaces (weak topologies) and of function and distribution spaces (Sobolev spaces) which is not demanded of the reader (Aubin 1979a). If the latter has a knowledge of the basic tools of convex analysis, non-regular analysis and nonlinear analysis, he will be well equipped to tackle these theories effectively.

It remains to wish the reader (in fact, the explorer) deserved success in mastering this exciting area of mathematics, nonlinear analysis.

Part I

Nonlinear Analysis: Theory

1. Minimisation Problems: General Theorems

1.1 Introduction

The aim of this chapter is to show that a minimisation problem:

$$\text{find } \bar{x} \in K \text{ such that } f(\bar{x}) \leq f(x) \ \forall x \in K$$

has a solution when the set K is compact and the function f from K into $\mathbb{R}$ is lower semi-continuous.

This leads us to define semi-continuous functions and to describe some of their properties.

1.2 Definitions

First, we shall study minimisation problems in a general framework: we assume we have
- a subset K of X
- a function f from K to R

and we seek a solution $\bar{x}$ of the problem

$$(i) \qquad\qquad \bar{x} \in K$$
$$(ii) \qquad\qquad f(\bar{x}) = \inf_{x \in K} f(x). \qquad\qquad (1)$$

For ease of notation, we begin by introducing a convenient method which avoids explicit mention of the subset K on which the function f is defined. We set

$$f_K(x) := \begin{cases} f(x) & \text{if } x \in K \\ +\infty & \text{if } x \notin K \end{cases} \qquad\qquad (2)$$

where f_K is no longer a real-valued function but a function from X to $\mathbb{R} \cup \{+\infty\}$ such that

$$K = \{x \in X \,|\, f_K(x) < +\infty\}. \qquad\qquad (3)$$

Moreover, any solution of (1) is a solution of the problem

$$f_K(\bar{x}) = \inf_{x \in X} f_K(x) \qquad (4)$$

and conversely.

We are thus led to introduce the class of functions f from X to $\mathbb{R} \cup \{+\infty\}$ and to associate them with their *domain*

$$\mathrm{Dom}\, f := \{x \in X | f(x) < +\infty\}. \qquad (5)$$

Equation (3) may thus be written as $K = \mathrm{Dom}\,(f_K)$. In order to exclude the degenerate case in which $\mathrm{Dom}\, f = \emptyset$, that is to say where f is the constant function equal to $+\infty$, we shall use the following definition.

Definition 1.1. *We shall say that a function f from X to $\mathbb{R} \cup \{+\infty\}$ is **non-trivial** if its domain is non-empty, that is to say if f is finite at at least one point.*

We shall often use the *indicator function* of a set, which characterises the set in the same way as characteristic functions in other areas of mathematics.

Definition 1.2. *Let K be a subset of X. We shall say that the function $\psi_K : X \to \mathbb{R} \cup \{+\infty\}$ defined by*

$$\psi_K(x) = \begin{cases} 0 & \text{if } x \in K \\ +\infty & \text{if } x \notin K \end{cases} \qquad (6)$$

*is the **indicator function** of K.*

Note that the sum $f + \psi_K$ of a function f and the indicator function of a subset K may be identified with the *restriction of f to K* and that the minimisation problem (1) is equivalent to the problem

$$f(\bar{x}) + \psi_K(\bar{x}) = \inf_{x \in K} (f(x) + \psi_K(x)). \qquad (7)$$

We shall see that this new formulation of the problem will enable us to derive interesting properties of its possible solutions in a convenient and fast way.

1.3 Epigraph

We may characterise a function f from X to $\mathbb{R} \cup \{+\infty\}$ by its epigraph, which is a subset of $X \times R$.

Definition 1.3. *Let f be a function from X to $\mathbb{R} \cup \{+\infty\}$. We shall call the subset*

$$\mathrm{Ep}\,(f) := \{(x, \lambda) \in X \times \mathbb{R} | f(x) \leq \lambda\} \qquad (8)$$

*the **epigraph** of f.*

The epigraph of f is non-empty if and only if f is nontrivial.

The following property of epigraphs will be useful.

Proposition 1.1. *Consider a family of functions f_i from X to $\mathbb{R} \cup \{+\infty\}$ and its upper envelope $\sup_{i \in I} f_i$. Then*

$$\text{Ep}\left(\sup_{i \in I} f_i\right) = \bigcap_{i \in I} \text{Ep}(f_i). \tag{9}$$

Proof. Exercise. □

1.4 Lower Sections

Definition 1.4. *Let f be a function from X to $\mathbb{R} \cup \{+\infty\}$. The sets*

$$S(f, \lambda) := \{x \in X \mid f(x) \leq \lambda\} \tag{10}$$

are called sections (lower, wide) of f.

Let $\alpha := \inf_{x \in X} f(x)$. By the verry definition of the infimum of a function, the set M of solutions of problem (1) may be written in the form

$$M = \bigcap_{\lambda > \alpha} S(f_K, \lambda).$$

Thus, the set of solutions M 'inherits' the properties of the sections of f which are 'stable with respect to intersection' (for example, closed, compact, convex, etc.).

Proposition 1.2. *Consider a family of functions f_i from X to $\mathbb{R} \cup \{+\infty\}$ and its upper envelope $\sup_{i \in I} f_i$. Then*

$$S\left(\sup_{i \in I} f_i, \lambda\right) = \bigcap_{i \in I} S(f_i, \lambda). \tag{11}$$

Proof. Exercise. □

1.5 Lower Semi-continuous Functions

Let X be a metric space.

We recall that a function f from X to $\mathbb{R} \cup \{+\infty\}$ is continuous at a point x_0 (which necessarily belongs to the domain of f) if, for all $\varepsilon > 0$, there exists $\eta > 0$ such that $\forall x \in B(x_0, \eta)$ we have both $\lambda := f(x_0) - \varepsilon \leq f(x)$ and $f(x) \leq f(x_0) + \varepsilon_0$. Demanding only one of these properties leads to a notion of semi-continuity introduced by René Baire.

Definition 1.5. *We shall say that a function f from X to $\mathbb{R} \cup \{+\infty\}$ is **lower semi-continuous** at x_0 if for all $\lambda < f(x_0)$, there exists $\eta > 0$ such that*

$$\forall x \in B(x_0, \eta), \quad \lambda \leq f(x). \tag{12}$$

We shall say that f is **lower semi-continuous** *if it is lower semi-continuous at every point of* X. *A function is* **upper semi-continuous** *if* $-f$ *is lower semi-continuous.*

We begin by proving the characteristic properties. We recall that, by definition,

$$\liminf_{x \to x_0} f(x) := \sup_{\eta > 0} \inf_{x \in B(x_0, \eta)} f(x). \tag{13}$$

Proposition 1.3. *A function f from X to $\mathbb{R} \cup \{+\infty\}$ is lower semi-continuous at x_0 if and only if*

$$f(x_0) \leq \liminf_{x \to x_0} f(x). \tag{14}$$

Proof.
a) Suppose that f is lower semi-continuous at x_0. For all $\lambda < f(x_0)$, there exists η such that

$$\lambda \leq \inf_{x \in B(x_0, \eta)} f(x) \leq \liminf_{x \to x_0} f(x).$$

Inequality (14) now follows.

b) Conversely, given any $\lambda < \sup_{\eta > 0} \inf_{x \in B(x_0, \eta)} f(x)$, by definition of the supremum, there exists $\eta > 0$ such that $\lambda \leq \inf_{x \in B(x_0, \eta)} f(x)$. Thus, condition (14) implies that f is lower semi-continuous at x_0. □

Proposition 1.4. *Let f be a function from X to $\mathbb{R} \cup \{+\infty\}$. The following assertions are equivalent*
a) f is lower semi-continuous;
b) the epigraph of f is closed;
c) all sections $S(f, \lambda)$ of f are closed.

Proof.
a) We assume that f is lower semi-continuous and show that its epigraph is closed. For this, we take a sequence of elements $(x_n, \lambda_n) \in \mathrm{Ep}(f)$ converging to (x, λ) and show that (x, λ) belongs $\mathrm{Ep}(f)$, whence that $f(x) \leq \lambda$. But Proposition 1.3 then implies that

$$f(x) \leq \liminf_{n \to \infty} f(x_n) \leq \liminf_{n \to \infty} \lambda_n = \lim_{n \to \infty} \lambda_n = \lambda,$$

since $f(x_n) \leq \lambda_n$ for all n.

b) We now suppose that $\mathrm{Ep}(f)$ is closed and show that an arbitrary section $S(f, \lambda)$ is also closed. For this, we consider a sequence of elements $x_n \in S(f, \lambda)$ converging to x and show that $x \in S(f, \lambda)$, whence that $(x, \lambda) \in \mathrm{Ep}(f)$. But this is a result of the fact that the sequence of elements (x_n, λ) of the epigraph of f, which is closed, converges to (x, λ).

c) We suppose that all the sections of f are closed. We take $x_0 \in X$ and $\lambda < f(x_0)$. Then (x_0, λ) does not belong to $S(f, \lambda)$, which is a closed set. Thus, there exists $\eta > 0$ such that $B(x_0, \eta) \cap S(f, \lambda) = \emptyset$, that is to say that $\lambda \leq f(x)$ for all $x \in B(x_0, \eta)$. Thus, f is lower semi-continuous at x_0.   □

Remark. If a function f is not lower semicontinuous, one can associate with it the function $\overline{f}$ the epigraph of which is the closure of the epigraph of f:$\mathrm{Ep}(\overline{f}) :=$ $\overline{\mathrm{Ep}(f)}$. It is the largest lower semicontinuous function smaller than or equal to f.

We deduce the following corollary

Corollary 1.1. *A subset K of X is closed if and only if its indicator function is lower semi-continuous.*

Proof. In fact, $\mathrm{Ep}(\psi_K) = K \times \mathbb{R}_+$ is closed if and only if K is closed.   □

Proposition 1.5. *The functions f, g, f_i from X to $\mathbb{R} \cup \{+\infty\}$ are assumed to be lower semi-continuous. Then*

a) $f + g$ is lower semi-continuous;
b) if $\alpha > 0$, then αf is lower semi-continuous;
c) $\inf(f, g)$ is lower semi-continuous;
d) if A is a continuous mapping from Y to X then $f \circ A$ is lower semi-continuous;
e) $\sup_{i \in I} f_i$ is lower semi-continuous.

Proof. The proof of the first four assertions is elementary. The fifth results from the fact that $\mathrm{Ep}(\sup_{i \in I} f_i) = \bigcap_{i \in I} \mathrm{Ep}(f_i)$ is closed (see Proposition 1.1).   □

We shall see how to generalise the third assertion (see Proposition 1.7).

Remark. If $f : X \to \mathbb{R} \cup \{+\infty\}$ is lower semi-continuous, the same is true of the restriction to $\mathrm{Dom}\, f$, $f_0 : \mathrm{Dom}\, f \to \mathbb{R}$, when $\mathrm{Dom}\, f$ has the induced metric.

There is no exact converse. Only the following theorem holds.

Proposition 1.6. *Suppose that K is a **closed** subset of X and that f is a lower semi-continuous function from the metric subspace K to $\mathbb{R}$. Then the function f_K from X to $\mathbb{R} \cup \{+\infty\}$ is lower semi-continuous.*

Proof. In fact, the sections $S(f_K, \lambda)$ and $S(f, \lambda)$ are identical. Since $S(f, \lambda)$ is closed in K, and since K is closed in X, it follows that $S(f_K, \lambda) = S(f, \lambda)$ is closed in X.   □

1.6 Lower Semi-compact Functions

Study of the minimisation problem suggests that we should distinguish the following class of functions.

Definition 1.6. *We shall say that a function f from X to $\mathbb{R} \cup \{+\infty\}$ is lower semi-compact (or inf-compact) if all its lower sections are relatively compact.*

We then have the following theorem.

Theorem 1.1. *Suppose that a nontrivial function f from X to $\mathbb{R} \cup \{+\infty\}$ is both lower semi-continuous and lower semi-compact. Then the set M of elements at which f attains its minimum is non-empty and compact.*

Proof. Let $\alpha = \inf_{x \in X} f(x) \in \mathbb{R} \cup \{+\infty\}$ and $\lambda_0 > \alpha$. For all $\lambda \in]\alpha, \lambda_0]$, there exists $x_\lambda \in S(f, \lambda) \subset S(f, \lambda_0)$. Since the set $S(f, \lambda_0)$ is *compact*, a subsequence of elements $x_{\lambda'}$ converges to an element $\bar{x}$ of $S(f, \lambda_0)$. Since f is lower semi-continuous, we deduce that

$$f(\bar{x}) \leq \liminf_{x_{\lambda'} \to x_0} f(x_{\lambda'}) \leq \liminf_{\lambda > \alpha} \lambda = \alpha \leq f(\bar{x}).$$

Thus, $f(\bar{x}) = \alpha$, which implies that α is finite. Moreover, $M = \bigcap_{\alpha < \lambda \leq \lambda_0} S(f, \lambda)$ being an intersection of compact sets, is compact. $\qquad \square$

Corollary 1.2. *Any lower semi-continuous function from a compact subset $K \subset X$ to $\mathbb{R}$ is bounded below and attains its minimum.*

Proof. We apply Theorem 1.1 to the function f_K defined by $f_K(x) = f(x)$ if $x \in K$ and $f_K(x) = \infty$ if $x \notin K$, noting that f_K is lower semi-continuous (since K is closed and f is lower semi-continuous) and that f_K is lower semi-compact, K being relatively compact. $\qquad \square$

Remark. This very simple theorem is a rare general theorem for the existence of solutions of an optimisation problem.

The difficulty essentially arises in the verification of the assumptions. For instance, when the vector space E is infinite dimensional, we can supply it with topologies which are not equivalent, contrary to the case of finite dimensional vector spaces (supplied with topologies for which the addition and the multiplication by scalars are continuous) are all equivalent. In this case, since compact subsets remain compact when the topology is weaker, supplying E with weaker topologies increases the possibilities of having f lower semicompact. But continuous or lower semicontinuous functions remain continuous or lower semicontinuous respectively whenever the topology of E is stronger, so that strengthening the topology of E is advantageous. Hence, for applying Theorem 1.1, we have to construct topologies on E satisfying opposite requirements.

We shall see another existence result which does not use compactness, but instead requires stronger assumptions on the regularity of the function to be minimised.

Proposition 1.7. *Suppose that K is a **compact** topological space and that g is a lower semi-continuous function from $X \times K$ to $\mathbb{R} \cup \{+\infty\}$. Then the function $f : X \to \mathbb{R} \cup \{+\infty\}$ defined by*

$$\forall x \in X, \qquad f(x) := \inf_{y \in K} g(x, y) \tag{15}$$

is also lower semi-continuous.

Proof. We take $\lambda \in \mathbb{R}$ and consider a sequence of elements $x_n \in S(f, \lambda)$ converging to an element x_0. We shall prove that $x_0 \in S(f, \lambda)$. Because

$y \to f(x_n, y)$ is lower semi-continuous, and since K is compact, there exists $y_n \in K$ such that $f(x_n) = g(x_n, y_n)$ (Corollary 1.2). Thus, the sequence y_n contains a subsequence of elements $y_{n'}$ which converges to an element y_0 of K. Then, the sequence of pairs $(x_{n'}, y_{n'})$ of $S(g, \lambda)$ converges to (x_0, y_0), which belongs to $S(g, \lambda)$ since g is a lower semi-continuous function. Consequently, $x_0 \in S(f, \lambda)$, since $f(x_0) \le g(x_0, y_0) \le \lambda$. □

Finally, we note the following interesting result.

Proposition 1.8. *Consider n lower semi-continuous functions f_i from X to $\mathbb{R} \cup \{+\infty\}$ and suppose that one of these is lower semi-compact. We associate them with the mapping F from $K := \bigcap_{i=1}^{m} \mathrm{Dom}\, f_i$ to $\mathbb{R}^n$ defined by*

$$\forall x \in K, \qquad F(x) := (f_1(x), \ldots, f_n(x)). \tag{16}$$

Then
$$\text{the set} \quad F(K) + \mathbb{R}_+^n \quad \text{is closed in } \mathbb{R}^n. \tag{17}$$

Proof. We consider a sequence of elements $x_n \in K$ and elements $u_n \in \mathbb{R}_+^n$ such that the sequence of elements $y_n := F(x_n) + u_n$ converges to an element y of $\mathbb{R}^n$, and show that y belongs to $F(K) + \mathbb{R}_+^n$.

Let f_{i_0} be the function which is both lower semi-continuous and lower semi-compact. Since $f_{i_0}(x_n) + u_{n i_0}$ converges to y_{i_0}, there exists n_0 such that $|y_{i_0} - f_{i_0}(x_n) - u_{n i_0}| \le 1$ whenever $n \ge n_0$. Since $f_{i_0}(x_n) \le y_{i_0} - u_{n i_0} + 1 \le y_{i_0} + 1$, we deduce that for $n \ge n_0$, the x_n belong to $S(f_{i_0}, y_{i_0} + 1)$, which is compact. Thus, there exists a subsequence of elements $x_{n'}$ which converges to an element $\bar{x}$. We take an index $i = 1, \ldots, n$. Since f_i is lower semi-continuous, we deduce that

$$f_i(\bar{x}) \le \liminf_{n \to \infty} f_i(x_n) = \liminf_{n \to \infty} (y_{n'_i} - u_{n'_i}) \le \liminf_{n \to \infty} y_{n'_i} = y_i.$$

Thus, setting $u_i := y_i - f_i(\bar{x})$, which is positive or zero, we have shown that $y = F(\bar{x}) + u$ where $\bar{x} \in K$ and $u \in \mathbb{R}_+^n$. □

1.7 Approximate Minimisation of Lower Semi-continuous Functions on a Complete Space

In the statement of Theorem 1.1, and its Corollary 1.2 on the existence of a solution to a minimisation problem, compactness plays a crucial role. However, it is remarkable that simply with the condition that the set over which f is minimised is complete, we nonetheless obtain an existence result for an approximate minimisation problem.

Theorem 1.2 (Ekeland). *Suppose that E is a **complete** metric space and that $f : E \to \mathbb{R}_+ \cup \{+\infty\}$ is nontrivial, positive and lower semi-continuous. Consider $x_0 \in \mathrm{Dom}\,(f)$ and $\varepsilon > 0$. There exists $\bar{x} \in E$ such that*

(i) $$f(\bar{x}) + \varepsilon d(x_0, \bar{x}) \le f(x_0)$$
(ii) $$\forall x \ne \bar{x}, \qquad f(\bar{x}) < f(x) + \varepsilon d(x, \bar{x}). \tag{18}$$

The first property is a *localization* property stating that $\bar{x}$ belongs to a ball centered around x_0 and of radius at least equal to $\frac{f(x_0)}{\varepsilon}$. The second property states that $\bar{x}$ minimizes the function $x \mapsto f(x) + \varepsilon d(x, \bar{x})$ (which depends upon the unknown solution $\bar{x}$!)

Before proving this theorem, we state a corollary which clarifies the notion of approximate solution.

Corollary 1.3. *The assumptions are as in Theorem 1.2. Suppose $\varepsilon, \lambda > 0$ and that x_0 is a point with $f(x_0) \le \inf f(x) + \varepsilon\lambda$. Then there exists $\bar{x} \in E$ such that*

(i) $$f(\bar{x}) \le f(x_0)$$
(ii) $$d(x_0, \bar{x}) \le \lambda$$
(iii) $$\forall x \in E, \qquad f(\bar{x}) \le f(x) + \varepsilon d(x, \bar{x}). \tag{19}$$

Proof of Theorem 1.2. We may naturally take $\varepsilon = 1$.

We shall associate the function f with the correspondence F of E into itself which associates a point x with the set $F(x)$ defined by

$$F(x) := \{y | f(y) + d(x, y) \le f(x)\}. \tag{20}$$

The sets $F(x)$ are closed and the correspondence F has the following property:

(i) $\qquad\qquad y \in F(y)$ $\qquad\qquad$ (reflexivity)
(ii) $\qquad$ if $y \in F(x)$, then $F(y) \subset F(x)$ $\qquad$ (transitivity). $\tag{21}$

Condition (21)(ii) is evident if $x \notin \mathrm{Dom}\, f$, since in this case $F(x) = E$.

Thus, we suppose that $f(x)$ is finite. Take $y \in F(x)$ and $z \in F(y)$. Adding the inequalities:

$$f(z) + d(y, z) \le f(y) \quad \text{and} \quad f(y) + d(x, y) \le f(x)$$

and using the triangle inequality, we obtain the inequality

$$f(z) + d(x, z) \le f(x),$$

which implies that $z \in F(x)$.

We associate the function f with the function v defined on $\mathrm{Dom}\, f$ by

$$v(y) := \inf_{z \in F(y)} f(z). \tag{22}$$

It is clear that

$$\forall y \in F(x), \qquad d(x, y) \le f(x) - v(x), \tag{23}$$

which implies the following upper bound on the diameter of $F(x)$

$$\text{Diam } (F(x)) \le 2(f(x) - v(x)). \tag{24}$$

Next, we define the following sequence beginning with x_0: we take x_{n+1} in $F(x_n)$ such that $f(x_{n+1}) \le v(x_n) + 2^{-n}$ (this is possible by definition of the infimum). Since $F(x_{n+1}) \subset F(x_n)$, by virtue of (21)(ii), we have

$$v(x_n) \le v(x_{n+1}). \tag{25}$$

On the other hand, since we always have $v(y) \le f(y)$, we obtain the inequalities

$$v(x_{n+1}) \le f(x_{n+1}) \le v(x_n) + 2^{-n} \le v(x_{n+1}) + 2^{-n} \tag{26}$$

and thus the inequalities

$$0 \le f(x_{n+1}) - v(x_{n+1}) \le 2^{-n}. \tag{27}$$

Consequently, formula (24) implies that the diameter of the closed sets $F(x_n)$ converges to 0. As these closed sets are nested and since the space is complete, it follows that

$$\bigcap_{n \ge 0} F(x_n) = \{\bar{x}\}. \tag{28}$$

Since $\bar{x}$ belongs to $F(x_0)$, the inequality (18)(i) is satisfied. On the other hand, $\bar{x}$ belongs to all the $F(x_n)$; it follows that $F(\bar{x}) \subset F(x_n)$ and consequently that

$$F(\bar{x}) = \{\bar{x}\}. \tag{29}$$

Thus, we deduce that if $x \ne \bar{x}$ then $x \notin F(\bar{x})$, whence $f(x) + d(\bar{x}, x) > f(\bar{x})$. Thus, we have proved (18)(ii). □

1.8 Application to Fixed-point Theorems

If G is a correspondence of E into itself, a solution $\bar{x}$ of the inclusion

$$\bar{x} \in G(\bar{x}) \tag{30}$$

is called a *fixed point* of G.

Theorem 1.3 (Caristi). *Let G be a nontrivial correspondence of a complete metric space E into itself. We suppose that there exists a proper, positive, lower semi-continuous function f from E to $\mathbb{R}_+ \cup \{+\infty\}$ such that*

$$\forall x \in E, \ \exists y \in G(x) \ \text{such that} \ f(y) + d(x, y) \le f(x). \tag{31}$$

*Then the correspondence G has a **fixed point**.*
If f is linked to G by the stronger relationship

$$\forall x \in E, \quad \forall y \in G(x), \quad f(y) + d(y, x) \leq f(x), \tag{32}$$

then there exists $\bar{x} \in E$ such that $G(\bar{x}) = \{\bar{x}\}$.

Proof. Suppose that $\bar{x}$ satisfies (18)(ii), with $\varepsilon < 1$ and that $\bar{y} \in G(\bar{x})$ satisfies $f(\bar{y}) + d(\bar{x}, y) \leq f(\bar{x})$. If $\bar{y}$ is not equal to $\bar{x}$, inequality (18)(ii) with $x := \bar{y}$ implies that $d(\bar{x}, \bar{y}) \leq \varepsilon d(\bar{x}, \bar{y})$, which is impossible since $\varepsilon < 1$. Thus, $\bar{y}$ is equal to $\bar{x}$. There is at least one such if condition (31) is satisfied, whilst all the $\bar{y} \in G(\bar{x})$ are equal to $\bar{x}$ if condition (32) is satisfied. $\square$

Since we are discussing fixed-point theorems, we shall prove another result in which f is no longer assumed to be lower semi-continuous; however the correspondence G must have a closed graph. The graph of a correspondence G from E to F is defined by

$$\text{Graph}\,(G) := \{(x, y) \in E \times F | y \in G(x)\}. \tag{33}$$

Theorem 1.4. *Let E be a complete metric space. We consider a correspondence G from E to E with a closed graph. If there exists a nontrivial positive function f from E to $\mathbb{R}_+ \cup \{+\infty\}$ satisfying condition (31), then the correspondence G has a fixed point.*

Proof. We take a point $x_0 \in \text{Dom}\, f$ and use a recurrence to calculate a sequence of elements $x_n \in E$ such that, by virtue of condition (31), we have

$$x_{n+1} \in G(x_n), \qquad d(x_{n+1}, x_n) \leq f(x_n) - f(x_{n+1}). \tag{34}$$

This implies that the sequence of positive numbers $f(x_n)$ is decreasing; thus, it converges to a number α. Adding the inequalities (34) from $n = p$ to $n = q - 1$, the triangle inequality implies that

$$d(x_p, x_q) \leq \sum_{n=p}^{q-1} d(x_{n+1}, x_n) \leq f(x_p) - f(x_q). \tag{35}$$

Since the term on the right tends to $\alpha - \alpha = 0$ as p and q tend to infinity, we deduce that the sequence of the x_n is a Cauchy sequence which thus converges to an element $\bar{x} \in E$ since the space is complete.

Since the pairs (x_n, x_{n+1}) belong to the graph of G, which is closed, and converge to the pair $(\bar{x}, \bar{x})$ which thus belongs to the graph of G, the limit $\bar{x}$ is a fixed point of G. $\square$

As a corollary we obtain the Banach–Picard fixed point theorem for contractions.

Theorem 1.5 (Banach–Picard). *Suppose that E is a complete metric space and that $g : E \to E$ is a* **contraction***:*

$$\exists k \in]0, 1[\quad such\ that\ \forall x, y \in E, \qquad d(g(x), g(y)) \leq kd(x, y). \tag{36}$$

Then g has a unique fixed point $\bar{x}$.

Proof. We associate g with the function f from E to $\mathbb{R}_+$ defined by

$$f(x) := \sum_{n=0}^{\infty} d(g^n(x), g^{n+1}(x)). \tag{37}$$

Condition (36) implies that

$$d(g^n(x), g^{n+1}(x)) \leq kd(g^{n-1}(x), g^n(x)) \leq k^n d(x, g(x)).$$

Thus, the function f satisfies the condition:

$$0 \leq f(x) \leq \frac{1}{1-k} d(x, g(x)) < +\infty. \tag{38}$$

On the other hand, note that

$$f(x) = d(x, g(x)) + \sum_{n=1}^{\infty} d(g^n(x), g^{n+1}(x)) = d(x, g(x)) + f(g(x)).$$

Thus, the assumptions of Theorem 1.4 are satisfied, and so there exists a fixed point for the contraction g. Moreover, we also have uniqueness; if $\bar{x}$ and $\bar{y}$ are fixed points of g, the inequality

$$d(\bar{x}, \bar{y}) = d(g(\bar{x}), g(\bar{y})) \leq kd(\bar{x}, \bar{y})$$

implies that $d(\bar{x}, \bar{y}) = 0$ since $k < 1$, whence that $\bar{x} = \bar{y}$. $\qquad \square$

2. Convex Functions and Proximation, Projection and Separation Theorems

2.1 Introduction

Convexity plays a crucial role in the study of minimisation problems. After defining convex functions and describing their elementary properties, we show that continuous convex functions are locally Lipschitz (Lipschitz in a suitable neighbourhood of each point). We then prove the theorem for the existence and uniqueness of a solution of the minimisation problem

$$\frac{1}{2}\|\bar{x} - x_0\|^2 + f(\bar{x}) = \inf_{x \in X} \left(\frac{1}{2}\|x - x_0\|^2 + f(x) \right)$$

when f is a nontrivial convex lower semi-continuous function from X to $\mathbb{R} \cup \{+\infty\}$.

As a particular case, we derive the theorem for the best approximation of x_0 by elements of a convex closed set. It is known that this theorem has very important consequences. Amongst these, we mention the separation theorems which we shall use to prove the fundamental theorems of duality theory in convex analysis.

2.2 Definitions

Let X be a vector space.

Definition 2.1. *We shall say that a function f from X to $\mathbb{R} \cup \{+\infty\}$ is **convex** if for any convex combination $x = \sum_{i=1}^{n} \lambda_i x_i$ of elements $x_i \in X$ we have the inequality*

$$f\left(\sum_{i=1}^{n} \lambda_i x_i\right) \leq \sum_{i=1}^{n} \lambda_i f(x_i). \tag{1}$$

*We shall say that f is **concave** if $-f$ is convex, and that f is **affine** if f is both convex and concave.*

We begin by characterising convex functions.

Proposition 2.1. *Let f be a function from X to $\mathbb{R} \cup \{+\infty\}$. The following conditions are equivalent*
a) f is convex
b) $\forall x, y \in X, \forall \alpha \in]0, 1[$

$$f(\alpha x + (1 - \alpha)y) \leq \alpha f(x) + (1 - \alpha)f(y)$$

c) the epigraph of f is convex.

Proof. Clearly a) implies b).

We show that b) implies c). We let (x, λ) and (y, μ) be two points of the epigraph of f and $\alpha \in]0, 1[$ and show that

$$\alpha(x, \lambda) + (1 - \alpha)(y, \mu) = (\alpha x + (1 - \alpha)y, \alpha\lambda + (1 - \alpha)\mu)$$

belongs to this epigraph. In fact, the inequalities $f(x) \leq \lambda$ and $f(y) \leq \mu$ imply that $\alpha f(x) + (1 - \alpha)f(y) \leq \alpha\lambda + (1 - \alpha)\mu$, since α and $(1 - \alpha)$ are positive. Consequently, $f(\alpha x + (1 - \alpha)y) \leq \alpha\lambda + (1 - \alpha)\mu$, from b).

Lastly, we show that c) implies a). Since the 2-tuples $(x_i, f(x_i))$ belong to $\mathrm{Ep}(f)$, which is convex, then $\sum_{i=1}^{n} \lambda_i(x_i, f(x_i)) = (\sum_{i=1}^{n} \lambda_i x_i, \sum_{i=1}^{n} \lambda_i f(x_i))$ belongs to $\mathrm{Ep}(f)$, which means that $f\left(\sum_{i=1}^{n} \lambda_i x_i\right) \leq \sum_{i=1}^{n} \lambda_i f(x_i)$. $\qquad\square$

We deduce the following corollary.

Corollary 2.1. *A subset K of X is convex if and only if its indicator function is convex.*

Proof. In fact, $\mathrm{Ep}(\psi_K) = K \times \mathbb{R}_+$ is convex if and only if K is convex. $\qquad\square$

Proposition 2.2. *We suppose that the functions f, g, f_i from X to $\mathbb{R} \cup \{+\infty\}$ are convex. Then*
a) $f + g$ is convex;
b) if $\alpha > 0$ then αf is convex;
c) if A is a linear mapping from a vector space Y to X, then $f \circ A$ is convex;
d) if $\phi : \mathbb{R} \to \mathbb{R}$ is convex and increasing then $\phi \circ f$ is convex;
e) $\sup_{i \in I} f_i$ is convex.

Proof. The first four assertions are evident, whilst the last one results from the equality $\mathrm{Ep}(\sup_{i \in I} f_i) = \cap_{i \in I} \mathrm{Ep}(f_i)$. $\qquad\square$

We mention the following obvious property.

Proposition 2.3. *If f is a convex function from X to $\mathbb{R} \cup \{+\infty\}$, then its sections $S(f, \lambda)$ are convex.*

Remark. The converse is not true. A function all of whose sections are convex is said to be *quasi-convex*.

Definition 2.2. *A nontrivial function $f : X \to \mathbb{R} \cup \{+\infty\}$ is **strictly convex** if for any two distinct points x and $y \in \mathrm{Dom}\, f$*

$$f\left(\frac{x+y}{2}\right) < \frac{f(x)+f(y)}{2}. \tag{2}$$

This condition enables us to give a sufficient condition for the uniqueness of a solution of an optimisation problem.

Proposition 2.4. *Let f be a nontrivial convex function from X to $\mathbb{R} \cup \{+\infty\}$. Then the set M of solutions $\bar{x} \in X$ of the problem $f(\bar{x}) = \inf_{x \in X} f(x)$ is convex. If f is strictly convex then M contains at most one point.*

Proof. Let $\alpha := \inf_{x \in X} f(x)$. The first assertion follows from the equality $M = \cap_{\lambda > \alpha} S(f, \alpha)$, which implies that M is an intersection of convex sets. If f is strictly convex and if x_1 and x_2 are two solutions of the problem $\alpha = \inf_{x \in X} f(x)$, we would have

$$\alpha = f\left(\frac{x_1 + x_2}{2}\right) < \frac{f(x_1) + f(x_2)}{2} = \alpha$$

which is impossible. $\qquad\square$

Proposition 2.5. *Let g be a convex function from $X \times Y$ to $\mathbb{R} \cup \{+\infty\}$. Then the function f from X to $\mathbb{R} \cup \{+\infty\}$ defined by*

$$f(x) := \inf_{y \in Y} g(x, y) \tag{3}$$

is convex.

Proof. Fix $\varepsilon > 0, \lambda \in]0, 1[$ and $x_i (i = 1, 2)$ in X. Equality (3) is true when at least one of the x_i does not belong to the domain of f. Consider the case in which x_1 and x_2 belong to Dom f. Then there exist y_1 and y_2 such that

$$g(x_i, y_i) \le f(x_i) + \varepsilon \qquad (i = 1, 2). \tag{4}$$

Since g is convex, we deduce that

$$g(\alpha x_1 + (1 - \alpha)x_2, \alpha y_1 + (1 - \alpha)y_2) \le \alpha f(x_1) + (1 - \alpha)f(x_2) + \varepsilon.$$

But $f(\alpha x_1 + (1-\alpha)x_2)$ is less than or equal to $g(\alpha x_1 + (1-\alpha)x_2, \alpha y_1 + (1-\alpha)y_2)$. Whence

$$f(\alpha x_1 + (1 - \alpha)x_2) \le \alpha f(x_1) + (1 - \alpha)f(x_2) + \varepsilon$$

and simply letting ε tend to 0 completes the proof. $\qquad\square$

Proposition 2.6. *Consider n convex functions f_i from X to $\mathbb{R} \cup \{+\infty\}$. Then the mapping F from $K := \cap_{i=1}^{n} \text{Dom } f_i$ to $\mathbb{R} \cup \{+\infty\}$ defined by*

$$\forall x \in K, \qquad F(x) := (f_1(x), \dots, f_n(x)) \tag{5}$$

satisfies the following properties:

$$\text{the sets } F(K) + \mathbb{R}^n_+ \text{ and } F(K) + \mathring{\mathbb{R}}^n_+ \text{ are convex.} \tag{6}$$

Proof. We prove only the second assertion. The cone $\mathring{\mathbb{R}}^n_+$, the interior of the cone $\mathbb{R}^n_+$, is formed from vectors u with strictly positive components u_j.

Fix two elements $y_i = F(x_i) + u_i$ $(i = 1, 2)$ of $F(K) + \mathring{\mathbb{R}}^n_+$, where $x_i \in K$ and $u_i \in \mathring{\mathbb{R}}^n_+$. If $\alpha \in]0, 1[$, we may write

$$y = \alpha y_1 + (1 - \alpha)y_2 = F(x) + u$$

where $x = \alpha x_1 + (1 - \alpha)y_2$ and

$$u = \alpha u_1 + (1 - \alpha)u_2 + \alpha F(x_1) + (1 - \alpha)F(x_2) - F(\alpha x_1 + (1 - \alpha)x_2).$$

The convexity of the functions f_i then implies that the components u_i of this vector u are strictly positive. Thus y belongs to $F(K) + \mathring{\mathbb{R}}^n_+$. $\qquad\square$

2.3 Examples of Convex Functions

The norms and seminorms on a vector space are convex functions.

More generally, any subadditive positively homogeneous function is a positively homogeneous convex function and conversely.

Let $((x, y))$ be a scalar semiproduct on the vector space X and set

$$f(x) := \frac{1}{2}((x, x)) = \frac{1}{2}\|x\|^2 \tag{7}$$

where $\|x\| := \sqrt{(x, x)}$ is the seminorm associated with this scalar semiproduct. Then f is convex and strictly convex if $\|x\|$ is a norm. If we now take $\alpha, \beta \in [0, 1], \beta = 1 - \alpha$, then

$$\|x - \alpha y - \beta z\|^2 = \alpha^2 \|x - y\|^2 + \beta \|x - z\|^2 - 2\alpha\beta\|y - z\|^2. \tag{8}$$

In fact, the member on the left may be written as

$$\|\alpha(x - y) + \beta(x - z)\|^2 = \alpha^2 \|x - y\|^2 + \beta^2 \|x - z\|^2 + 2\alpha\beta((x - y, x - z)).$$

Multiplying the equality

$$\|y - z\|^2 = \|y - x + x - z\|^2 = \|x - y\|^2 + \|y - z\|^2 - 2((x - y, x - z))$$

by $\alpha\beta$ and adding it to the previous equality, we obtain the desired result.

Taking $x = 0$, we obtain

$$f(\alpha y + \beta z) = \alpha f(y) + \beta f(z) - \alpha\beta\|y - z\|^2 \le \alpha f(y) + \beta f(z)$$

and, if $\alpha = \frac{1}{2}$ and if $\|.\|$ is a norm, then

$$f\left(\frac{1}{2}(y+z)\right) \le \frac{1}{2}(f(y)+f(z)) - \frac{1}{4}\|y-z\|^2 < \frac{1}{2}(f(y)+f(z))$$

when $y \ne z$.   □

We recall that a continuous scalar semiproduct $((x,y))$ on X corresponds to a continuous linear operator L from X to X^* which satisfies

(i) $L = L^*$ (L is self-conjugate)
(ii) $\forall x \in X, \quad \langle Lx, x \rangle \ge 0$ (L is positive semi-definite)   (9)

It is defined by the formula

$$\forall x, y \in X \qquad \langle Lx, y \rangle = ((x, y)). \tag{10}$$

2.4 Continuous Convex Functions

We shall show that a convex function continuous at a point is actually Lipschitz in a neighbourhood of that point.

Definition 2.3. *A function f from an open subset Ω to $\mathbb{R}$ is locally Lipschitz if for each point $x \in \Omega$ there exists a neighbourhood of x on which f is Lipschitz.*

Theorem 2.1. *Let $f : X \to \mathbb{R} \cup \{+\infty\}$ be a nontrivial convex function. The following conditions are equivalent*
a) f is bounded above on an open subset (necessarily contained in Dom f).
b) f is locally Lipschitz on the interior of Dom f.

Proof. a). Clearly condition b) implies condition a).

b). Suppose then that f is bounded on a ball $x_0 + \eta B \subset \text{Dom } f$ by a constant $a < +\infty$. We associate with each $x \in X$ the element $y := \dfrac{x_0 - (1-\theta)x}{\theta}$ where $\theta := \dfrac{\|x - x_0\|}{\eta + \|x - x_0\|} < 1$. Then $\|y - x_0\| = \eta$ and consequently, $f(y) \le \alpha$. The convexity of f implies that

$$f(x_0) = f(\theta y + (1-\theta)x) \le \theta a + (1-\theta)f(x).$$

Whence

$$f(x_0) \le \frac{\theta}{1-\theta}(a - f(x_0)) + f(x)$$

and consequently, replacing θ by its value

$$\forall x \in X, \qquad f(x_0) - f(x) \le \frac{a - f(x_0)}{\eta}\|x - x_0\|. \tag{11}$$

Now take $x \in x_0 + \eta B$ and $y := \dfrac{x - (1-\theta)x_0}{\theta}$ where $\theta := \dfrac{\|x - x_0\|}{\eta} \le 1$. Then $\|y - x_0\| \le \eta$ and consequently, $f(y) \le a$. The convexity of f implies that

$$f(x) = f(\theta y + (1 - \theta)x_0) \leq \theta a + (1 - \theta)f(x_0)$$

and consequently, replacing θ by its value, that

$$\forall x \in x_0 + \eta B, \qquad f(x) - f(x_0) \leq \frac{a - f(x_0)}{\eta}\|x - x_0\|. \tag{12}$$

Inequalities (11) and (12) imply that

$$\forall x \in x_0 + \eta B, \qquad f(x) - f(x_0) \leq \frac{a - f(x_0)}{\eta}\|x - x_0\| \tag{13}$$

and consequently, that f is continuous at x_0.

c) We now prove that f is Lipschitz on the ball $x_0 + \beta B$ where $\beta < \eta$. Fix an integer n larger than $\dfrac{\|x_1 - x_0\|}{\eta - \beta}$. Take x_1 and x_2 in the ball $x_0 + \beta B$ and divide the segment from x_1 to x_2 into n parts, using the points $y_j := x_1 + \dfrac{j}{n}(x_1 - x_2)$, $(j = 0, \ldots, n)$. Note that $y_0 = x_1$, $y_n = x_2$ and that the points y_j belong to the ball $x_0 + \beta B$. It is clear that the balls $y_j + (\eta - \beta)B$ are then contained in the ball $x_0 + \eta B$, so that f is bounded by a on the balls of radius $y_j + (\eta - \beta)B$. Inequality (13), with x_0 replaced by y_j and η replaced by $\eta - \beta$, implies that

$$|f(y_{j+1}) - f(y_j)| \leq \frac{a - f(y_j)}{\eta - \beta}\|y_{j+1} - y_j\|$$

since $\|y_{j+1} - y_j\| = \dfrac{\|x_1 - x_0\|}{n} \leq \eta - \beta$.

On the other hand, inequality (13) implies that

$$f(x_0) - f(y_j) \leq \frac{a - f(x_0)}{\eta}\|y_j - x_0\| \leq a - f(x_0).$$

Then

$$|f(y_{j+1}) - f(y_j)| \leq \frac{2(a - f(x_0))}{\eta - \beta}\|y_{j+1} - y_j\|.$$

Since $\|x_1 - x_0\| = \sum_{i=1}^{n-1}\|y_{j+1} - y_j\|$, we now have

$$|f(x_1) - f(x_2)| \leq \sum_{j=1}^{n-1}|f(y_{j+1}) - f(y_j)| \leq \frac{2(a - f(x_0))}{\eta - \beta}\|x_1 - x_2\|.$$

Thus, f is Lipschitz on the ball $x_0 + \beta B$.

d). Lastly, we shall show that f is Lipschitz on a suitable neighbourhood of each point x_1 in the interior of the domain of f. By virtue of the above, it is sufficient to show that f is bounded above on a neighbourhood of x_1. Let $\gamma > 0$ be such that $x_1 + \gamma B$ is contained in Dom f. Set $\lambda = \dfrac{\gamma}{\gamma + \|x_1 - x_0\|}$, which is strictly less than 1. It is easy to see that the element

$$x_2 := x_0 + \frac{1}{(1-\lambda)}(x_1 - x_0) = \frac{x_1 - \lambda x_0}{1-\lambda} \tag{14}$$

belongs to $x_1 + \gamma B$, and that f is bounded above on the ball $x_1 + \lambda \eta B$ by $\lambda a + (\lambda - 1)f(x_2)$. In fact, if y belongs to the ball $x_1 + \lambda \eta B$, then the element $z := \frac{1}{\lambda}(y - (1-\lambda)x_2)$ belongs to the ball $x_0 + \eta B$. Then $f(z) \le a$ and, by convexity,

$$f(y) = f(\lambda z + (1-\lambda)x_2)) \le \lambda f(z) + (1-\lambda)f(x_2) \le \lambda a + (1-\lambda)f(x_2).$$

This completes the proof of the theorem.   □

Corollary 2.2. *If the interior of the domain of a convex function f from $\mathbb{R}^n$ to $\mathbb{R} \cup \{+\infty\}$ is non-empty, then f is locally Lipschitz on Int Dom f.*

Proof. Consider a ball $x_0 + \eta B$ contained in the domain of f. We may then find n points $x_i \in x_0 + \eta B$ such that the vectors $x_i - x_0$ are linearly independent. Thus, the set S of convex combinations $\sum_{i=1}^n \lambda_i x_i$, where $\lambda_i > 0$ for all i, is open and contained in the domain of f. Consequently, f is bounded above on the open set S by $\max_{i=1,\dots,n} f(x_i)$ since

$$f\left(\sum_{i=1}^n \lambda_i x_i\right) \le \sum_{i=1}^n \lambda_i x_i \le \max_{i=1,\dots,n} f(x_i).$$

Theorem 2.1 now applies.   □

Remark. Baire's theorem (see (Aubin 1977) page 189) implies the following corollary.

Corollary 2.3. *If the interior of the domain of a convex lower semi-continuous function f from a Hilbert space X to $\mathbb{R} \cup \{+\infty\}$ is non-empty, then f is locally Lipschitz on Int Dom f.*

Proof. Baire's theorem now implies that any lower semi-continuous function defined on an open set (here, Int Dom f) is bounded above on a non-empty open set. Theorem 2.1 then applies.   □

2.5 The Proximation Theorem

We shall consider minimisation problems of the form

$$f_\lambda(x) := \inf_{y \in X}\left[f(y) + \frac{1}{2\lambda}\|y - x\|^2\right] \tag{15}$$

where f is a function from a Hilbert space X to $\mathbb{R} \cup \{+\infty\}$, $\|\cdot\|$ is the Hilbert norm of X and λ is a positive parameter.

Theorem 2.2. *Suppose that $f : X \to \mathbb{R} \cup \{+\infty\}$ is a nontrivial, convex, lower* *semi-continuous function from a Hilbert space X to $\mathbb{R} \cup \{+\infty\}$. There exists a* **unique** *solution (denoted by $J_\lambda(x)$) of the problem (15):*

$$f_\lambda(x) = f(J_\lambda x) + \frac{1}{2\lambda}\|J_\lambda x - x\|^2. \tag{16}$$

This solution is characterised by the following variational inequalities

$$\forall y \in X, \qquad \frac{1}{\lambda}\langle J_\lambda x - x, J_\lambda x - y \rangle + f(J_\lambda x) - f(y) \leq 0. \tag{17}$$

Before proving this theorem, we shall apply it to the case where $f = \psi_K$ is the characteristic function of a set. This leads to the projection theorem, since in this case

$$f_\lambda(x) = \frac{1}{2\lambda}d(x, K)^2$$

where $d(x, K) := \inf_{y \in K}\|x - y\|$ is the distance from x to K.

Theorem 2.3 (Best Approximation). *Let K be a* **closed convex** *subset of a Hilbert space X. The minimisation problem*

(i) $\qquad\qquad\qquad\qquad J_x \in K$
(ii) $\qquad\qquad\qquad\qquad \|x - Jx\| = d(x, K)$ $\qquad\qquad$ (18)

has a unique solution J_x which is characterised by the variational inequalities

(i) $\qquad\qquad\qquad\qquad J_x \in K$
(ii) $\qquad\quad y \in K, \qquad \langle Jx - x, Jx - y \rangle \leq 0.$ $\qquad\qquad$ (19)

Definition 2.4. *The mapping J of X onto K is called the* **projector of best approximation** *of X onto K.*

Proof. a) If f is positive or zero, then f_λ is also positive or zero. This is the case, for example, of the Best Approximation Theorem (where $f = \psi_K$). If f is not positive, then we use a consequence of the projection theorem (Theorem 3.1) which implies that f is bounded below by an affine function: there exist $p \in X^*$ and $a \in \mathbb{R}$ such that:

$$\forall y \in X, \qquad f(y) \geq \langle p, y \rangle + a.$$

Since the Cauchy–Schwarz inequality implies that

$$\langle p, x - y \rangle \leq \frac{1}{\lambda}\|\lambda p\| \, \|x - y\| \leq \frac{\lambda}{2}\|p\|^2 + \frac{1}{2\lambda}\|y - x\|^2,$$

this inequality implies that

$$
\begin{aligned}
f(y) + \frac{1}{2\lambda}\|y - x\|^2 &\geq \langle p, y - x \rangle + a + \langle p, x \rangle + \frac{1}{2\lambda}\|y - x\|^2 \\
&\geq a + \langle p, x \rangle - \frac{\lambda}{2}\|p\|^2
\end{aligned}
$$

and thus that

$$f_\lambda(x) \geq a + \langle p, x \rangle - \frac{\lambda}{2} \|p\|^2 > -\infty.$$

b) We show that any solution $\bar{x}$ of the problem

$$f_\lambda(x) = f(\bar{x}) + \frac{1}{2\lambda} \|\bar{x} - x\|^2 \tag{20}$$

satisfies

$$\forall y \in X, \qquad \frac{1}{\lambda} \langle \bar{x} - x, \bar{x} - y \rangle + f(\bar{x}) - f(y) \leq 0. \tag{21}$$

We take $z = \bar{x} + \theta(y - \bar{x}) = \theta y + (1 - \theta)\bar{x}$ where $\theta \in]0, 1[$. Then we obtain the inequality

$$\begin{aligned} f_\lambda(\bar{x}) + \frac{1}{2\lambda} \|\bar{x} - x\|^2 &\leq f(\bar{x} + \theta(y - \bar{x})) + \frac{1}{2\lambda} \|\bar{x} + \theta(y - \bar{x}) - x\|^2 \\ &\leq (1 - \theta)f(\bar{x}) + \theta f(y) + \frac{1}{2\lambda} \|\bar{x} - x\|^2 \\ &\quad + \frac{\theta}{\lambda} \langle \bar{x} - x, y - \bar{x} \rangle + \frac{\theta^2}{2\lambda} \|y - \bar{x}\|^2 \end{aligned}$$

which implies that

$$f_\lambda(\bar{x}) - f(y) + \frac{1}{\lambda} \langle \bar{x} - x, \bar{x} - y \rangle \leq \frac{\theta}{2\lambda} \|y - \bar{x}\|^2.$$

It is now sufficient to let θ tend to 0.

c) Suppose, conversely, that $\bar{x}$ satisfies the variational inequalities (21). We recall that

$$\frac{1}{2} \|\bar{x} - x\|^2 - \frac{1}{2} \|y - x\|^2 \leq \langle \bar{x} - x, \bar{x} - y \rangle$$

and that consequently

$$\begin{aligned} f(\bar{x}) + \frac{1}{2\lambda} \|\bar{x} - x\|^2 - f(y) - \frac{1}{2\lambda} \|y - x\|^2 &\leq f(\bar{x}) - f(y) + \frac{1}{\lambda} \langle \bar{x} - x, \bar{x} - y \rangle \\ &\leq 0 \end{aligned}$$

for all $y \in X$.

d) There exists a solution $\bar{x}$ of the problem $f_\lambda(x)$. To prove this, we consider a minimising sequence of elements $x_n \in X$ satisfying

$$f(x_n) + \frac{1}{2\lambda} \|x_n - x\|^2 \leq f_\lambda(x) + \frac{1}{n}. \tag{22}$$

We shall show that this is a Cauchy sequence. In fact, the so-called median formula implies that

$$\|x_n - x_m\|^2 = 2\|x_n - x\|^2 + 2\|x_m - x\|^2 - 4\left\|\frac{x_n + x_m}{2} - x\right\|^2. \qquad (23)$$

Consequently, by virtue of (22) and (23), we have

$$
\begin{aligned}
\|x_n - x_m\|^2 &\leq 4\lambda\left(\frac{1}{n} + \frac{1}{m} + 2f_\lambda(x) - f(x_m) - f(x_n)\right) \\
&\quad + 8\lambda\left(f\left(\frac{x_n + x_m}{2}\right) - f_\lambda(x)\right) \\
&= 4\lambda\left(\frac{1}{n} + \frac{1}{m} + 2f\left(\frac{x_n + x_m}{2}\right) - f(x_n) - f(x_m)\right) \\
&\leq 4\lambda\left(\frac{1}{n} + \frac{1}{m}\right)
\end{aligned}
$$

since f is convex.

Thus, x_n converges to an element $\bar{x}$ of X, since X is complete. The lower semi-continuity of f implies that

$$
\begin{aligned}
f(\bar{x}) + \frac{1}{2\lambda}\|\bar{x} - x\|^2 &\leq \liminf_{x_n \to \bar{x}}\left(f(x_n) + \frac{1}{2\lambda}\|x_n - \bar{x}\|^2\right) \\
&\leq f_\lambda(x).
\end{aligned}
$$

Whence $f_\lambda(x) = f(\bar{x}) + \frac{1}{2\lambda}\|\bar{x} - x\|^2$.

e) We now prove the uniqueness. If $\bar{x}$ and $\bar{\bar{x}}$ are two solutions to the problem of minimising $f_\lambda(x)$, we deduce from the variational inequality (21) that $f(\bar{x}) - f(\bar{\bar{x}}) + \frac{1}{\lambda}\langle\bar{x} - x, \bar{x} - \bar{\bar{x}}\rangle \leq 0$. Interchanging the roles of $\bar{x}$ and $\bar{\bar{x}}$ we obtain the inequality $f(\bar{\bar{x}}) - f(\bar{x}) + \frac{1}{\lambda}\langle\bar{\bar{x}} - x, \bar{\bar{x}} - \bar{x}\rangle \leq 0$. Adding these inequalities, we deduce that $\frac{1}{\lambda}\|\bar{x} - \bar{\bar{x}}\|^2 \leq 0$. Whence $\bar{x} = \bar{\bar{x}}$. $\qquad\square$

We note that the mappings J_λ and $1 - J_\lambda$ are both continuous, indeed Lipschitz with constant 1.

Proposition 2.7. *The mappings J_λ and $1 - J_\lambda$ are Lipschitz with constant 1 (independent of λ) and 'monotone' in the sense that:*

(i) $\qquad\qquad \langle J_\lambda x - J_\lambda y, x - y\rangle \geq \|J_\lambda x - J_\lambda y\|^2$

(ii) $\quad \langle(1 - J_\lambda)x - (1 - J_\lambda)y, x - y\rangle \geq \|(1 - J_\lambda)x - (1 - J_\lambda)y\|^2. \qquad (24)$

Proof. The variational inequality which characterises $J_\lambda x$ implies that

$$f(J_\lambda x) - f(J_\lambda y) + \frac{1}{\lambda}\langle J_\lambda x - x, J_\lambda x - J_\lambda y\rangle \leq 0.$$

Switching the roles of x and y, we have

$$f(J_\lambda y) - f(J_\lambda x) + \frac{1}{\lambda}\langle J_\lambda y - y, J_\lambda y - J_\lambda x\rangle \leq 0.$$

Adding these two inequalities, we find that

$$\langle J_\lambda x - J_\lambda y - (x - y), J_\lambda x - J_\lambda y\rangle \leq 0. \qquad (25)$$

The inequalities (24)(i) and (ii) follow from this inequality.

This being so, we write

$$
\begin{aligned}
\|x - y\|^2 &= \|x - J_\lambda x - (y - J_\lambda y) + (J_\lambda x - J_\lambda y)\|^2 \\
&= \|(1 - J_\lambda)x - (1 - J_\lambda)y\|^2 + \|J_\lambda x - J_\lambda y\|^2 \\
&\quad + 2\langle (1 - J_\lambda)x - (1 - J_\lambda)y, J_\lambda x - J_\lambda y \rangle.
\end{aligned}
$$

Following (25), we deduce that

$$
\|x - y\|^2 \geq \|(1 - J_\lambda)x - (1 - J_\lambda)y\|^2 + \|J_\lambda x - J_\lambda y\|^2.
$$

This completes the proof of Proposition 2.7.   □

Remark. We shall study this question further to show, amongst other things, that

$$
\begin{aligned}
\lim_{\lambda \to 0} J_\lambda x &= x && \text{if } x \in \mathrm{Dom}\, f \\
\lim_{\lambda \to 0} f_\lambda(x) &= f(x) \\
\lim_{\lambda \to \infty} f_\lambda(x) &= \inf_{x \in X} f(x)
\end{aligned}
$$

and that f_λ is a convex differentiable function with $\nabla f_\lambda(x) = \dfrac{x - J_\lambda x}{\lambda}$. This is the reason why f_λ is called the *Moreau approximation* of f.

2.6 Separation Theorems

We shall use the Best Approximation Theorem to deduce one of the most useful analytical results, known as the Separation Theorem .

Theorem 2.4 (Separation Theorem). *Consider a non-empty, convex, closed subset K of a Hilbert space X. If x_0 does not belong to K, there exist a continuous linear form $p \in X^*$ and $\varepsilon > 0$ such that*

$$
\sup_{y \in K} \langle p, y \rangle \leq \langle p, x_0 \rangle - \varepsilon. \tag{26}
$$

Proof. We consider the projection $J x_0$ of best approximation of x_0 onto K. The variational inequality which characterises $J x_0$ implies that

$$
\langle J x_0 - x_0, J x_0 - y \rangle \leq 0 \qquad \forall y \in K.
$$

We deduce that

$$
\|J x_0 - x_0\|^2 \leq \langle x_0 - J x_0, x_0 - y \rangle \qquad \forall y \in K.
$$

Since $x_0 \notin K$, $\|J x_0 - x_0\|^2$ is strictly positive, and the linear form $p = J x_0 - x_0$ satisfies the conclusion of the theorem (26).   □

Remark. Set $a := \langle p, x_0 \rangle - \sup_{y \in K} \langle p, y \rangle$ and $b = \langle p, x_0 \rangle - \dfrac{a}{2} = \sup_{y \in K} \langle p, y \rangle + \dfrac{a}{2}$.
The hyperplane

$$H = \{x \in X | \langle p, x \rangle = b\}$$

separates x_0 *from* K, *since* $\langle p, x_0 \rangle > b$ *and* $\langle p, y \rangle \leq b$ *for all* $y \in K$.

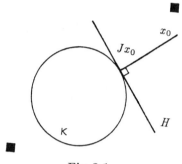

Fig. 2.1.

If X is a finite-dimensional space, we obtain a Large Separation Theorem, without assuming that the set K is closed. This is very useful, since it is often difficult to prove that a set is closed.

Theorem 2.5 (Large Separation). *Let K be a non-empty convex subset of a* **finite-dimensional** *space X. If x_0 does not belong to K, there exists a linear form $p \in X^*$ such that*

$$p \neq 0 \quad and \quad \sup_{y \in K} \langle p, y \rangle \leq \langle p, x_0 \rangle. \tag{27}$$

Proof. Although K is not closed, all the convex hulls of finite families of points of K are however closed convex subsets of K which we may separate from x_0 by virtue of the above theorem. We shall use this idea.

Next, with any $x \in K$ we associate the subset F_x of the unit sphere defined by

$$F_x := \{p \in X^* \,|\, \|p\|_* = 1 \text{ and } \langle p, x \rangle \leq \langle p, x_0 \rangle\}. \tag{28}$$

We note that the set of the solutions (of norm one) of (27) is the intersection $\cap_{x \in K} F_x$ of the sets F_x. Thus, we need to show that this intersection is non-empty. For this we use the fact the unit sphere $S := \{p \in X^* \,|\, \|p\|_* = 1\}$ is *compact, since X is finite-dimensional.* As the subsets F_x are clearly closed, it is sufficient to show that they satisfy the finite-intersection property: for any family $x_1, \ldots, x_n$, the intersection $\cap_{i=1}^n F_{x_i} \neq \emptyset$. To prove this, we consider the convex hull of the x_i, $M := \{\sum_{i=1}^n \lambda_i x_i | \lambda_i \geq 0, \sum_{i=1}^n \lambda_i = 1\}$. Since K is convex, M is contained in K, and consequently $x_0 \notin M$. On the other hand, M is convex and closed (compact even). Thus, the Separation Theorem implies that there exists a linear form (which may always be taken to have norm 1) such that $\sup_{y \in M} \langle p, y \rangle < \langle p, x_0 \rangle$. Since the x_i belong to M, then $\langle p, x_i \rangle \leq \langle p, x_0 \rangle$, whence p belongs to F_{x_i} for each $i = 1, \ldots, n$.

Thus, the finite-intersection property is satisfied, whence the set $\cap_{x \in K} F_x$, being non-empty, contains a linear form p which therefore satisfies (27). $\square$

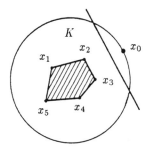

Fig. 2.2.

There are numerous corollaries to the separation theorems but we find it preferable to use one or other of the previous results. We shall, however, show how these results may be used to 'separate' two disjunct subsets M and N. To say that two subsets M and N of a vector space are disjunct is equivalent to the statement that 0 does not belong to $M - N$.

$$M \cap N = \emptyset \Leftrightarrow 0 \notin M - N. \tag{29}$$

We also note that

$$\sup_{z \in M-N} \langle p, z \rangle = \sup_{x \in M} \langle p, x \rangle - \inf_{y \in N} \langle p, y \rangle. \tag{30}$$

Having established these two remarks, we obtain the following corollary.

Corollary 2.4. *Consider two non-empty, **disjunct** subsets of a Hilbert space X.*
a) If we assume that

$$\text{the set } M - N \text{ is convex and closed,} \tag{31}$$

then there exist a continuous linear form $p \in X^*$ and $\varepsilon > 0$ such that

$$\sup_{x \in M} \langle p, x \rangle \leq \inf_{y \in N} \langle p, y \rangle - \varepsilon. \tag{32}$$

b) If we assume that

$$X \text{ is finite dimensional and } M - N \text{ is convex,} \tag{33}$$

then there exists a linear form $p \in X^*$ such that

$$p \neq 0 \quad and \quad \sup_{x \in M} \langle p, x \rangle \leq \inf_{y \in N} \langle p, y \rangle. \tag{34}$$

Proof. It is sufficient to apply the Separation Theorem and the Large Separation Theorem to the case in which $K = M - N$ with $x_0 = 0$, using properties (29) and (30). □

We may now give examples of properties implying the assumptions of this corollary.

For example, we recall that

$$\text{if } M \text{ and } N \text{ are convex, then } M - N \text{ is convex} \tag{35}$$

and that

$$\text{if } M \text{ is compact and } N \text{ is closed, then } M - N \text{ is closed.} \tag{36}$$

But, be warned, we shall use examples in which $M - N$ is convex and M is not and in which $M - N$ is closed although neither M nor N is compact (see Propositions 2.6 and 1.8).

The first separation theorems and the foundations of what was to become functional analysis are due to the mathematician Minkowski (1910). The extension of these theorems to Banach spaces and the equivalence to the problem of extending continuous linear forms is due independently to Hahn (the founder of the famous Vienna Circle) and Banach. The Hahn–Banach theorem is one of the three fundamental theorems of linear functional analysis all of which carry the name of Banach (the two other theorems, the Banach Closed Graph Theorem and the Banach-Steinhauss Theorem, deal with continuous linear operator and are based on the Baire Theorem.). In 1922, Banach published his first discoveries about *"les opérations dans les ensembles abstraits et leurs applications aux équations intégrales"*. In 1932, he published his masterpiece, the monograph *Théorie des Opérateurs Linéaires*, which had and continues to have a determining influence on the course of the history of mathematics.

Note that, whilst integral equations were the principal motivation which drove Banach, Hilbert and other mathematicians at the beginning of this century to build the foundations of functional analysis, the latter has been applied in very different areas of mathematics and, by ricochet, in numerous disciplines.

It is this universality of mathematical results, having their origin in one discipline and finding applications in others, which makes mathematics so fascinating.

3. Conjugate Functions and Convex Minimisation Problems

3.1 Introduction

The power and the beauty of convex analysis stem from the existence of a one-to-one correspondence between the convex lower semi-continuous functions on X and those on its dual X^*. This correspondence plays a role analogous to that of transposition, which is also a one-to-one correspondence between the continuous linear operators from X to Y and those from Y^* to X^*. In associating a convex lower semi-continuous function with its conjugate, we in some way double the number of properties since we will have the option of using the properties of the function or its conjugate.

This transformation also shows that the cone of convex lower semi-continuous functions, which is stable on passage to the upper envelope, is in fact obtained by saturation of the space of affine functions continuous under this operation. In simple terms, this means that *any convex lower semi-continuous function is the upper envelope of the continuous affine functions which minorise it*. To make this more precise, let us consider the minimisation of a nontrivial function f from X to $\mathbb{R} \cup \{+\infty\}$. In fact, since the function f is never known exactly, it is wise to study not only the minimisation of the function f, but also that of a family of perturbed functions.

For simplicity and efficiency, we restrict ourselves to simple perturbations. In our given context, this means that we shall perturb f by continuous linear functions and study the *family* of minimisation problems

$$-f^*(p) := \inf_{x \in X} [f(x) - \langle p, x \rangle] \qquad (*)$$

and the variation of this infimum as a function of p. In particular,

$$-f^*(0) := \inf_{x \in X} f(x).$$

The formula $(*)$ may be rewritten in the form

$$f^*(p) = \sup_{x \in X} [\langle p, x \rangle - f(x)]$$

which immediately shows that the function $f^* : p \in X^* \to f^*(p) \in \mathbb{R} \cup \{+\infty\}$, which is the upper envelope of the continuous affine functions $p \to \langle p, x \rangle - f(x)$ on X^*, is a convex lower semi-continuous function.

The result mentioned above is explained by this assertion: *a function f is convex, lower semi-continuous and nontrivial if and only if f is equal to its biconjugate* $(f^*)^*$.

The second important result is known as Fenchel's Theorem. We consider

(i) two Hilbert spaces (or reflexive Banach spaces) X and Y;

(ii) a continuous linear operator A from X to Y;

(iii) two nontrivial, convex, lower semi-continuous functions
$f : X \to \mathbb{R} \cup \{+\infty\}$ and $g : Y \to \mathbb{R} \cup \{+\infty\}$ satisfying

    (a) $0 \in \mathrm{Int}\,(A\,\mathrm{Dom}f - \mathrm{Dom}\,g)$

    (b) $0 \in \mathrm{Int}\,(A^*\,\mathrm{Dom}f^* + \mathrm{Dom}\,g^*)$

We shall prove that *there exist solutions* $\bar{x} \in X$ *and* $\bar{q} \in Y^*$ *of the minimisation problems*

$$v := \inf_{x \in X}[f(x) + g(Ax)] = f(\bar{x}) + g(A\bar{x})$$

and

$$v_* := \inf_{q \in Y^*}[f^*(-A^*q) + g^*(q)] = f^*(-A^*\bar{q}) + g^*(\bar{q})$$

and that, in addition, *the two minimisation problems are linked by the equation*

$$v + v_* = 0.$$

In the next section, we shall establish the connections between the solutions of the v problem and those of the v_* problem (known as the *dual* of the v problem).

We shall formulate a calculus of conjugate functions which will enrich the field of applications of these two theorems. Since a closed convex subset K of X is characterised by its indicator function ψ_K, which is convex and lower semi-continuous, it is consequently equivalently characterised by the conjugate function ψ_K^* of ψ_K defined on X^* by

$$\sigma_K(p) := \psi_K^*(p) = \sup_{x \in K}\langle p, x\rangle.$$

This function, called the *support function* of the subset K, is very useful in that it enables us to replace the manipulation of closed convex subsets by the more familiar manipulation of convex lower semi-continuous functions. The discovery and use of this fact is due to Minkowski.

This will lead us naturally to the notion of polarity between closed convex cones of X and of X^*. If K is a closed convex cone we denote its (negative) polar cone by

$$K^- := \{p \in X^* | \forall x \in K, \langle p, x\rangle \le 0\}.$$

This is also a closed convex cone. We shall prove that $K = (K^-)^-$.

As discovered by Steinitz from 1912, this relationship extends the orthogonality relationships between vector subspaces to closed convex cones.

3.2 Characterisation of Convex Lower Semi-continuous Functions

Following the Danish mathematician Fenchel, who introduced this concept in 1949, after a long history beginning with Young's inequality in 1912, we introduce the following definition

Definition 3.1. *Let f be a nontrivial function from X to $\mathbb{R} \cup \{+\infty\}$. Then the function f^* from X^* to $\mathbb{R} \cup \{+\infty\}$ defined by*

$$\forall p \in X^*, \qquad f^*(p) := \sup_{x \in X}[\langle p, x \rangle - f(x)] \in \mathbb{R} \cup \{+\infty\} \tag{1}$$

is called the (Fenchel) **conjugate** *of f and the function $f^{**} : X \to \overline{\mathbb{R}}$ defined by*

$$\forall x \in X, \qquad f^{**}(x) := \sup_{p \in X}[\langle p, x \rangle - f^*(p)] \tag{2}$$

is the **biconjugate** *of f.*

Note that the so-called Fenchel inequality

$$\forall x \in X, \ \forall p \in X^*, \qquad \langle p, x \rangle \le f(x) + f^*(p) \tag{3}$$

always holds and that

$$\forall x \in X, \qquad f^{**}(x) \le f(x). \tag{4}$$

Remark. If we interpret the vector space X as a space of commodities, its dual X^* as the space of prices (continuous linear functions associating a commodity with its value) and f as a cost function, then $\langle p, x \rangle - f(x)$ is a profit and the conjugate function is the maximum-profit function, which associates every price p with the maximum profit which it may obtain.

If a function f coincides with its biconjugate, then f is necessarily convex and lower semi-continuous. The converse is also true.

Theorem 3.1. *A nontrivial function $f : X \to \mathbb{R} \cup \{+\infty\}$ is convex and lower semi-continuous if and only if $f = f^{**}$. In this case, f^* is also nontrivial.*

Remark. Since in this case

$$f(x) = \sup_{p \in X^*}[\langle p, x \rangle - f^*(p)], \tag{5}$$

we deduce that any nontrivial convex lower semi-continuous function is the upper envelope of the affine functions which minorise it.

Proof. The idea of the proof is very simple. Since the epigraph of f is a closed convex set, any point (x, a) which does not belong to it is separated from $\mathrm{Ep}(f)$ by a hyperplane which is the graph of a continuous affine function minorising f.

We shall now substantiate this idea.

a) Suppose that $a < f(x)$. Since the pair (x, a) does not belong to $\mathrm{Ep}(f)$, which is convex and closed, there exist a continuous linear form $(p, b) \in X^* \times \mathbb{R}$ and $\varepsilon > 0$ such that

$$\forall y \in \mathrm{Dom} f, \ \forall \lambda \geq f(y), \qquad \langle p, y \rangle - b\lambda \leq \langle p, x \rangle - ba - \varepsilon \qquad (6)$$

by virtue of the *Separation Theorem* (Theorem 2.4).

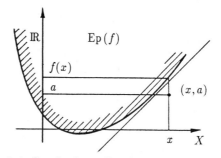

Fig. 3.1. Graph of an affine function minorising f.

b) We note that $b \geq 0$. If not, we take y in the domain of f and $\lambda = f(y) + \mu$. We would have

$$-b\mu \leq \langle p, x - y \rangle + b(f(y) - a) - \varepsilon < +\infty.$$

Then we obtain a contradiction if we let μ tend to $+\infty$.

c) We show that if $b > 0$, then $a < f^{**}(x)$. In fact, we may divide the inequality (6) by b; whence, setting $\bar{p} = p/b$ and taking $\lambda = f(y)$, we obtain

$$\forall y \in \mathrm{Dom} f, \qquad \langle \bar{p}, y \rangle - f(y) \leq \langle \bar{p}, x \rangle - a - \varepsilon/b.$$

Then, taking the supremum with respect to y, we have

$$f^*(\bar{p}) < \langle \bar{p}, x \rangle - a.$$

This implies that

(i) $\bar{p}$ belongs to the domain of f^*

(ii) $a < \langle \bar{p}, x \rangle - f^*(\bar{p}) \leq f^{**}(x).$ \qquad (7)

d) We consider the case in which x belongs to the domain of f. In this case, b is always strictly positive. To see this, it is sufficient to take $y = x$ and $\lambda = f(x)$ in formula (6) to show that

$$b \geq \varepsilon/(f(x) - a)$$

since $f(x) - a$ is a strictly positive real number. Then, from part b), we deduce the existence of $\bar{p} \in \mathrm{Dom}\, f^*$ and that $a \leq f^{**}(x) \leq f(x)$ for all $a < f(x)$. Thus, $f^{**}(x)$ is equal to $f(x)$.

e) We consider the case in which $f(x) = +\infty$ and a is an arbitrarily large number. Either b is strictly positive, in which case part b) implies that $a < f^{**}(x)$, or $b = 0$. In the latter case, (6) implies that

$$\forall y \in \mathrm{Dom}\, f, \qquad \langle p, y - x \rangle + \varepsilon \leq 0. \tag{8}$$

Let us take $\bar{p}$ in the domain of f^* (we have shown that such an element exists, since $\mathrm{Dom}\, f$ is non-empty). Fenchel's inequality implies that

$$\langle \bar{p}, y \rangle - f^*(\bar{p}) - f(y) \leq 0. \tag{9}$$

We take $\mu > 0$, multiply the inequality (8) by μ and add it to the inequality (9) to obtain

$$\langle \bar{p} + \mu p, y \rangle - f(y) \leq f^*(\bar{p}) + \mu \langle p, x \rangle - \mu \varepsilon.$$

Taking the supremum with respect to y, we obtain:

$$f^*(\bar{p} + \mu p) \leq f^*(\bar{p}) + \mu \langle p, x \rangle - \mu \varepsilon$$

which may be written in the form

$$\langle p, x \rangle + \mu \varepsilon - f^*(\bar{p}) \leq \langle \bar{p} + \mu p, x \rangle - f^*(\bar{p} + \mu p) \leq f^{**}(x).$$

Taking $\mu = \frac{a + f^*(\bar{p}) - \langle \bar{p}, x \rangle}{\varepsilon}$, which is strictly positive, we have again proved that $a \leq f^{**}(x)$. Thus, since $f^{**}(x)$ is greater than an arbitrary finite number, we deduce that $f^{**}(x) = +\infty$.

3.3 Fenchel's Theorem

We shall now prove Fenchel's duality theorem which, in conjunction with the previous theorem, provides the framework for convex analysis.

Suppose we have two Hilbert spaces (or reflexive Banach spaces) X and Y, together with

(i)   a continuous linear operator $A \in L(X, Y)$;
(ii)   two nontrivial, convex, lower semi-continuous functions
        $f : X \to \mathbb{R} \cup \{+\infty\}$ and $g : Y \to \mathbb{R} \cup \{+\infty\}$. (10)

We shall study the minimisation problem

$$v := \inf_{x \in X}[f(x) + g(Ax)]. \tag{11}$$

Note that the function $f + g \circ A$ which we propose to minimise is only nontrivial if $A\, \mathrm{Dom}\, f \cap \mathrm{Dom}\, g \neq \emptyset$, that is to say, if

$$0 \in A \operatorname{Dom} f - \operatorname{Dom} g. \tag{12}$$

In this case, we have $v < +\infty$.

Now we introduce the dual minimisation problem

$$v_* := \inf_{q \in Y_*} [f^*(-A^*q) + g^*(q)] \tag{13}$$

where $A^* \in L(Y^*, X^*)$ is the transpose of A, $f^* : X^* \to \mathbb{R} \cup \{+\infty\}$ is the conjugate of f and $g^* : Y^* \to \mathbb{R} \cup \{+\infty\}$ is the conjugate of g. This only makes sense if we assume that

$$0 \in A^* \operatorname{Dom} g^* + \operatorname{Dom} f^* \tag{14}$$

and in this case, $v_* < +\infty$.

Note that we still have the inequality

$$v + v_* \geq 0 \tag{15}$$

since, by virtue of Fenchel's inequality,

$$f(x) + g(Ax) + f^*(-Aq) + g^*(q) \geq \langle -A^*q, x \rangle + \langle q, Ax \rangle = 0.$$

Consequently, conditions (12) and (14) imply that v and v_* are finite.

By way of a slight reinforcement of condition (12), guaranteeing that the function $f + g \circ A$ is nontrivial, we shall show that $v + v_*$ is equal to zero and that the dual problem has a solution.

Theorem 3.2 (Fenchel). *Suppose that X and Y are Hilbert spaces (or reflexive Banach spaces), that $A \in L(X,Y)$ is a continuous linear operator from X to Y and that $f : X \to \mathbb{R} \cup \{+\infty\}$ and $g : Y \to \mathbb{R} \cup \{+\infty\}$ are nontrivial, convex, lower semi-continuous functions. We consider the case in which $0 \in A \operatorname{Dom} f - \operatorname{Dom} g$ and $0 \in A^* \operatorname{Dom} g^* + \operatorname{Dom} f^*$ (which is equivalent to the assumption that v and v_* are finite).*

If we suppose that

$$0 \in \operatorname{Int}(A \operatorname{Dom} f - \operatorname{Dom} g), \tag{16}$$

then

(i) $\qquad\qquad\qquad v + v_* = 0$

(ii) $\qquad \exists \bar{q} \in Y^* \text{ such that } f^*(-A^*\bar{q}) + g^*(\bar{q}) = v_*. \tag{17}$

If we suppose that

$$0 \in \operatorname{Int}(A^* \operatorname{Dom} g^* + \operatorname{Dom} f^*), \tag{18}$$

then

(i) $\qquad\qquad\qquad v + v_* = 0;$

(ii) $\qquad \exists \bar{x} \in X \text{ such that } f(\bar{x}) + g(A\bar{x}) = v. \tag{19}$

Proof.

a) We shall begin by proving the theorem for the case in which the space Y is finite dimensional.

We introduce the mapping ϕ from $\mathrm{Dom}\, f \times \mathrm{Dom}\, g$ to $Y \times \mathbb{R}$ defined by

$$\phi(x, y) = \{Ax - y, f(x) + g(y)\} \tag{20}$$

together with

(i)   the vector $(0, v) \in Y \times \mathbb{R}$
(ii) the cone $Q = \{0\} \times]0, \infty[\subset Y \times \mathbb{R}$ $\qquad\qquad$ (21)

In a proof analogous to that of Proposition 2.6, it is easy to show that the linearity of A and the convexity of the functions f and g imply that

$$\phi(\mathrm{Dom}\, f \times \mathrm{Dom}\, g) + Q \text{ is a convex subset of } Y \times \mathbb{R}. \tag{22}$$

Furthermore, if we suppose that $(0, v)$ belongs to $\phi(\mathrm{Dom} f \times \mathrm{Dom}\, g) + Q$, we may deduce the existence of $x \in \mathrm{Dom}\, f$ and $y \in \mathrm{Dom}\, g$ such that $Ax - y = 0$ and $v > f(x) + g(y) = f(x) + g(Ax)$, which would contradict the definition of v. Thus,

$$(0, v) \notin \phi(\mathrm{Dom}\, f \times \mathrm{Dom}\, g) + Q. \tag{23}$$

Since Y is a finite-dimensional space, we may use the Large Separation Theorem to show that there exists a linear form $(p, a) \in Y^* \times \mathbb{R}$ such that

(i) $\qquad (p, a) \neq 0$

(ii) $\qquad av = \langle (p, a), (0, v) \rangle$
$$\leq \inf_{\substack{x \in \mathrm{Dom} f \\ y \in \mathrm{Dom} g}} [a(f(x) + g(y)) + \langle p, Ax - y \rangle] + \inf_{\theta > 0} a\theta. \tag{24}$$

Since the number $\inf_{\theta>0} a\theta$ is bounded below, we deduce that it is zero and that a is positive or zero. We cannot have $a = 0$, since in that case, the inequality (24)(ii) would imply that

$$0 \leq \inf_{\substack{x \in \mathrm{Dom} f \\ y \in \mathrm{Dom} g}} \langle p, Ax - y \rangle = \inf_{z \in A\mathrm{Dom} f - \mathrm{Dom} g} \langle p, z \rangle. \tag{25}$$

Since the set $A\,\mathrm{Dom} f - \mathrm{Dom}\, g$ contains a ball of radius η and centre 0, by virtue of (16), we deduce that $0 \leq -\eta \|p\|$ and thus that $p = 0$. This contradicts (24)(i).

Consequently, a is strictly positive. Dividing the inequality (24)(ii) by a and taking $\bar{p} = p/a$, we obtain

$$\begin{aligned}
v &\leq \inf_{\substack{x \in \mathrm{Dom} f \\ y \in \mathrm{Dom} g}} [\langle A^* \bar{p}, x \rangle - \langle \bar{p}, y \rangle + f(x) + g(y)] \\
&= -\sup_{\substack{x \in X \\ y \in Y}} [\langle -A^* \bar{p}, x \rangle + \langle \bar{p}, y \rangle - f(x) - g(y)] \\
&= -f^*(-A^* \bar{p}) - g^*(\bar{p}).
\end{aligned}$$

Whence, $f^*(-A^*\bar{p}) + g^*(\bar{p}) = -v \le v_*$, which proves that $\bar{p}$ is a solution of the dual problem and that $v_* = -v$.

The second assertion is proved by replacing f by g^*, g by f^* and A by $-A^*$.

b) We now give a proof in the case of infinite-dimensional spaces.

In this case, we consider the mapping ψ from $\text{Dom} f^* \times \text{Dom} g^*$ to $\mathbb{R} \times X^*$ defined by

$$\psi(p, q) = (f^*(p) + g^*(q), p + A^*q) \tag{26}$$

together with the set

$$\psi(\text{Dom} f^* \times \text{Dom} g^*) + \mathbb{R}_+ \times \{0\}. \tag{27}$$

α) It is easy to prove that this set is convex. We show that it is closed. For this, we consider a sequence of elements (v_n, r_n), belonging to this set, converging to (v_*, r_*) in $\mathbb{R} \times X^*$. Thus, there exist elements $p_n \in X^*$ and $q_n \in Y^*$ such that

$$v_n \ge f^*(p_n) + g^*(q_n) \qquad r_n = p_n + A^*q_n. \tag{28}$$

We shall deduce from the assumption (16) that the sequence of elements q_n is weakly bounded.

In fact, the assumption (16) implies the existence of a ball of radius $\gamma > 0$ contained in $\text{Dom} g - A \text{Dom} f$. Thus, for all $z \in Y$, there exist $x \in \text{Dom} f$ and $y \in \text{Dom} g$ such that $\dfrac{\gamma}{\|z\|} z = y - Ax$. Consequently,

$$\begin{aligned}
\frac{\gamma}{\|z\|}\langle q_n, z \rangle &= \langle q_n, y \rangle - \langle A^*q_n, x \rangle \\
&= \langle q_n, y \rangle + \langle p_n, x \rangle - \langle r_n, x \rangle \\
&\le g(q_n) + f^*(p_n) + g(y) + f(x) - \langle r_n, x \rangle \\
&\le g(y) + f(x) + v_n - \langle r_n, x \rangle.
\end{aligned}$$

Since the sequences v_n and $\langle r_n, x \rangle$ are convergent, they are bounded and thus we have shown that

$$\forall z \in Y, \qquad \sup_{n \ge 0} \langle q_n, z \rangle < +\infty. \tag{29}$$

The Banach–Steinhauss theorem then implies that the sequence of elements q_n is weakly compact; whence, it has a subsequence $q_{n'}$ which converges weakly to an element q_* of Y^* and consequently, the subsequence $p_{n'} - r_{n'} - A^*q_n$ converges weakly to $p_* = r - A^*q_*$ (see Schwartz 1970).

Since the functions f^* and g^* are weakly lower semi-continuous, we deduce that

$$\begin{aligned}
f^*(p_*) + g^*(q_*) &\le \liminf_{n \to \infty} f^*(p_n) + \liminf_{n \to \infty} g^*(q_n) \\
&\le \liminf_{n \to \infty}(f^*(p_n) + g^*(q_n)) \le \lim_{n \to \infty} v_n = v.
\end{aligned}$$

Thus, we have shown that

$$v \geq f^*(p_*) + g^*(q_*), \qquad r = p_* + A^*q_*,$$

whence, that (v, r) belongs to $\psi(\mathrm{Dom}\, f^* \times \mathrm{Dom}\, g^*) + \mathbb{R}_+ \times \{0\}$.

β) Next we shall show that

$$(-v, 0) \in \psi(\mathrm{Dom}\, f^* \times \mathrm{Dom}\, g^*) + \mathbb{R}_+ \times \{0\}. \tag{30}$$

This assertion implies the theorem, since there exists $\bar{q} \in \mathrm{Dom}\, g^*$ such that $-A\bar{q} \in \mathrm{Dom} f^*$ and $-v \geq f^*(-A^*\bar{q}) + g^*(\bar{q}) \geq v_* \geq -v$.

Consequently,

$$-v = v_* = f^*(-A^*\bar{q}) + g^*(\bar{q}).$$

We shall now suppose that assertion (30) is false. Since the set $\psi(\mathrm{Dom}\, f^* \times \mathrm{Dom}\, g^*) + \mathbb{R}_+ \times \{0\}$ is convex and closed and since $\mathbb{R} \times X$ is the dual of $\mathbb{R} \times X^*$, the pair $(-v, 0)$ may be strictly separated from this set; thus, there exist $(\alpha, -\bar{x}) \in \mathbb{R} \times X$ and $\varepsilon > 0$ such that

$$-\alpha v \leq \inf_{(p,q)} [\alpha(f^*(p) + g^*(q)) + \langle p + A^*q, \bar{x}\rangle] + \inf_{\theta > 0} \alpha\theta - \varepsilon.$$

Since $\inf_{\theta \geq 0} \alpha\theta$ is bounded below, it follows that $\inf_{\theta \geq 0} \alpha\theta = 0$ and that α is positive or zero. It cannot be zero, for in that case we would have

$$0 \leq \inf_{(p,q) \in \mathrm{Dom} f^* \times \mathrm{Dom}\, g^*} \langle p + A^*q, -\bar{x}\rangle - \varepsilon.$$

Since (14) implies the existence of $p \in \mathrm{Dom} f^*$ and $q \in \mathrm{Dom}\, g^*$ such that $p + A^*q = 0$, we would have $0 \leq -\varepsilon$, which is impossible.

Dividing by $\alpha > 0$ and setting $x := \bar{x}/\alpha$ and $\eta = \varepsilon/\alpha$, we obtain

$$\begin{aligned}
-v &\leq \inf_{(p,q)} [f^*(p) + g^*(q) - \langle p, x\rangle - \langle q, Ax\rangle] - \eta \\
&= -\sup_{(p,q)} [\langle p, x\rangle + \langle q, Ax\rangle - f^*(p) - g^*(q)] - \eta \\
&= -(f(x) + g(Ax)) - \varepsilon \leq -v - \eta.
\end{aligned}$$

This is impossible. Thus, assertion (30) is true and the proof of the theorem is complete. $\qquad\square$

3.4 Properties of Conjugate Functions

Firstly, we note the following elementary propositions.

Proposition 3.1.
a) If $f \leq g$, then $g^* \leq f^*$.
b) If $A \in L(X, X)$ is an isomorphism, then

$$(f \circ A)^* = f^* \circ A^{*-1}.$$

c) If $g(x) := f(x - x_0) + \langle p_0, x \rangle + a$, then

$$g^*(p) = f^*(p - p_0) + \langle p, x_0 \rangle - (a + \langle p_0, x_0 \rangle).$$

d) If $g(x) := f(\lambda x)$, then $g^*(p) = f^*\left(\frac{p}{\lambda}\right)$ and if $h(x) := \lambda f(x)$, then $h^*(p) = \lambda f^*\left(\frac{p}{\lambda}\right)$.

Proof. The first assertion is evident. The second assertion may be proved by showing that

$$\sup_{x \in X}[\langle p, x \rangle - f(Ax)] = \sup_{y \in X}[\langle A^{*-1}p, y \rangle - f(y)] = f^*(A^{*-1}p).$$

For the third assertion, we observe that

$$
\begin{aligned}
\sup_{x \in X}[\langle p, x \rangle - g(x)] &= \sup_{x \in X}[\langle p - p_0, x \rangle - f(x - x_0)] - a \\
&= \sup_{x \in X}[\langle p - p_0, y \rangle - f(y)] - a + \langle p - p_0, x_0 \rangle \\
&= f^*(p - p_0) + \langle p, x_0 \rangle - a - \langle p_0, x_0 \rangle.
\end{aligned}
$$
□

Proposition 3.2. *Suppose that X and Y are two Hilbert spaces and that f is a nontrivial convex function from $X \times Y$ to $\mathbb{R} \cup \{+\infty\}$. Set $g(y) := \inf_{x \in X} f(x, y)$. Then*

$$g^*(q) = f^*(0, q). \tag{31}$$

Proof.

$$
\begin{aligned}
g^*(q) &= \sup_{y \in Y}[\langle q, y \rangle - \inf_{x \in X} f(x, y)] \\
&= \sup_{y \in Y} \sup_{x \in X}[\langle 0, x \rangle + \langle q, y \rangle - f(x, y)] = f^*(0, q).
\end{aligned}
$$
□

Proposition 3.3. *Suppose that X and Y are two Hilbert spaces, that $B \in L(Y, X)$ is a continuous linear operator from Y to X and that $f : X \to \mathbb{R} \cup \{+\infty\}$ and $g : Y \to \mathbb{R} \cup \{+\infty\}$ are two nontrivial functions. Set $h(x) := \inf_{y \in Y}(f(x - By) + g(y))$. Then*

$$h^*(p) = f^*(p) + g^*(B^*p). \tag{32}$$

Proof.

$$
\begin{aligned}
\sup_{x \in X}[\langle p, x \rangle - \inf_{y \in Y}(f(x - By) + g(y))] &= \sup_{\substack{x \in X \\ y \in Y}}[\langle p, x \rangle - f(x - By) - g(y)] \\
&= \sup_{\substack{x \in X \\ y \in Y}}[\langle p, x + By \rangle - f(x) - g(y)] \\
&= f^*(p) + g^*(B^*p).
\end{aligned}
$$
□

When $X = Y$ and $B = \mathbf{I}$ is the identity, the function $h := f \oplus_\uparrow g$ defined by $h(x) := \inf_{y \in Y}(f(x - y) + g(y))$ is called the *inf-convolution* of the functions f and g. The above proposition states that the conjugate of the inf-convolution of two functions is the sum of the conjugates.

Next we shall calculate the function conjugate to $f^* + g^* \circ B^*$. We shall not recover the function h, since we do not know if the latter is lower semi-continuous. For this, we need a slightly more restrictive assumption, namely

$$0 \in \mathrm{Int}\,(B^* \,\mathrm{Dom}\, f^* - \mathrm{Dom}\, g^*). \tag{33}$$

In fact, this is a consequence of the following proposition:

Proposition 3.4. *Suppose that X and Y are two Hilbert spaces, that $A \in L(X,Y)$ is a continuous linear operator and that $f : X \to \mathbb{R} \cup \{+\infty\}$ and $g : Y \to \mathbb{R} \cup \{+\infty\}$ are two nontrivial, convex, lower semi-continuous functions. Suppose further that*

$$0 \in \mathrm{Int}\,(A \,\mathrm{Dom}\, f - \mathrm{Dom}\, g). \tag{34}$$

Then, for all $p \in A^* \,\mathrm{Dom}\, g^* + \mathrm{Dom}\, f^*$, there exists $\bar{q} \in Y^*$ such that

$$
\begin{aligned}
(f + g \circ A)^*(p) &= f^*(p - A^*\bar{q}) + g^*(\bar{q}) \\
&= \inf_{q \in Y^*} (f^*(p - A^*q) + g^*(q)).
\end{aligned} \tag{35}
$$

Proof. We may write

$$\sup_{x \in X}[\langle p, x \rangle - f(x) - g(Ax)] = -\inf[f(x) - \langle p, x \rangle + g(Ax)].$$

We apply Fenchel's Theorem with f replaced by $f(\cdot) - \langle p, \cdot \rangle$, the domain of which coincides with that of f and the conjugate function of which is equal to $q \to f^*(q + p)$. Thus, there exists $\bar{q} \in Y^*$ such that

$$
\begin{aligned}
\sup_{x \in X}[\langle p, x \rangle - f(x) - g(Ax)] &= f^*(p - A^*\bar{q}) + g^*(\bar{q}) \\
&= \inf_{q \in Y}[f^*(p - A^*q) + g^*(q)]. \qquad \square
\end{aligned}
$$

It is useful to state the following consequence explicitly:

Proposition 3.5. *Suppose that X and Y are two Hilbert spaces, that $A \in L(X,Y)$ is a continuous linear operator from X to Y and that $g : Y \to \mathbb{R} \cup \{+\infty\}$ is a nontrivial, convex, lower semi-continuous function. We suppose further that*

$$0 \in \mathrm{Int}\,(\mathrm{Im}\, A - \mathrm{Dom}\, g) \tag{36}$$

Then, for all $p \in A^* \,\mathrm{Dom}\, g^*$, there exists $\bar{q} \in \mathrm{Dom}\, g^*$ satisfying

$$A^*\bar{q} = p \quad and \quad (g \circ A)^*(p) = g^*(\bar{q}) = \min_{A^*q=p} g^*(q).$$

Proof. We apply the previous proposition with $f = 0$, where the domain is the whole space X. Its conjugate function f^* is defined by $f^*(p) = \{0\}$ if $p = 0$ and $f^*(p) = +\infty$ otherwise. Consequently, $f^*(p - A^*q)$ is finite (and equal to 0) if and only if $p = A^*q$.   □

The following result will be used later; in the meantime, it may be considered as an exercise.

Proposition 3.6. *Suppose X and Y are two Hilbert spaces, that $A \in L(X,Y)$ is a continuous linear operator from X to Y and that $f : X \to \mathbb{R} \cup \{+\infty\}$ and $g : Y \to \mathbb{R} \cup \{+\infty\}$ are two nontrivial, convex, lower semi-continuous functions. We suppose further that*

$$0 \in \mathrm{Int}\,(\mathrm{Dom}\,g - A\,\mathrm{Dom}\,f) \tag{37}$$

We set $e(x, y) := f(x) + g(Ax + y)$. Then, for all $(p, q) \in X^* \times Y^*$

$$e^*(p, q) = f^*(p - A^*q) + g^*(q). \tag{38}$$

Proof. We may write

$$e(x, y) = f(x) + g(Ax + y) = h(C(x, y)) \tag{39}$$

where h is a function from $X \times Y$ to $\mathbb{R} \cup \{+\infty\}$ given by $h(x, y) = f(x) + g(y)$ with domain $\mathrm{Dom}\,h = \mathrm{Dom}\,f \times \mathrm{Dom}\,g$ and where $C \in L(X \times Y, X \times Y)$ is defined by $C(x, y) = (x, Ax + y)$. Its transpose $C^* \in L(X^* \times Y^*, X^* \times Y^*)$ is defined by $C^*(p, q) = (p + A^*q, q)$. We shall apply Proposition 3.1 to calculate the function conjugate to $h \circ C$, since the operator C is clearly an isomorphism of $X \times Y$ onto itself.   □

Corollary 3.1. *The assumptions are as in Proposition 3.6, above. We set $h(y) := \inf_{x \in X}(f(x) + g(Ax + y))$. Then*

$$h^*(q) = f^*(-A^*q) + g^*(q). \tag{40}$$

Proof. We apply Propositions 3.2 and 3.6   □

Example. Conjugate functions of quadratic functions.

Proposition 3.7. *Suppose that X is a Hilbert space and that L is a continuous linear operator from X to X^* satisfying*

(i) $L = L^*$
(ii) $\langle Lx, x \rangle \geq 0 \qquad \forall x \in X$
(iii) $\mathrm{Im}\,L$ *is closed in X^*.* $\qquad\qquad\qquad\qquad\qquad\qquad\qquad\qquad$ (41)

Let f be the function from X to $\mathbb{R}_+$ defined by

$$f(x) = \frac{1}{2}\langle Lx, x \rangle. \tag{42}$$

Then its conjugate function is equal to

$$f^*(p) = \begin{cases} \frac{1}{2}\langle p, x \rangle & \text{where } x \in L^{-1}(p) & \text{when } p \in \operatorname{Im} L \\ +\infty & & \text{when } p \notin \operatorname{Im} L \end{cases} \tag{43}$$

Proof.
a) First we take $p \notin \operatorname{Im} L$. Since the image of L is closed, it follows that $\operatorname{Im} L = (\operatorname{Ker} L)^\perp$

(see Theorem 3.4, below) and thus there exists an element $x_0 \in \operatorname{Ker} L$ such that $\langle p, x_0 \rangle$ is strictly positive. Whence,

$$f^*(p) \geq \sup_{\lambda > 0}(\langle p, \lambda x_0 \rangle - f(\lambda x_0)) = \left(\sup_{\lambda > 0} \lambda\right)\langle p, x_0 \rangle = +\infty$$

since $f(\lambda x_0) = \frac{1}{2}\langle \lambda L(x_0), \lambda x_0 \rangle = 0$.

b) Now we take $p \in \operatorname{Im} L$ with $\bar{x}$ a solution of $p = L\bar{x}$. Then $l(x, y) = \langle Lx, y \rangle$ is a scalar semiproduct and the Cauchy–Schwarz inequality implies that

$$\langle L\bar{x}, y \rangle \leq \sqrt{\langle L\bar{x}, \bar{x} \rangle}\sqrt{\langle Ly, y \rangle} \leq \frac{1}{2}\langle L\bar{x}, \bar{x} \rangle + \frac{1}{2}\langle Ly, y \rangle.$$

Whence

$$f^*(\bar{p}) = \sup_y \left(\langle L\bar{x}, y \rangle - \frac{1}{2}\langle Ly, y \rangle\right) \leq \frac{1}{2}\langle L\bar{x}, \bar{x} \rangle = \frac{1}{2}\langle p, \bar{x} \rangle.$$

On the other hand,

$$\frac{1}{2}\langle p, \bar{x} \rangle = \langle p, \bar{x} \rangle - \frac{1}{2}\langle L\bar{x}, \bar{x} \rangle \leq f^*(p).$$

Thus, we have shown that $f^*(p) = \langle p, \bar{x} \rangle$ for all solutions $\bar{x}$ of the equation $L\bar{x} = p$. □

Corollary 3.2. *Let X be a Hilbert space and $L \in L(X, X^*)$ the duality operator. The conjugate function of the function f defined by $f(x) = \frac{1}{2}\|x\|^2$ is the function f^* defined by*

$$f^*(p) = \frac{1}{2}\|p\|_*^2 \quad \text{where } \|p\|_* = \sup_{x \in X} \frac{\langle p, x \rangle}{\|x\|} = \sqrt{\langle L^{-1}p, p \rangle}. \tag{44}$$

Proof. The duality operator L satisfies the properties (41), is surjective and is associated with the norms $\|x\|$ and $\|p\|_*$ by the relationships $\|x\|^2 = \langle Lx, x \rangle$ and $\|p\|_*^2 = \langle L^{-1}p, p \rangle$. □

Corollary 3.3. *Let X be a Hilbert space and $f : X \to \mathbb{R} \cup \{+\infty\}$ a nontrivial, convex, lower semi-continuous function. Then, for all $\lambda > 0$,*

$$\inf_{x \in X} \left(f(x) + \frac{1}{2\lambda} \|x\|^2 \right) + \inf_{p \in X^*} \left(f^*(p) + \frac{\lambda}{2} \|p\|_*^2 \right) = 0. \tag{45}$$

Proof. We apply Theorem 3.2 with $g(x) = \frac{1}{2\lambda} \|x\|^2$, where the conjugate function is defined by $g^*(p) = \frac{\lambda}{2} \|p\|_*^2$. □

3.5 Support Functions

We have already mentioned that it is possible to characterise a subset $K \subset X$ by its characteristic function ψ_K defined by $\psi_K(x) = 0$ if $x \in K$ and $\psi_K(x) = +\infty$ otherwise.

Its conjugate function is defined by

$$\psi_K^*(p) = \sup_{x \in K} \langle p, x \rangle. \tag{46}$$

Definition 3.2. *The conjugate function ψ_K^* of the indicator function of a subset K is called the **support function** of K and is often denoted by*

$$\sigma_K(p) := \sigma(K, p) := \sup_{x \in K} \langle p, x \rangle. \tag{47}$$

*The domain of $\sigma_K(.)$ is called the **barrier cone** K and is often denoted by $b(K) := \operatorname{Dom} \sigma_K$.*

Examples.
a) If $K = \{x_0\}$ then $\sigma_K(p) = \langle p, x_0 \rangle$.
b) If $K = B$ then $\sigma_K(p) = \|p\|_*$.
c) If K is a cone then

$$\sigma_K(p) = \psi_{K^-}(p) \text{ and } b(K) = K^- \tag{48}$$

where

$$K^- = \{p \in X^* | \forall x \in K, \ \langle p, x \rangle \le 0\} \text{ is the } negative \ polar \ cone \text{ of } K. \tag{49}$$

d) If K is a vector subspace then

$$\sigma_K(p) = \psi_{K^\perp}(p) \text{ and } b(K) = K^\perp \tag{50}$$

where $K^\perp = \{p \in X^* | \forall x \in K, \ \langle p, x \rangle = 0\}$ is the *orthogonal subspace* corresponding to K (the *orthogonal* for short).

We note that

$$\text{if } 0 \in K \text{ then } \sigma_K \geq 0 \tag{51}$$

and that

$$\text{if } K \text{ is symmetric then } \sigma_K \text{ is even.} \tag{52}$$

Proposition 3.8. *Any support function σ_K of a non-empty subset $K \subset X$ is a convex, lower semi-continuous, positively homogeneous function from X^* to $\mathbb{R} \cup \{+\infty\}$.*

Conversely, any function $\sigma : X^* \to \mathbb{R} \cup \{+\infty\}$ which is convex, lower semi-continuous and positively homogeneous is the support function of the set

$$K_\sigma := \{x \in X | \forall p \in X^*, \ \langle p, x \rangle \leq \sigma(p)\}. \tag{53}$$

Proof. The first assertion is evident. To establish the second assertion, we calculate the conjugate function of σ.

If x belongs to K_σ, then $\sigma^*(x) = 0$, since

$$\sigma^*(x) = \sup_{p \in X}(\langle p, x \rangle - \sigma(p)) \leq 0 = \langle 0, x \rangle - \sigma(0) \leq \sigma^*(x)$$

If x does not belong to K_σ, then there exists p_0 with $\langle p_0, x \rangle - \sigma(p_0) > 0$. Thus,

$$\sigma^*(x) \geq \sup_{\lambda > 0}(\langle \lambda p_0, x \rangle - \sigma(\lambda p_0)) \geq \sup_{\lambda > 0} \lambda(\langle p_0, x \rangle - \sigma(p_0)) = +\infty.$$

Thus, we have proved that σ^* is the support function of K_σ. □

Theorem 3.3. *If K is a convex closed subset of X, then*

$$K = \{x \in X | \forall p \in X^*, \ \langle p, x \rangle \leq \sigma_K(p)\}. \tag{54}$$

If K is a closed convex cone then

$$K = (K^-)^-. \tag{55}$$

If K is a closed vector subspace then

$$K = (K^\perp)^\perp. \tag{56}$$

Proof. If K is convex and closed then ψ_K is convex and lower semi-continuous and consequently $\psi_K = (\psi_K^*)^* = \sigma_K^*$. Thus, ψ_K is the indicator function of the set K_{σ_K}, which is nothing other than the right-hand side of formula (54).

Formulae (55) and (56) follow from the above together with the fact that $\sigma_K = \psi_{K^-}$ if K is a cone and $\sigma_K = \psi_{K^\perp}$ if K is a vector subspace. □

The following result is known as the bipolars theorem.

Theorem 3.4. *Suppose that $A \in L(X, Y)$ is a continuous linear operator and that K is a subset of X. Then*

$$A(K)^- = A^{*-1}(K^-) \tag{57}$$

and if $A(K)$ is a closed convex cone, then $A(K) = (A^{*-1}(K^-))^-$.
  In particular,

$$\operatorname{Ker} A^* = (\operatorname{Im} A)^\perp \tag{58}$$

and if $\operatorname{Im} A$ is closed, $\operatorname{Im} A = (\operatorname{Ker} A^*)^\perp$. **Proof.** In fact, p belongs to $A(K)^-$ if and only if

$$\forall x \in K, \qquad \langle p, Ax \rangle = \langle A^* p, x \rangle \le 0,$$

that is to say, if and only of $A^* p$ belongs to K^-. The second assertion follows from Theorem 3.3. The equality (58) is the particular case in which $K = X$, $K^- = \{0\}$. □

Since the restriction of a function f to a subset K is the sum of f and the indicator function of K, we obtain the following formula.

Proposition 3.9. *Let f be a nontrivial, convex, lower semi-continuous function from X to $\mathbb{R} \cup \{+\infty\}$ and let K be a closed, convex subset of X. If $0 \in \operatorname{Int}(\operatorname{Dom} f - K)$ and $p \in \operatorname{Dom} f^* + b(K)$, then there exists $\bar{q} \in b(K)$ such that*

$$(f|_K)^*(p) = f^*(p - \bar{q}) + \sigma_K(\bar{q}). \tag{59}$$

Since the barrier cone is the domain of the support function of K, which is convex and positively homogeneous, it is a convex cone, *which is not necessarily closed.*

It is clear that K is simply bounded if and only if $b(K) = X$, since to say that K is simply bounded is equivalent to the statement that

$$\forall p \in X^*, \quad \sigma_K(p) = \sup_{x \in K} \langle p, x \rangle < +\infty. \tag{60}$$

The 'uniform-boundedness' theorem says that, in fact, the simply bounded sets are the bounded sets (as simple as that!).

It follows that barrier cones in some way measure the 'lack of boundedness' of sets. The smaller the barrier cone of a set, the more 'unbounded' this set, if we dare to use this ill-sounding neologism.

Proposition 3.10. *Let K be a closed convex subset. Then, for all $x_0 \in K$,*

$$b(K)^- = \bigcap_{\lambda > 0} \lambda(K - x_0). \tag{61}$$

Definition 3.3. *The negative polar cone of the barrier cone of K is called the* **asymptotic cone** *of K.*

Proof of Proposition 3.10. Provisionally, we set $L := \cap_{\lambda>0}\lambda(K - x_0)$.
a) We take $x \in L$. For all $\lambda > 0$, there exists $y_\lambda \in K$ such that $x = \lambda(y_\lambda - x_0)$. Thus, $\langle p, x \rangle = \lambda(\langle p, y_\lambda \rangle - \langle p, x_0 \rangle) \le \lambda(\sigma_K(p) - \langle p, x_0 \rangle) < +\infty$ if p belongs to the barrier cone. It suffices to make λ tend to zero to see that $\langle p, x \rangle \le 0$ for all $p \in b(K)$, that is to say that L is contained in $b(K)^-$.
b) Conversely, we take x in $b(K)^-$ and $\lambda > 0$. Since $\frac{x}{\lambda}$ belongs to $b(K)^-$, we deduce that for all $p \in b(K)$,

$$\left\langle p, x_0 + \frac{x}{\lambda} \right\rangle \le \langle p, x_0 \rangle + \left\langle p, \frac{x}{\lambda} \right\rangle \le \langle p, x_0 \rangle \le \sigma_K(p).$$

Since K is convex and closed, Theorem 3.3 implies that $\frac{x}{\lambda} + x_0$ belongs to K, whence that x belongs to L.

Formulae relating to support functions and barrier cones

of The following formulae relating to support functions and barrier cones may be deduced from the properties of conjugate functions.

Remark. If f is a proper, convex, lower semi-continuous function, then $\sigma_{\text{Ep}(f)}(p, -1) = f^*(p)$.

- If $K \subset L$ then
$$b(L) \subset b(K) \text{ and } \sigma_K \le \sigma_L. \tag{62}$$

- If $K_i \subset X_i$ $(i = 1, \dots, n)$, then

$$b\left(\prod_{i=1}^n K_i\right) = \prod_{i=0}^n b(K_i) \text{ and } \sigma_K(p_1, \dots, p_n) = \sum_{i=1}^n \sigma_{K_i}(p_i). \tag{63}$$

- $b\left(\overline{co}\bigcup_{i\in I} K_i\right) \subset \bigcap_{i\in I} b(K_i)$ and $\sigma\left(\overline{co}\left(\bigcup_{i\in I}\sigma_{K_i}\right)(p)\right) = \sup_{i\in I}\sigma_{K_i}(p). \tag{64}$

- If $B \in L(X, Y)$, then
$$b(\overline{B(K)}) = B^{*-1}b(K) \text{ and } \sigma_{\overline{B(K)}}(p) = \sigma_K(B^*p). \tag{65}$$

- $b(K_1 + K_2) = b(K_1) \cap b(K_2)$ and $\sigma_{K_1+K_2}(p) = \sigma_{K_1}(p) + \sigma_{K_2}(p). \tag{66}$

- If P is a convex closed cone then

$$b(K + P) = b(K) \cap P^- \quad \text{and} \quad \sigma_{K+P}(x) = \begin{cases} \sigma_K(p) & \text{if } p \in P^- \\ +\infty & \text{otherwise} \end{cases} \qquad (67)$$

-
$$b(K + \{x_0\}) = b(K) \quad \text{and} \quad \sigma_{K+x_0}(p) = \sigma_K(p) + \langle p, x_0 \rangle. \qquad (68)$$

- If $A \in L(X, Y)$, if $L \subset X$ and $M \subset Y$ are closed convex subsets and if $0 \in \text{Int}(A(L) - M)$ then

$$b(L \cap A^{-1}(M)) = b(L) + A^* b(M)$$

and $\forall p \in b(K)$, $\exists \bar{q} \in b(M)$ such that

$$\sigma_{L \cap A^{-1}(M)}(p) = \sigma_L(p - A^* \bar{q}) + \sigma_M(\bar{q}) = \inf_{q \in Y_*}(\sigma_L(p - A^* q) + \sigma_M(q)). \qquad (69)$$

- If $A \in L(X, Y)$, if $M \subset Y$ is convex and closed and if $0 \in \text{Int}(\text{Im}(A) - M)$, then

$$b(A^{-1}(M)) = A^* b(M)$$

and $\forall p \in b(A^{-1}(M))$, $\exists \bar{q} \in b(M)$ satisfying

$$A^* \bar{q} = p \quad \text{and} \quad \sigma_{A^{-1}(M)}(p) = \sigma_M(\bar{q}) = \inf_{A^* q = p} \sigma_M(q). \qquad (70)$$

- If K_1 and K_2 are convex closed subsets of X such that $0 \in \text{Int}(K_1 - K_2)$, then $b(K_1 \cap K_2) = b(K_1) + b(K_2)$ and for all $p \in b(K_1 \cap K_2)$, there exist $\bar{p}_i \in b(K_i)$ $(i = 1, 2)$ such that $p = \bar{p}_1 + \bar{p}_2$ and

$$\sigma_{K_1 \cap K_2}(p) = \sigma_{K_1}(\bar{p}_1) + \sigma_{K_2}(\bar{p}_2) = \inf_{p = p_1 + p_2}(\sigma_{K_1}(p_1) + \sigma_{K_2}(p_2)). \qquad (71)$$

3.6 The Cramèr Transform

The Cramèr transform C associates with any nonnegative measure $d\mu$ on a finite dimensional vector space $\mathbb{R}^n$ the nonnegative extended function $C_\mu : \mathbb{R}^n \mapsto \mathbb{R}_+ \cup \{+\infty\}$ defined on $\mathbb{R}^n$ (identified with its dual) by :

$$C_\mu(p) := \sup_{x \in \mathbb{R}^n} \left(\langle p, x \rangle - \log \left(\int_{\mathbb{R}^n} e^{\langle x, y \rangle} d\mu(y) \right) \right)$$

In other words, it is the product of the *Laplace transform* $\mu \mapsto \int_{\mathbb{R}^n} e^{\langle x, y \rangle} d\mu(y)$, of the logarithm and of the *Fenchel transform* (conjugate functions) $g(\cdot) \mapsto g^\star(\cdot)$. This Cramèr transform plays an important role in statistics, and in particular, in the field of large deviations. Since C_μ is the supremum of affine functions with respect to p, this is a lower semicontinuous convex function. It satisfies

$$C_\mu(p) \geq \langle p, 0 \rangle - \log \left(\int_{\mathbb{R}^n} e^{\langle 0, y \rangle} d\mu(y) \right) = -\log \left(\int_{\mathbb{R}^n} d\mu(y) \right)$$

so that when $d\mu$ is a probability measure, its Cramèr transform C_μ is nonnegative.

We may regard nontrivial nonnegative extended functions as *membership cost functions* of "toll sets", following a suggestion of Dubois and Prades. Indeed, they provide another implementation of the idea underlying "fuzzy sets" exposed in chapter 13, since the set $[0, \infty]^E$ of nonnegative extended functions f from E to $\mathbb{R}_+ \cup \{+\infty\}$ is the closed convex hull of the set $\{0, \infty\}^E$ of indicators:

Definition 3.4 *We shall regard an extended nonnegative function* $f : X \mapsto \mathbb{R}_+ \cup \{+\infty\}$ *as a toll set. Its* domain *is the domain of* f, *i.e., the set of elements* x *such that* $f(x)$ *is finite, and the* core *of* f *is the set of elements* x *such that* $f(x) = 0$. *The* complement *of the toll set* f *is the complement of its domain and the complement of its core is called the* toll *boundary.*

We shall say that the toll set f *is* convex *(respectively* closed, *a* cone*) if the extended function* f *is convex (respectively lower semicontinuous, positively homogeneous).*

We observe that the membership function of the empty set is the constant function equal to $+\infty$.

The Cramèr transform provides a mathematical reason for which toll sets furnish a sensible mathematical representation of the concept of randomness, but different from the representation by probabilities. This is justified by the following observations.

The indicators $\psi_{\{a\}}$ of singleta a are images of *Dirac measures* δ_a: Indeed, if δ_a is the Dirac measure at the point $a \in \mathbb{R}^n$, then

$$C_{\delta_a}(p) = \sup_{x \in \mathbb{R}^n} (\langle p, x \rangle - \langle a, x \rangle) = \begin{cases} 0 & \text{if } p = a \\ +\infty & \text{if } p \neq a \end{cases} = \psi_a(p)$$

The Cramèr transform of the Gaussian with mean m and variance σ is the quadratic function $G_{\sigma, m}$ defined by

$$G_{\sigma, m}(x) := \frac{1}{2} \left\| \frac{x - m}{\sigma} \right\|^2$$

which we can regard as a Gaussian toll set with mean m and variance σ. Such toll sets play the role of Gaussians in probability theory.

The function $x \mapsto \log \left(\int_{\mathbb{R}^n} e^{\langle x, y \rangle} d\mu(y) \right)$ is

1. *convex*

    Indeed, applying Hölder inequality with exponents $\frac{1}{\alpha_i}$, we obtain

$$\int_{\mathbb{R}^n} e^{\langle \alpha_1 x_1 + \alpha_2 x_2, y \rangle} d\mu(y) = \int_{\mathbb{R}^n} \left(e^{\langle x_1, y \rangle} \right)^{\alpha_1} \left(e^{\langle x_2, y \rangle} \right)^{\alpha_2} d\mu(y)$$

$$\leq \left(\int_{\mathbb{R}^n} e^{\langle x_1, y \rangle} d\mu y \right)^{\alpha_1} \left(\int_{\mathbb{R}^n} e^{\langle x_2, y \rangle} d\mu y \right)^{\alpha_2}$$

    By taking the logarithms, we get the convexity of this function with respect to x.

2. and *lower semicontinuous*

    Since the measure $d\mu$ is nonnegative, Fatou's Lemma implies that if x_p converges to x, then

$$\int_{\mathbb{R}^n} e^{\langle x, y \rangle} d\mu(y) \leq \liminf_{p \to \infty} \int_{\mathbb{R}^n} e^{\langle x^p, y \rangle} d\mu(y)$$

    Hence the lower semicontinuity of the Laplace transform of $d\mu$ is established. Since the logarithm is increasing and continuous, it is continuous and nondecreasing.

Therefore

$$C_\mu^\star(x) = \log \left(\int_{\mathbb{R}^n} e^{\langle x, y \rangle} d\mu(y) \right)$$

It is actually differentiable and its gradient is equal to

$$\nabla C_\mu^\star(x) = \frac{\int_{\mathbb{R}^n} y e^{\langle x, y \rangle} d\mu(y)}{\int_{\mathbb{R}^n} e^{\langle x, y \rangle} d\mu(y)}$$

When $d\mu$ is the probability law of a random variable, then its mean is equal to $\nabla C_\mu^\star(0)$, which is centered if and only if its Cramèr transform vanishes at 0.

Inf-convolution plays the role of the usual convolution product of two integrable functions f and g defined by

$$(f \star g)(x) := \int_{\mathbb{R}^n} f(x - y) g(y) dy$$

We thus deduce that the Laplace transform of a convolution product is the product of the Laplace transforms because

$$\int_{\mathbb{R}^n} e^{\langle x, y \rangle} \int_{\mathbb{R}^n} f(y - z) g(z) dy dz = \int_{\mathbb{R}^n} \int_{\mathbb{R}^n} e^{\langle x, z \rangle} g(z) e^{\langle x, y - z \rangle} g(y - z) dy dz$$

$$= \int_{\mathbb{R}^n} e^{\langle x, z \rangle} g(z) dz \int_{\mathbb{R}^n} e^{\langle x, u \rangle} g(u) du$$

Therefore, taking the logarithm, we obtain

$$\begin{cases} \log \left(\int_{\mathbb{R}^n} e^{\langle x, y \rangle} (f \star g)(y) dy \right) \\ = \log \left(\int_{\mathbb{R}^n} e^{\langle x, y \rangle} f(y) dy \right) + \log \left(\int_{\mathbb{R}^n} e^{\langle x, y \rangle} g(y) dy \right) \end{cases}$$

The Fenchel conjugate of this sum being the inf-convolution of the Fenchel conjugates, we infer that the Cramèr transform of a convolution product is the inf-convolution of the Cramèr transforms:

$$C_{f \star g} \;=\; C_f \oplus_\uparrow C_g$$

The Proximation Theorem implies that inf-convolution by a quadratic function maps a lower semicontinuous convex function to a continuously differentiable convex function, in the same way that the convolution product by a Gaussian maps a function to an indefinitely differentiable function:

$$f_\sigma(x) := \inf_{y \in X}\left[f(y) + \frac{1}{2}\left\| \frac{x-y}{\sigma} \right\|^2 \right]$$

The Cramèr transform thus maps the convolution by a Gaussian into inf-convolution by quadratic functions.

The quadratic functions

$$G_{\sigma,\,m}(x) \;:=\; \frac{1}{2}\left\| \frac{x-m}{\sigma} \right\|^2$$

are regarded as *Gaussian toll sets* with mean m and variance σ. They form a class stable by inf-convolution:

Proposition 3.11 *The Gaussian toll sets are stable under inf-convolution:*

$$\left(G_{\sigma_1,\,m_1} \oplus G_{\sigma_2,\,m_2} \right)(x) \;=\; G_{\sqrt{\sigma_1^2+\sigma_2^2},\, m_1+m_2}$$

Proof — One must compute the solution to the minimization problem

$$\inf_y \left(\frac{1}{2}\left\| \frac{x-y-m_1}{\sigma_1} \right\|^2 + \frac{1}{2}\left\| \frac{y-m_2}{\sigma_2} \right\|^2 \right)$$

From Fermat's Rule, this problem achieves its minimum at

$$\bar{y} := \frac{\sigma_2^2(x-m_1) + \sigma_1^2 m_2}{\sigma_1^2 + \sigma_2^2}$$

Consequently,

$$\left\{ \begin{aligned} \left(G_{\sigma_1,\,m_1} \oplus G_{\sigma_2,\,m_2} \right)(x) &= \frac{1}{2}\left\| \frac{x-\bar{y}-m_1}{\sigma_1} \right\|^2 + \frac{1}{2}\left\| \frac{\bar{y}-m_2}{\sigma_2} \right\|^2 \\ &= \frac{1}{2}\left\| \frac{x-(m_1+m_2)}{\sqrt{\sigma_1^2+\sigma_2^2}} \right\|^2 = G_{\sqrt{\sigma_1^2+\sigma_2^2},\, m_1+m_2} \quad \Box \end{aligned} \right.$$

Remark. The Cramèr transform justifies a striking formal analogy between optimization and probability theory. We shall only sketch it without entering details which may lead us too far. When f is a nonnegative extended function from X to $\mathbb{R} \cup \{+\infty\}$, we can regard the "set-defined map" $K \mapsto M_f(K) := \inf_{y \in f}(y)$ as a *Maslov measure* whose "*density*" is f on the family $\mathcal{F}(X)$ of closed subsets, i.e., a "set-defined map" satisfying

$$\begin{cases} i) & M_f(X) = \inf_{y \in X} f(y) \\ ii) & M_f(\emptyset) = +\infty \\ iii) & M_f(K \cup L) = \min(M_f(K), M_f(L)) \end{cases}$$

Maslov measures are analogous to usual nonnegative measures, which are set-defined maps from the σ-algebra $\mathcal{A}$ on a measured space Ω to $\mathbb{R}_+$. *Maslov probabilities* are those satisfying

$$M_f(X) = \inf_{y \in X} f(y) = 0$$

To the integral

$$x(\cdot) \mapsto \int_\Omega x(\omega) d\mu(\omega)$$

of a nonnegative measurable function defined on a measured space $(\Omega, \mathcal{A}, d\mu)$ corresponds the *infimum of a lower semicontinuous function $g : E \mapsto \mathbb{R} \cup \{+\infty\}$ on a metric space E* defined by

$$g(\cdot) \mapsto \inf_{x \in E}(g(x) + f(x))$$

To the Dirac measure $\delta_a : x(\cdot) \mapsto x(a)$ corresponds the indicator ψ_a because

$$g(\cdot) \mapsto \inf_{x \in g}(g(x) + \psi_a(x)) = g(a)$$

To the integral $\int_A d\mu(\omega)$ of the characteristic function of a measurable set $A \in \mathcal{A}$ providing the measure $d\mu$ of a subset A corresponds the minimization problem of a function $g(\cdot)$ on the closed subset A

$$g(\cdot) \mapsto \inf_{x \in E}(\psi_A(x) + g(x)) = \inf_{x \in A} g(x)$$

Consequently, to the measure $d\mu$, which is a function from the σ-algebra $\mathcal{A}$ to the half-line $\mathbb{R}_+$ supplied with the operations $+$ and $\times$, corresponds the *Maslov measure* M_f, function from the family of compact subsets of E to $[0, +\infty]$.

The analogy then becomes algebraic, because $(\mathbb{R}_+, +, \times)$ supplied with the usual addition and multiplication and neutral elements 0 and 1 on one hand, and $(\mathbb{R}_+, \inf, +)$ supplied with the infimum and the usual addition and the neutral elements $+\infty$ and 0 on the other hand, are two instances of "dioids, which are kind of rings supplied with two operations which do not have inverses.

4. Subdifferentials of Convex Functions

4.1 Introduction

The crucial discovery of the concept of differential calculus is due to Pierre de Fermat (1601–1655), who was one of the most important innovators in the history of mathematics. It is to him that we owe a rule for determining extrema, described, without proof, in a short treatise *Methodus ad disquirendam Maximam et Minimam* written in 1637. The importance of his discoveries in number theory has eclipsed the contributions which this exceptional and modest man made to other areas of mathematics. Fermat also was the first to discover the "principle of least time" in optics, the prototype of the variational principles governing so many physical and mechanical laws. He shared independently with Descartes the invention of analytic geometry and with Pascal the creation of the mathematical theory of probability. His achievements in number theory overshadowed his other contributions, as the Last Fermat Theorem which remained a challenge for such a long time, and still is a challenge if indeed the simple proof of Piere de Fermat did exist. Not to mention his compositions in French, Latin, Italian and Spanish verse and his Grecian erudition. It is also notable that he was able to find time for these occupations in the midst of his duties as counsellor to the parliament of Toulouse (even taking into account Fermat's genius, this makes us reflect on the leisure activities offered by a lawyer's career).

But Fermat never knew the concept of the derivative which was only formulated later by Newton (in 1671) and by Leibniz in his publication on differential calculus entitled *Nova methodus pro maximis et minimis* in 1684. However, Newton himself recognized explicitly that he got the hint of the differential calculus from Fermat's method of building tangents devised half a century earlier.

Fermat was also the one who discovered that the derivative of a (polynomial) function vanishes when it reaches an extremum. (This is Fermat's Rule, which remains the main strategy for obtaining necessary conditions of optimality, from mathematical programming to calculus of variations to optimal control).

The analogy between Fermat's method (restricted to algebraic functions) and that of Leibniz is remarkable, since, as you know, this rule involves searching for the extrema of a function f among the solutions of the equation $f'(x) = 0$, a problem much more familiar to mathematicians.

This rule has been applied, justified, improved, adapted and generalised in the course of three centuries of work on optimisation theory, the theory of

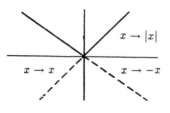

Fig. 4.1.

the calculus of variations and (now) optimal control theory. Three centuries of intensive work by numerous mathematicians, punctuated by important stages and bearing the seal of the works of Euler (XVIIIth century), Lagrange, Jacobi (XIXth century), Poincaré and Hilbert (at the dawn of this century) are continued even today, since the results which we shall describe are recent (which does not necessarily mean complicated, since scientific progress also involves simplification).

The concept of functions of several differentiable variables has been known since Jacobi and that of differentiable functions on normed spaces since Fréchet and Gâteaux. The rule due to Fermat and Leibniz remains valid. If the function f attains its minimum with respect to x, the gradient of f is zero at that point. There are many reasons why we should not stop there.

Firstly, we may seek to minimise so-called nondifferentiable functions. Optimisation theory, game theory, etc., involve such functions since the operations of supremum and infimum destroy the usual differentiability properties; for example, we mention the function $x \to |x|$, which is not differentiable at the point $x = 0$, but which is obtained as the upper envelope of the differentiable functions $x \to \alpha x$ when α ranges over $[-1, +1]$.

We may wonder (like others before us) why there should be so much fuss in the case of nondifferentiability at a single point. All the more so since we shall see that any convex lower semi-continuous function may be approximated by differentiable functions; for example, the function $x \to |x|$ may be approximated by the functions f_λ defined by:

$$f_\lambda(y) = \begin{cases} -y - \frac{\lambda}{2} & \text{if } y \leq -\lambda \\ \frac{y^2}{2\lambda} & \text{if } |y| \leq \lambda \\ y - \frac{\lambda}{2} & \text{if } y \geq \lambda \end{cases}$$

However, if we are interested in the minimum of $x \to |x|$ which is attained at 0, we note that it is at this point that the function is not differentiable and thus that Fermat's rule cannot be applied. What can we do? In fact, we may retain Fermat's rule, modifying the concept of gradient and generalising it appropriately. Examination of the function $x \to |x|$ may put us on the right track. Since $x \to |x|$ is the upper envelope of the functions $x \to \alpha x$ the derivatives of which at 0 are α, when α ranges over $[-1, +1]$ why not consider the set $[+1, -1]$

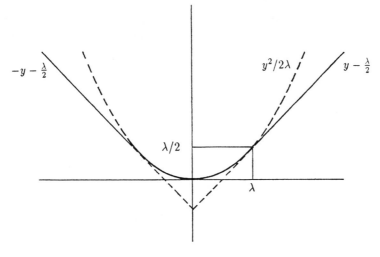

Fig. 4.2.

of these derivatives as a candidate? Clearly, we must overcome our hesitation at the multi-valued nature of this solution, which simply results from a lack of familiarity (and our conservatism). But, to convince ourselves of the importance of this stroke of daring, we need only note that Fermat's rule remains true for this example, since

$$0 \text{ belongs to } [-1, +1].$$

Since we have seen that any convex lower semi-continuous function is the upper envelope of the continuous affine functions $x \rightarrow \langle p, x \rangle - f^*(x)$ which minorise it, we thus consider the set of the gradients p of those affine functions which pass through the point $(x_0, f(x_0))$, in other words, the set of p such that:

$$\langle p, x_0 \rangle - f^*(p) = f(x_0).$$

In the context of this theory, we shall choose this set (convex, closed, possibly empty), called the *subdifferential* $\partial f(x_0)$ of f at x_0 as a candidate for a generalisation of the concept of gradient. In the context of other theories (for example, partial differential equations), other strategies such as distribution theory will be more appropriate.

If there is only one affine function (the tangent) then $\partial f(x_0)$ reduces to the usual gradient of f at x_0: $\partial f(x_0) = \{\nabla f(x_0)\}$.

We shall show that Fermat's rule remains true: $\bar{x}$ minimises a nontrivial, convex, lower semi-continuous function f if and only if $0 \in \partial f(\bar{x})$.

In order to exploit this result, we need to develop a subdifferential calculus, analogous to the usual differential calculus. We shall establish conditions under which formulae such as

$$\partial(f+g)(x_0) = \partial f(x_0) + \partial g(x_0)$$
$$\partial(f \circ A)(x_0) = A^* \partial f(Ax_0)$$
$$\partial\left(\sup_{i=1,\dots,n} f_i\right)(x_0) = \overline{\mathrm{co}}\left(\bigcup_{i \in I(x_0)} \partial f_i(x_0)\right),$$

where $I(x_0) := \{i = 1, \dots, n | f_i(x_0) = \sup_{i=1,\dots,n} f_i(x_0)\}$ and '$\overline{\mathrm{co}}$' denotes the closed convex hull, are true.

One important class of convex nondifferentiable functions consists of the restrictions $f_K = f + \psi_K$ of convex functions to closed convex subsets K. When the interior of K is empty, we cannot talk of either the derivative or the gradient in the usual sense.

However, we can apply the formula

$$\partial f_K(x) = \partial f(x) + \partial \psi_K(x)$$

which, when f is differentiable, gives

$$\partial f_K(x) = \nabla f(x) + \partial \psi_K(x).$$

A simple calculation shows that the subdifferential $\partial \psi_K$ of the indicator function of K is the closed convex cone

$$\partial \psi_K(x) = \{p \in X^* | \forall y \in K, \ \langle p, y - x \rangle \leq 0\}.$$

The elements $p \in X^*$ of this set play the role of normals to K at x. This is why $\partial \psi_K(x)$ is called the *normal cone* to K at x and is denoted by $N_K(x)$.

Since we have already replaced the notion of orthogonality for vector subspaces by the notion of polarity for cones, it is natural to consider the negative polar cone $T_K(x) := N_K(x)^-$ of the normal cone to K at x as the *tangent cone* to K at x. This will be all the more justified when it is shown that

$$T_K(x) = \text{closure}\left(\bigcup_{h>0} \frac{1}{h}(K - x)\right).$$

In fact, this formula shows that a vector is tangent to K at x if it is the limit of vectors $v \in X$ such that $x + tv$ belongs to K for all $t \in [0, h_0]$. Such vectors are the derivatives (right) of the curves $t \to x + tv$ passing through x and lying in K.

4.2 Definitions

We shall begin by exhibiting an important property of convex functions from X to $\mathbb{R} \cup \{+\infty\}$.

Proposition 4.1. *Let f be a **nontrivial convex** function from X to $\mathbb{R} \cup \{+\infty\}$. Suppose $x_0 \in \mathrm{Dom} f$ and $v \in X$. Then the limit*

$$Df(x_0)(v) = \lim_{h \to 0_+} \frac{f(x_0 + hv) - f(x_0)}{h} \tag{1}$$

exists in $\overline{\mathbb{R}}$ $(:= \{-\infty\} \cup \mathbb{R} \cup \{+\infty\})$ and satisfies

$$f(x_0) - f(x_0 - v) \le Df(x_0)(v) \le f(x_0 + v) - f(x_0). \tag{2}$$

Moreover,

$$v \to Df(x_0)(v) \text{ is convex and positively homogeneous.} \tag{3}$$

Proof.

a) The function $h \to \frac{f(x_0 + hv) - f(x_0)}{h}$ is increasing. In fact, if $h_1 \le h_2$, then

$$f(x_0 + h_1 v) - f(x_0) = f\left(\frac{h_1}{h_2}(x_0 + h_2 v) + \left(1 - \frac{h_1}{h_2}\right)x_0\right) - f(x_0).$$

Since f is convex and h_1/h_2 is less than one, it follows that

$$f(x_0 + h_1 v) - f(x_0) \le \frac{h_1}{h_2} f(x_0 + h_2 v) + \left(1 - \frac{h_1}{h_2}\right) f(x_0) - f(x_0);$$

whence, that

$$\frac{f(x_0 + h_1 v) - f(x_0)}{h_1} \le \frac{f(x_0 + h_2 v) - f(x_0)}{h_2}.$$

Thus, these differential quotients have a limit in $\overline{\mathbb{R}}$ as $h \to 0_+$:

$$Df(x_0)(v) = \inf_{h > 0} \frac{f(x_0 + hv) - f(x_0)}{h}. \tag{4}$$

b) Taking $h = 1$, equation (4) implies that

$$Df(x_0)(v) \le f(x_0 + v) - f(x_0).$$

Writing $x_0 = \frac{1}{1+h}(x_0 + hv) + \frac{h}{1+h}(x_0 - v)$ and using the convexity of f, we obtain

$$f(x_0) \le \frac{1}{1+h} f(x_0 + hv) + \frac{h}{1+h} f(x_0 - v).$$

This inequality implies that for all $h > 0$

$$f(x_0) - f(x_0 - v) \leq \frac{f(x_0 + hv) - f(x_0)}{h}$$

and consequently, by virtue of (4), that $f(x_0) - f(x_0 - v)$ is less than or equal to $Df(x_0)(v)$.

c) Clearly, $v \to Df(x_0)(v)$ is positively homogeneous. We show that it is convex:

$$
\begin{aligned}
f(x_0 &+ h(\lambda v_1 + (1 - \lambda)v_2)) - f(x_0) \\
&= f(\lambda(x_0 + hv_1) + (1 - \lambda)(x_0 + hv_2)) - \lambda f(x_0) - (1 - \lambda)f(x_0) \\
&\leq \lambda(f(x_0 + hv_1) - f(x_0)) + (1 - \lambda)(f(x_0 + hv_2) - f(x_0)).
\end{aligned}
$$

Dividing by $h > 0$, and letting h tend to 0_+, we deduce that

$$Df(x_0)(\lambda v_1 + (1 - \lambda)v_2) \leq \lambda Df(x_0)(v_1) + (1 - \lambda)Df(x_0)(v_2). \qquad \square$$

In general, $v \to Df(x_0)(v)$ is not lower semi-continuous.

Definition 4.1. *We shall say that $Df(x_0)(v)$ is the **right derivative** of f at x_0 in the direction v and that $v \to Df(x_0)(v)$ is the right derivative of f at x_0.*

*If $v \to Df(x_0)(v)$ is a continuous linear function, we say that f is **Gâteaux differentiable** at x_0, and the continuous linear form $\nabla f(x_0)$ defined by*

$$\forall v \in X, \qquad \langle \nabla f(x_0), v \rangle = Df(x_0)(v) \tag{5}$$

*is called the **gradient of f at x_0**.*

Whilst the right derivative is not necessarily linear and continuous, it is always convex and positively homogeneous. If it is nontrivial and lower semi-continuous, Proposition 3.8 tells us that the right derivative is the support function of the convex closed set

$$\{p \in X^* | \forall v \in X, \ \langle p, v \rangle \leq Df(x_0)(v)\}.$$

Nothing prevents us from considering this set in the general case.

Definition 4.2. *Let $f : X \to \mathbb{R} \cup \{+\infty\}$ be a nontrivial convex function. We call the subset $\partial f(x_0)$ defined by*

$$\partial f(x_0) := \{p \in X^* | \forall v \in X, \ \langle p, v \rangle \leq Df(x_0)(v)\} \tag{6}$$

*the **subdifferential** of f at x_0. The elements p of $\partial f(x_0)$ are often called **subgradients**.*

The subdifferential $\partial f(x_0)$ is *always a convex closed set* and may be empty (this is the case if $Df(x_0)(v) = -\infty$ for at least one direction v).

The concept of subdifferential generalises the notion of gradient in the sense that, when f is Gâteaux differentiable at x_0, the subdifferential reduces to the set consisting simply of the gradient $\nabla f(x_0)$ of f at x_0:

$$\partial f(x_0) = \{\nabla f(x_0)\} \quad \text{when} \quad \nabla f(x_0) \text{ exists.} \tag{7}$$

If the right derivative $Df(x_0)(\cdot)$ is nontrivial and lower semi-continuous, Proposition 3.8 implies that

$$Df(x_0)(v) = \sigma(\partial f(x_0), v). \tag{8}$$

We shall characterise the subgradient of f at x_0.

Proposition 4.2. *Let f be a nontrivial convex function from X to $\mathbb{R} \cup \{+\infty\}$. Suppose that $\partial f(x_0) \neq \emptyset$.*

The following assertions are equivalent

a) $\qquad\qquad\qquad p \in \partial f(x)$
b) $\qquad\qquad\qquad \langle p, x \rangle = f(x) + f^*(p)$
c) $\qquad f(x) - \langle p, x \rangle \leq \inf_{x \in X} (f(y) - \langle p, y \rangle) \qquad \forall y \in X. \tag{9}$

Proof. The inequality (2), where $v = y - x$ proves that a) implies c), whilst c) and b) are clearly equivalent. We show that c) implies a). Firstly, taking $y = x + hv$, c) implies that $\langle p, v \rangle \leq \frac{f(x+hv)-f(x)}{h}$ and consequently, that $\langle p, v \rangle \leq Df(x)(v)$ for all $v \in X$. Thus, p belongs to $\partial f(x)$. $\qquad\qquad\square$

Remark. Property b), which characterises the subdifferential using the conjugate function will be very useful, since it is very simple to use.

Moreover, it has the following consequence

Corollary 4.1. *Suppose $f : X \to \mathbb{R} \cup \{+\infty\}$ is a nontrivial, convex, lower semi-continuous function. Then*

$$p \in \partial f(x) \Leftrightarrow x \in \partial f^*(p). \tag{10}$$

This may be expressed in another way, by defining the inverse of the set-valued map $x \to \partial f(x)$ to be the set-valued map $p \to (\partial f)^{-1}(p)$ given by

$$x \in \partial f^{-1}(p) \Leftrightarrow p \in \partial f(x). \tag{11}$$

Then Corollary 4.1 states that: *the inverse of the subdifferential $x \to \partial f(x)$ is the subdifferential $p \to \partial f^*(p)$ of the conjugate function of f.* Whence, by abuse of terminology, it is again convenient to call the set-valued map $x \to \partial f(x)$ the subdifferential.

It was Fenchel who recognised the analogue of the *Legendre transformation* which associates a function f with a function g such that

$$p = \nabla f(x) \Leftrightarrow x = \nabla g(p).$$

It is easy to see the advantageous consequences of such a property.

Since, at that time, no one dared to talk of set-valued maps, it was assumed that the *mapping* $x \to \nabla f(x)$ *was a homeomorphism from an open subset* Ω_1 *of* $\mathbb{R}^n$ *onto an open subset* Ω_2 *of* $\mathbb{R}^n$ (that is, a bijective and bicontinuous function).

Then the solution is given by the function g defined on Ω_2 by

$$g(p) = \langle p, (\nabla f)^{-1}(p) \rangle - f((\nabla f)^{-1}(p)).$$

Setting $x = (\nabla f)^{-1}(p)$, we obtain the identity

$$\langle p, x \rangle = g(p) + f(x)$$

analogous to the property (9) b).

The function g is called the *Legendre transformation* of f. When f is also convex, g *coincides with the conjugate* f^*.

Next we suppose that $p \in \Omega_2$ and $x \in \partial f^*(p)$. Following (9) a), x maximises the function $y \to \langle p, y \rangle - f(y)$. Since f is differentiable at x, we deduce that $0 = \nabla(\langle p, \cdot \rangle - f(\cdot))(x) = p - \nabla f(x)$. Thus, $p = \nabla f(x)$ and

$$f^*(p) = \langle p, x \rangle - f(x) = \langle p, (\nabla f)^{-1}(p) \rangle - f((\nabla f)^{-1}p) = g(p).$$

In summary, in the context of convex analysis, the conjugate function of a convex function plays the same role as that played by the Legendre transformation in classical (regular) analysis.

4.3 Subdifferentiability of Convex Continuous Functions

Theorem 4.1. *Suppose that a convex function f is continuous on the interior of its domain. Then f is right differentiable on* Int Dom f *and satisfies*

$$Df(x)(u) = \limsup_{\substack{y \to x \\ h \to 0_+}} \frac{f(y + hu) - f(y)}{h}. \tag{12}$$

Moreover,

(i) $(x, u) \in$ Int Dom $f \times X \to Df(x)(u)$ *is upper semi-continuous*
(ii) $\exists c > 0$ *such that* $\forall u \in X,\ |Df(x)(u)| \leq c\|u\|$. $\tag{13}$

Proof. Since f is bounded above on a neighbourhood of x, there exists $\alpha > 0$ such that $x - \alpha u$ and $x + \alpha u$ belong to the domain of f. The inequalities (2) of Proposition 4.1 imply that $Df(x)(u)$ is finite. Thus f is right differentiable. Since f is Lipschitz on a neighbourhood of x by virtue of Theorem 2.1, there exists a constant $c > 0$ such that

$$Df(x)(u) \leq \frac{f(x + hu) - f(x)}{h} \leq c\|u\| \tag{14}$$

which implies that $Df(x)(\cdot)$ is Lipschitz, whence lower semi-continuous. Provisionally, we set

$$D_c f(x)(u) = \limsup_{\substack{y \to x \\ h \to 0+}} \frac{f(y + hu) - f(y)}{h}.$$

The inequality $Df(x)(u) \leq D_c f(x)(u)$ is clear and we show that the inverse inequality holds. Since the function

$$(h, y) \to \frac{f(y + hu) - f(y)}{h}$$

is continuous at (λ, x), there exists $\alpha > 0$ such that

$$\frac{f(y + hu) - f(y)}{h} \leq \frac{f(x + \lambda u) - f(x)}{\lambda} + \varepsilon$$

when $|h - \lambda| \leq \alpha$ and $\|y - x\| \leq \alpha$. This implies, in particular, thanks to the fact that $h \to \dfrac{(y + hu) - f(y)}{h}$ is increasing, that

$$\sup_{\|y-x\|\leq\alpha} \sup_{0<h\leq\lambda+\alpha} \frac{f(y + hu) - f(y)}{h} \leq \frac{f(x + \lambda u) - f(x)}{\lambda} + \varepsilon.$$

Taking the infimum with respect to λ and α, we obtain

$$D_c f(x)(u) \leq Df(x)(u) + \varepsilon.$$

Thus, it is sufficient to let ε tend to 0.

Finally, the function $(x, u) \to Df(x, u)$ is upper semi-continuous as the lower envelope of the continuous functions $(x, u) \to \dfrac{f(x + hu) - f(x)}{h}$. $\qquad\square$

We now state Theorem 4.1 in subdifferential terms.

Theorem 4.2. *Suppose that a convex function f is continuous on the interior of its domain. Then,*

$$\forall x \in \text{Int Dom } f, \ \partial f(x) \text{ is non-empty and bounded.} \tag{15}$$

Moreover,

$$(x, u) \in \text{Int Dom } f \times X \to \sigma(\partial f(x), u) \text{ is upper semi-continuous.} \tag{16}$$

Proof. Since the function $u \to Df(x)(u)$ is nontrivial and lower semi-continuous, it is the support function of the subdifferential

$$\partial f(x) : Df(x)(u) = \sigma(\partial f(x), u).$$

The inequality (14) may be written as $\sigma(\partial f(x), u) \leq c\|u\| = \sigma(cB, u)$, which implies that $\partial f(x)$ is contained in the ball cB of radius $c > 0$. $\qquad\square$

Corollary 4.2. *Suppose that a convex function f is continuous on the interior of its domain. Then f is Gâteaux differentiable at $x \in \text{Int Dom } f$ if and only if $\partial f(x)$ contains only one point (which is the gradient of f).*

4.4 Subdifferentiability of Convex Lower Semi-continuous Functions

When the convex function f is only lower semi-continuous we can nevertheless show that f is subdifferentiable on a sufficiently large set, since it is *dense* on the domain of f.

Theorem 4.3. *Let $f : X \rightarrow \mathbb{R} \cup \{+\infty\}$ be a nontrivial, convex, lower semi-continuous function. Then*
a) f is subdifferentiable on a dense subset of the domain of f;
b) for all $\lambda > 0$, the set-valued map $x \rightarrow x + \lambda \partial f(x)$ is surjective and its inverse $J_\lambda := (1 + \lambda \partial f(\cdot))^{-1}$ is a Lipschitz mapping with constant equal to 1. **Proof.**
a) This is a consequence of Theorem 2.2. We begin by proving the second assertion. For all $\lambda > 0$, the unique solution $J_\lambda x$ of the minimisation problem

$$f_\lambda(x) := \inf_{y \in X} \left[f(y) + \frac{1}{2\lambda} \|y - x\|^2 \right]$$

satisfies the inequality

$$\forall y \in X, \qquad f(J_\lambda x) - f(y) \le \left\langle \frac{1}{\lambda}(x - J_\lambda x), J_\lambda x - y \right\rangle$$

which says precisely that $J_\lambda x$ is the (*unique*) solution of the inclusion

$$x \in J_\lambda x + \lambda \partial f(J_\lambda x) = (1 + \lambda \partial f(\cdot))(J_\lambda x). \tag{17}$$

Thus, J_λ is the inverse of the set-valued map $1 + \lambda \partial f(\cdot)$ and Proposition 2.7 implies that J_λ is Lipschitz with constant 1. In particular, f is subdifferentiable at $J_\lambda x$.

b) For x belonging to the domain of f, we shall show that $J_\lambda x$ converges to x, which proves the first part of the theorem. We take p in the domain of f^* (which is non-empty, by virtue of Theorem 3.1). Since

$$\frac{1}{2\lambda} \|J_\lambda x - x\|^2 + f(J_\lambda x) = f_\lambda(x) \le f(x)$$

and since

$$-f(J_\lambda x) \le f^*(p) - \langle p, J_\lambda x \rangle$$

we deduce that

$$\begin{aligned}
\frac{1}{2\lambda} \|J_\lambda x - x\|^2 &\le f(x) + f^*(p) - \langle p, x \rangle + \langle p, x - J_\lambda x \rangle \\
&\le \frac{1}{4\lambda} \|J_\lambda x - x\|^2 + f(x) + f^*(p) - \langle p, x \rangle + \lambda \|p\|^2
\end{aligned}$$

(since $ab \le a^2/4\lambda + b^2\lambda$). Thus, since λ converges to 0,

$$\|J_\lambda x - x\|^2 \le 4\lambda(f(x) + f^*(p) - \langle p, x \rangle + \lambda\|p\|^2) \to 0. \qquad \square$$

Remark. The single-valued nonlinear operators $J_\lambda := (1 + \lambda\partial f(\cdot))^{-1}$ are often called the *Moreau-Yosida* approximation of the set-valued map $\partial f : x \to \partial f(x)$ for the following reason. When f is convex, *the subdifferential map ∂f is monotone*, i.e., that its graph is monotone in the sense that for all pairs (x, p) and (y, q) of Graph(∂f),

$$\langle p - q, x - y \rangle \ge 0.$$

Indeed, it is sufficient to add the inequalities

$$f(x) - f(y) \le \langle p, x - y \rangle$$

and

$$f(y) - f(x) \le \langle q, y - x \rangle$$

When f is convex and lower semicontinuous, one can prove that *the subdifferential is maximal monotone* in the sense that its graph is maximal among the monotone graphs. For maximal monotone set-valued maps $A : E \mapsto E$, one can prove that $J_\lambda := (1 + \lambda A)^{-1}$ i the *Yosida approximation of A.*

4.5 Subdifferential Calculus

Theorem 4.4. *We consider two Hilbert spaces X and Y, a continuous linear operator $A \in L(X, Y)$ and two nontrivial, convex, lower semi-continuous functions $f : X \to \mathbb{R} \cup \{+\infty\}$ and $g : Y \to \mathbb{R} \cup \{+\infty\}$.*
   We assume further that

$$0 \in \text{Int}(A \, \text{Dom} \, f - \text{Dom} \, g). \tag{18}$$

Then,

$$\partial(f + g \circ A)(x) = \partial f(x) + A^*\partial g(Ax). \tag{19}$$

Proof. It is easy to check that $\partial f(x) + A^*\partial g(Ax)$ is contained in $\partial(f + g\circ A)(x)$. The inverse inclusion follows from Proposition 3.4. We take $p \in \partial(f + g\circ A)(x)$. There exists $\bar{q} \in Y^*$ such that $(f + g \circ A)^*(p) = f^*(p - A^*\bar{q}) + g^*(\bar{q})$. Thus, from equation (9) b),

$$\langle p, x \rangle = f(x) + g(Ax) + (f + g \circ A)^*(p)$$
$$= (f(x) + f^*(p - A^*\bar{q})) + (g(Ax) + g^*(\bar{q})).$$

Consequently,

$$0 = ((\langle p - A^*\bar{q}, x \rangle - f(x) - f^*(p - A^*\bar{q})) + (\langle \bar{q}, Ax \rangle - g(Ax) - g^*(\bar{q})).$$

Since each of these two expressions is negative or zero, it follows that they are both zero, whence that $\bar{q} \in \partial g(Ax)$ and $p - A^*\bar{q} \in \partial f(x)$. Thus, we have shown that $p = p - A^*\bar{q} + A^*\bar{q} \in \partial f(x) + A^*\partial g(Ax)$. $\square$.

Corollary 4.3. *If f and g are two nontrivial, convex, lower semi-continuous functions from X to $\mathbb{R} \cup \{+\infty\}$ and if*

$$0 \in \text{Int} (\text{Dom } f - \text{Dom } g) \tag{20}$$

then

$$\partial(f + g)(x) = \partial f(x) + \partial g(x). \tag{21}$$

If g is a nontrivial, convex, lower semi-continuous function from Y to $\mathbb{R} \cup \{+\infty\}$ and if $A \in L(X, Y)$ satisfies

$$0 \in \text{Int} (\text{Im } A - \text{Dom } g) \tag{22}$$

then

$$\partial(g \circ A)(x) = A^*\partial g(Ax). \tag{23}$$

Proposition 4.3. *Let g be a nontrivial convex function from $X \times Y$ to $\mathbb{R} \cup \{+\infty\}$. Consider the function $h : Y \to \mathbb{R} \cup \{+\infty\}$ defined by*

$$h(y) := \inf_{x \in X} g(x, y). \tag{24}$$

If $\bar{x} \in X$ satisfies $h(y) = g(\bar{x}, y)$, then the following conditions are equivalent:

(a) $q \in \partial h(y)$
(b) $(0, q) \in \partial g(\bar{x}, y).$ (25)

Proof. Since $h^*(q) = g^*(0, q)$, following Proposition 3.2, we deduce that q belongs to $\partial h(y)$ if and only if $\langle q, y \rangle = h(y) + h^*(q) = g(\bar{x}, y) + g^*(0, q)$, that is, if and only if $(0, q) \in \partial g(\bar{x}, y)$. $\square$

Proposition 4.4. *We consider a family of **convex** functions $x \to f(x, p)$ indexed by a parameter p running over a set P. We assume that*
(i) P is compact
(ii) There exists a neighbourhood U of x such that, for all y in U,
  $p \to f(y, p)$ is upper semi-continuous.
(iii) $\forall p \in P, y \to f(y, p)$ is continuous at x. (26)

Consider the upper envelope k of the functions $f(\cdot, p)$, defined by $k(y) = \sup_{p \in P} f(y, p)$. Set

$$P(x) := \{p \in P | k(x) = f(x, p)\}. \tag{27}$$

Then

$$Dk(x)(v) = \sup_{p \in P(x)} Df(x, p)(v) \tag{28}$$

and

$$\partial k(x) = \overline{\mathrm{co}} \left(\bigcup_{p \in P(x)} \partial f(x, p) \right). \tag{29}$$

Proof. Since when p belongs to $P(x)$ we may write

$$\frac{f(x + hv, p) - f(x, p)}{h} \le \frac{k(x + hv) - k(x)}{h},$$

letting h tend to 0 we obtain

$$\sup_{p \in P(x)} Df(x, p)(v) \le Dk(x)(v). \tag{30}$$

We must establish the inverse inequality. Fix $\varepsilon > 0$; we shall show that there exists $p \in P(x)$ such that $Dk(x)(v) - \varepsilon \le Df(x, p)(v)$. Since the function k is convex we know that

$$Dk(x)(v) = \inf_{h>0} \frac{k(x + hv) - k(x)}{h}.$$

Then, for all $h > 0$, the set

$$B_h := \left\{ p \in P \left| \frac{f(x + hv, p) - k(x)}{h} \ge Dk(x)(v) - \varepsilon \right. \right\} \tag{31}$$

is non-empty. Consider the neighbourhood U mentioned in assumption (26)(ii). There exists $h_0 > 0$ such that $x + hv$ belongs to U for all $h \le h_0$. Since $p \to f(x + hv, p)$ is upper semi-continuous, the set B_h is closed. On the other hand, if $h_1 \le h_2$, then $B_{h_1} \subset B_{h_2}$; if p belongs to B_{h_1}, the convexity of f with respect to x implies that

$$Dk(x)(v) - \varepsilon \le \frac{1}{h_1} \left[\left(1 - \frac{h_1}{h_2} \right) (f(x, p) - k(x)) + \frac{h_1}{h_2} (f(x + h_2v, p) - k(x)) \right]$$

$$\le \frac{1}{h_2} (f(x + h_2v, p) - k(x))$$

since $x + h_1v = \left(1 - \frac{h_1}{h_2} \right) x + \frac{h_1}{h_2} (x + h_2v)$ and since $f(x, p) - k(x) \le 0$ for all p. Consequently, since P is compact, the intersection $\cap_{0 < h \le h_0} B_h$ is non-empty and all elements $\bar{p}$ of this intersection satisfy

$$h(Dk(x)(v) - \varepsilon) \le f(x + hv, \bar{p}) - k(x). \tag{32}$$

Letting h tend to 0, we deduce that $f(x, \bar{p}) - k(x) \ge 0$, whence $\bar{p}$ belongs to $P(x)$. Dividing the inequality (32) by $h > 0$, replacing $k(x)$ by $f(x, \bar{p})$ and letting h tend to 0, we obtain the inequality

$$Dk(x)(v) - \varepsilon \le Df(x, p)(v) \le \sup_{p \in P(x)} Df(x, p)(v).$$

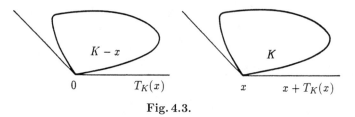

Fig. 4.3.

Thus, it is sufficient to let ε tend to 0.

Since $y \to f(y,p)$ is continuous at x, we know that $Df(x,p)(\cdot)$ is continuous for each p, whence that $Dk(x)(\cdot)$ is lower semi-continuous. Equation (28) may be written as

$$\sigma(\partial k(x), v) = \sup_{p \in P(x)} \sigma(\partial f(x,p), v)$$

which, by virtue of equation (64) of Chapter 3, implies equation (29).   □

Corollary 4.4. *Consider n convex functions f_i continuous at a point x. Then*

$$\partial \left(\sup_{i=1,\ldots,n} f_i \right)(x) = \overline{co} \left(\bigcup_{i \in I(x)} \partial f_i(x) \right) \tag{33}$$

where $I(x) = \{i = 1, \ldots, n | f_i(x) = \sup_{j=1,\ldots,n} f_j(x)\}$.

4.6 Tangent and Normal Cones

We consider a convex subset K. If $x \in K$, it is easy to check that

$$\partial \psi_K(x) = \{p \in X^* | \langle p, x \rangle = \sigma_K(p)\}. \tag{34}$$

Definition 4.3. *If K is a convex subset and if $x \in K$, $\partial \psi_K(x)$ is called the* **normal cone** *to K at x and is denoted by*

$$N_K(x) := \partial \psi_K(x) \tag{35}$$

The cone defined by

$$T_K(x) := \text{closure} \left(\bigcup_{h>0} \frac{1}{h}(K - x) \right) \tag{36}$$

is called the **tangent** *cone to K at x.*

These two cones are polar to each other.

Proposition 4.5. *The tangent and normal cones to K are linked by the relationship*

$$\forall x \in K, \qquad N_K(x) = T_K(x)^-. \qquad (37)$$

Proof. Since $K - x$ is contained in $T_K(x)$, when p belongs to $T_K(x)^-$, we have $\langle p, y - x \rangle \leq 0$ for all $y \in K$ and consequently p belongs to the normal cone to K at x.

Conversely, we fix p in $N_K(x)$, choose $v \in T_K(x)$ and show that $\langle p, x \rangle \leq 0$. But $v = \lim_{n \to \infty} \frac{1}{h_n}(y_n - x)$ where $h_n > 0$ and $y_n \in K$. Since $\langle p, y_n - x \rangle \leq 0$, we deduce that $\langle p, v \rangle$ is the limit of a sequence of negative or zero numbers $\frac{1}{h_n}\langle p, y_n - x \rangle$ and thus is itself negative or zero. $\square$

We note that

$$\text{if } x \in \text{Int } K, \text{ then } N_K(x) = \{0\} \text{ and } T_K(x) = X \qquad (38)$$

and that

$$\text{if } K = \{x_0\}, \text{ then } N_K(x_0) = X \text{ and } T_K(x_0) = \{0\}. \qquad (39)$$

It is also easy to show that if $K := B$ is the unit sphere, if $\|x\| = 1$ and if L denotes the duality operator, then

$$N_K(x) = \{\lambda L x\}_{\lambda \geq 0} \text{ and } T_K(x) = \{v \in X | \langle v, x \rangle \leq 0\}. \qquad (40)$$

This follows from the Cauchy–Schwarz inequality, which implies that

$$\langle \lambda L x, y - x \rangle \leq \lambda \|x\|(\|y\| - \|x\|) = \lambda(\|y\| - 1) \leq 0$$

when $\|y\| \leq 1$. Thus, $\lambda L x \in N_K(x)$. Conversely, if $p \in N_K(x)$, we deduce that $\|p\|\|x\| = \|p\| = \sup_{\|y\| \leq 1} \langle p, y \rangle \leq \langle p, x \rangle$ and consequently, that $p = \lambda L x$. Similarly, if $K := \mathbb{R}_+^n$ and if $x \in \mathbb{R}_+^n$, then

(i) $N_K(x) = \{p \in -\mathbb{R}_+^n | p_i = 0 \text{ when } x_i > 0\}$
(ii) $T_K(x) = \{v \in \mathbb{R}^n | v_i \geq 0 \text{ when } x_i = 0\}$ (41)

This follows from the fact that $\sup_{y \in \mathbb{R}_+^n} \langle p, y \rangle = \langle p, x \rangle = \sigma_{\mathbb{R}_+^n}(p) = \psi_{\mathbb{R}_-^n}(p)$. Thus, $p \in N_K(x)$ if and only if $p \in \mathbb{R}_-^n$ and if $\langle p, x \rangle = \sum_{i | x_i > 0} p_i x_i = 0$. The second formula is obtained by polarity.

We denote

$$M^n := \left\{ x \in \mathbb{R}_+^n \,\middle|\, \sum_{i=1}^{n} x_i = 1 \right\}. \qquad (42)$$

We shall deduce the formulae

$$N_{M^n}(x) = \{p \in \mathbb{R}^n | p_i = \max_{j=1,\ldots,n} p_j \text{ when } x_i > 0\} \qquad (43)$$

and

$$T_{M^n}(x) = \{v \in \mathbb{R}^n | \sum_{i=1}^{n} v_i = 0 \text{ and } v_i \geq 0 \text{ when } x_i = 0\} \qquad (44)$$

from formula (49), below.

Formulae Relating to Normal and Tangent Cones

By applying the subdifferential calculus to set indicator functions, we obtain a number of formulae which enable us to calculate normal and tangent cones.

- If $K \subset L$ and $x \in K$ then $T_K(x) \subset T_L(x)$ and $N_L(x) \subset N_K(x)$. $\qquad$ (45)
- If $K_i \subset X_i$ $(i = 1, \ldots, n)$ then

$$T_{\prod_{i=1}^n K_i}(x_1, \ldots, x_n) = \prod_{i=1}^n T_{K_i(x_i)} \qquad \text{and}$$

$$N_{\prod_{i=1}^n K_i}(x_1, \ldots, x_n) = \prod_{i=1}^n N_{K_i(x_i)} \qquad (46)$$

- If $B \in L(X, Y)$ then

$$T_{B(K)}(Bx) = \text{closure}(BT_K(x)) \quad \text{and} \quad N_{B(K)}(Bx) = B^{*-1}N_K(x). \qquad (47)$$

- $$\begin{aligned} T_{K_1+K_2}(x_1 + x_2) &= \text{closure}(T_{K_1}(x_1) + T_{K_2}(x_2)) \qquad \text{and} \\ N_{K_1+K_2}(x_1 + x_2) &= N_{K_1}(x_1) \cap N_{K_2}(x_2). \end{aligned} \qquad (48)$$

- If $A \in (L(X, Y))$ and if $L \subset X$ and $M \subset Y$ are convex closed subsets satisfying $0 \in \text{Int}(A(L) - M)$, then

$$\begin{aligned} T_{L \cap A^{-1}(M)}(x) &= T_L(x) \cap A^{-1}T_M(Ax) \qquad \text{and} \\ N_{L \cap A^{-1}(M)}(x) &= N_L(x) + A^* N_M(Ax). \end{aligned} \qquad (49)$$

- If $A \in L(X, Y)$ and if $M \subset Y$ is a convex closed subset satisfying $0 \in \text{Int}\,(\text{Im}\,A - M)$, then

$$T_{A^{-1}(M)}(x) = A^{-1}T_M(Ax) \quad \text{and} \quad N_{A^{-1}(M)}(x) = A^* N_M(Ax). \qquad (50)$$

- If K_1 and K_2 are convex closed subsets of X such that $0 \in \text{Int}\,(K_1 - K_2)$, then

$$\begin{aligned} T_{K_1 \cap K_2}(x) &= T_{K_1}(x) \cap T_{K_2}(x) \qquad \text{and} \\ N_{K_1 \cap K_2}(x) &= N_{K_1}(x) + N_{K_2}(x). \end{aligned} \qquad (51)$$

Proof of formulae (45)–(51). Formulae (45), (46), (47) and (48) are trivial to check for normal cones and follow by polarity for tangent cones.

Since the indicator function $\psi_{L \cap A^{-1}(M)}$ of $L \cap A^{-1}(M)$ is equal to $\psi_L + \psi_M \circ A$, formula (49) for normal cones follows from Theorem 4.4, and may be deduced for tangent cones by polarity. Formulae (50) and (51) are corollaries of formula (49).

Since $M^n := \{x \in \mathbb{R}^n_+ \mid \sum_{i=1}^n x_i = 1\}$, formula (49) may be applied with $X = \mathbb{R}^n_+$, $Y = \mathbb{R}$, $L = \mathbb{R}^n_+$, $M = \{1\}$ and A equal to the operator defined by $Ax = \sum_{i=1}^n x_i$. Thus, we derive formulae (43) and (44).   □

Remark. There are closed relations between tangent and normal cone to convex subsets and right-derivatives and subdifferentials of a convex function. First, the normal cone to K at x was defined by $N_K(x) := \partial\psi_K(x)$. One can check easily that

$$D\psi_K(x) = \psi_{T_K(x)}$$

so that the concepts of normal and tangent cones can be derived from the concept of subdifferential and right-derivative of a convex function. Conversely, one can also observe that

$$\forall\, x \in E, \ \mathrm{Ep}(Df(x)) = T_{\mathrm{Ep}(f)}(x, f(x))$$

and that $p \in \partial f(x)$ if and only if

$$(p, -1) \in N_{\mathrm{Ep}(f)}(x, f(x))$$

so that the concepts of right derivative and subdifferential of a convex function can be derived from the concepts of tangent and normal cones.

5. Marginal Properties of Solutions of Convex Minimisation Problems

5.1 Introduction

Fenchel's Theorem has already given us sufficient conditions for the existence of solutions of convex minimisation problems. As a consequence of Fermat's rule, suitably adapted, instead of searching for solutions of the minimisation problem,

$$-f^*(0) = \inf_{x \in X} f(x) \qquad (*)$$

we may seek to solve the *inclusion* (or set-valued equation)

$$0 \in \partial f(\bar{x}). \qquad (**)$$

Moreover, Fenchel's transformation shows that the set of solutions of the minimisation problem $(*)$ is the subdifferential

$$\partial f^*(0) = \{\bar{x} \in X | f(\bar{x}) = -f^*(0)\}$$

of the conjugate function f^* at 0.

We shall call this property (which is a very simple property in convex analysis) a marginal property of the solutions $\bar{x}$, to underline the use by neoclassical economists of adjective 'marginal' instead and in place of the adjective 'differential' used by mathematicians.

The subdifferential calculus which we described in the previous chapter will enable us to exploit this double characterisation for more specific minimisation problems. We have chosen a class of problems with a structure which is strong enough for us to acquire sufficiently useful information, but general enough to cover numerous examples. (This compromise is a matter of taste – that is, it is subjective.)

We shall consider a family of minimisation problems of the form

$$h(y) := \inf_{x \in X} [f(x) - \langle p, x \rangle + g(Ax + y)]$$

where

$$f : X \to \mathbb{R} \cup \{+\infty\} \text{ and } g : Y \to \mathbb{R} \cup \{+\infty\}$$

are nontrivial, convex and lower semi-continuous and where $A \in L(X,Y)$ is a continuous linear operator. We shall show that there exists a solution $\bar{x}$ of this problem under the assumption

$$p \in \text{Int} \left(\text{Dom } f^* + A^* \text{Dom } g^* \right)$$

which provides an additional justification for the introduction of conjugate functions.

We shall then show that if

$$y \in \text{Int} \left(\text{Dom } g - A \text{Dom } f \right),$$

then the set of such solutions is the set of solutions of the *inclusion* (or *set-valued equation*)

$$p \in \partial f(\bar{x}) + A^* \partial g(A\bar{x}. + y)$$

That is not all: we may associate the problem $h(y)$ with its dual problem

$$e_*(p) = \inf_{q \in Y^*} [f^*(p - A^*q) + g^*(q) - \langle q, y \rangle]$$

We shall then prove that the set of solutions $\bar{x}$ of the problem $h(y)$ is the subdifferential $\partial e_*(p)$ of the function $p \to e_*(p)$.

This more precise result (which describes the set of solutions of a minimisation problem as the subdifferential of the function e_*) plays a very prominent role in economic theory: *a minimal solution $\bar{x}$ of $h(y)$ measures the rate of variation of the marginal function e_* as the parameter p is varied.*

Moreover, the same assumptions also imply the same results for the dual problem $e_*(p)$: the set of solutions is non-empty, it is the subdifferential of the marginal function $h : y \to h(y)$ and it is the set of solutions of the inclusion

$$y \in \partial g^*(\bar{q}) - A\partial f^*(p - A^*\bar{q}).$$

5.2 Fermat's Rule

As previously mentioned, the set of solutions $\bar{x}$ of the minimisation problem

$$- f^*(0) := \inf_{x \in X} f(x) \tag{1}$$

is the set of solutions of $0 \in \partial f(\bar{x})$ since $\bar{x}$ belongs to $\partial f^*(0)$ if and only if

$$f(\bar{x}) = \langle 0, \bar{x} \rangle - f^*(0) = \inf_{x \in X} f(x).$$

Consequently, when f is nontrivial, convex and lower semi-continuous

$$\partial f^*(0) \text{ is the set of solutions of the minimisation problem (1).} \tag{2}$$

To exploit this result, we shall use the properties of conjugate functions and subdifferentials, as established above.

The following structures provide a framework which is sufficiently general to be the source of numerous examples and sufficiently specific that the technical difficulties are limited:

(i)   two Hilbert spaces X and Y;
(ii) two nontrivial, convex, lower semi-continuous functions
      $f : X \to \mathbb{R} \cup \{+\infty\}$ and $g : Y \to \mathbb{R} \cup \{+\infty\}$;
(iii) a continuous linear operator $A \in L(X, Y)$. $\hspace{2cm}$ (3)

We shall choose elements $y \in Y$ and $p \in X^*$ as parameters of the optimisation problems

$$h(y) := \inf_{x \in X} \left(f(x) - \langle p, x \rangle + g(Ax + y) \right) \tag{4}$$

and

$$e_*(p) := \inf_{q \in Y^*} \left(f^*(p - A^*q) + g^*(q) - \langle q, y \rangle \right) \tag{5}$$

which we shall solve simultaneously.

We shall say the minimisation problems $h(y)$ and $e_*(p)$ are *dual* and that the (convex) functions $h : y \to h(y)$ and $e_* : p \to e_*(p)$ are the *marginal* functions which describe the variation of the optimal values as a function of the parameters $y \in Y$ and $p \in X^*$.

The study of these marginal functions and above all of the properties of their gradients (or failing that their subgradients) is a subject of interest to economists.

Theorem 5.1. *a) We suppose that the conditions (3) are satisfied. If*

$$p \in \mathrm{Int}\, (\mathrm{Dom}\, f^* + A^* \mathrm{Dom}\, g^*), \tag{6}$$

then there exists a solution $\bar{x}$ of the problem $h(y)$ and

$$h(y) + e_*(p) = 0. \tag{7}$$

b) If we suppose further that

$$y \in \mathrm{Int}\, (\mathrm{Dom}\, g - A\, \mathrm{Dom}\, f) \tag{8}$$

then the following conditions are equivalent

(i)   *$\bar{x}$ is a solution of the problem $h(y)$;*
(ii) *$\bar{x}$ belongs to the subdifferential $\partial e_*(p)$ of the marginal function e_*;*
(iii) *$\bar{x}$ is a solution of the inclusion $p \in \partial f(\bar{x}) + A^* \partial g(A\bar{x} + y)$.* $\hspace{1cm}$ (9)

c) Similarly, assumption (8) implies that there exists a solution $\bar{q}$ of the problem $e_*(p)$ and the two assumptions imply that the following conditions are equivalent:

(i)   *$\bar{q}$ is a solution of the problem $e_*(p)$;*
(ii)   *$\bar{q}$ belongs to the subdifferential $\partial h(y)$ of the marginal function h;*
(iii)   *$\bar{q}$ is a solution of the inclusion $y \in \partial g^*(\bar{q}) - A\partial f^*(p - A^*\bar{q})$.*   (10)

d) The two assumptions imply that the solutions $\bar{x}$ and $\bar{q}$ of the problems $h(y)$ and $e_*(p)$ are solutions of the system of inclusions

(i) $$p \in \partial f(\bar{x}) + A^*(\bar{q})$$
(ii) $$y \in -A\bar{x} + \partial g^*(\bar{q}).$$ (11)

Remark. An optimal solution of one of the minimisation problems $h(y)$ or $e_*(p)$ is usually called a *Lagrange (or Kuhn–Tucker) multiplier*, the inclusion (9)(iii) is usually called the *Euler–Lagrange inclusion* and the inclusion (10)(iii) is the *Euler–Lagrange dual inclusion*. The system of inclusions (11) is usually called the *Hamiltonian system*.

The mapping $(x, q) \rightarrow (\partial f(x) + A^*q) \times (-Ax + \partial g^*(q))$ from $X \times Y^*$ to its dual $X^* \times Y$ may be written symbolically in matrix form by

$$\begin{pmatrix} \partial f & A^* \\ -A & \partial g^* \end{pmatrix}.$$ (12)

The set of solutions $(\bar{x}, \bar{q})$ of the minimisation problems $h(y)$ and $e_*(p)$ may be written in the form

$$\begin{pmatrix} \partial f & A^* \\ -A & \partial g^* \end{pmatrix}^{-1} \begin{pmatrix} p \\ y \end{pmatrix}.$$

This notation highlights the variation of the set of solutions as a function of the parameters $p \in X^*$ and $y \in Y$.

Proof of Theorem 5.1. a) The existence of solutions of the problems $h(y)$ and $e_*(p)$ and the equality $h(y) + e_*(p) = 0$ follows from Theorem 3.2 (Fenchel's Theorem) with f replaced by $x \rightarrow f(x) - \langle p, x \rangle$ and g replaced by $z \rightarrow g(z + y)$, since in this case $v = h(y)$ and $v_* = e_*(p)$.

b) We may write

$$h(y) = \inf_{x \in X} \varphi(C(x, y))$$

where

$$\begin{aligned} \varphi(x, y) &:= f(x) - \langle p, x \rangle + g(y) \\ C(x, y) &:= (x, Ax + y). \end{aligned}$$

Since the operator C is clearly an isomorphism of $X \times Y$ onto itself, Proposition 3.1 implies that

$$
\begin{aligned}
\partial(\varphi \circ C)(x, y) &= C^* \partial \varphi(C(x, y)) \\
&= C^*((\partial f(x) - p) \times \partial g(Ax + y)) \\
&= \{(\partial f(x) - p + A^* q) \times \{q\}\}_{q \in \partial g(Ax+y)}.
\end{aligned}
$$

Proposition 4.3 implies that if $\bar{x}$ is a solution of the problem $h(y)$, then $\bar{q}$ belongs to $\partial h(y)$ if and only if $(0, \bar{q})$ belongs to $\partial(\varphi \circ C)(x, y)$, in other words, if $\bar{q} \in \partial g(Ax + y)$ and $0 \in \partial f(\bar{x}) - p + A^* \bar{q}$, by virtue of the previous formula. Thus, we have shown that if $\bar{x}$ is a solution of $h(y)$, then the following conditions are equivalent

(i) $\qquad\qquad\qquad \bar{q} \in \partial h(y)$

(ii) $\qquad\quad 0 \in \partial f(\bar{x}) - p + A^* \bar{q}$ and $q \in \partial g(A\bar{x} + y).$ $\qquad\qquad$ (13)

Eliminating $\bar{q}$ from these two inclusions, we find that

$$
p \in \partial f(\bar{x}) + A^* \partial g(A\bar{x} + y). \tag{14}
$$

This last property implies that

$$
0 \in \partial(f(\cdot) - \langle p, \cdot \rangle + g(A(\cdot) + y))(\bar{x}) \tag{15}
$$

which shows that any solution $\bar{x}$ of the inclusion (14) is a solution of the minimisation problem $h(y)$. Conversely, assumption (8) and Theorem 4.4 imply that any solution of $h(y)$ which is a solution of the inclusion (15) is a solution of the inclusion (14). This latter implies that there exists $\bar{q} \in \partial g(A\bar{x} + y)$ such that $p \in \partial f(\bar{x}) + A^* \bar{q}$, in other words, such that $\bar{q} \in \partial h(y)$. Thus, we have proved that properties (9)(i) and (9)(iii) are equivalent. Similarly, replacing the functions f and g and the operator A by g^*, f^* and $-A^*$, respectively, properties (10)(i) and (10)(iii) may be shown to be equivalent.

The system of inclusions (11) is clearly equivalent to the systems of inclusions (9)(iii) and (10)(iii): this proves the last part of the theorem. The equivalence of (13)(i) and (13)(ii) then implies the equivalence of (9)(i) and (9)(ii) and, replacing f, g and A by g^*, f^* and $-A^*$, the equivalence of (10)(i) and (10)(ii). $\qquad\qquad\qquad\qquad\qquad\qquad\qquad\qquad\qquad\qquad\qquad\qquad$ $\square$

Remarks. When assumptions (6) and (8) of Theorem 5.1 are satisfied, solution of the problem $h(y)$ is equivalent to solution of the inclusion (set-valued equation)

$$
p \in \partial f(\bar{x}) + A^* \partial g(A\bar{x} + y). \tag{16}
$$

Theorem 5.1 indicates another way of solving this problem. This involves first solving the inclusion

$$
y \in \partial g^*(\bar{q}) - A \partial f^*(p - A^* \bar{q}) \tag{17}
$$

and then choosing $\bar{x}$ in the set

$$\partial f^*(p - A^*\bar{q}) \cap A^{-1}(\partial g^*(\bar{q}) - y). \tag{18}$$

This procedure is only sensible if the second inclusion is easier to solve than the first. This clearly depends on the functions f and g. If g is differentiable, it may be better to solve the inclusion (16). If, moreover, f^* is differentiable it may be easier to solve the inclusion (17) which, in this case, may be written as

$$A\nabla f^*(p - A^*\bar{q}) + y \in \partial g^*(\bar{q}) \tag{19}$$

or as

$$\forall q \in Y, \qquad \langle -A\nabla f^*(p - A^*\bar{q}) - y, \bar{q} - q \rangle + g^*(\bar{q}) - g^*(q) \leq 0. \tag{20}$$

5.3 Minimisation Problems with Constraints

Let us consider

(i)   two Hilbert spaces X and Y;
(ii)   a continuous linear operator $A \in L(X, Y)$;
(iii)   a convex closed subset $M \subset Y$;
(iv)   a nontrivial, convex, lower semi-continuous function
        $f : X \to \mathbb{R} \cup \{+\infty\}$ and two elements $y \in Y$ and $p \in X^*$. (21)

We consider the minimisation problem

$$h(y) := \inf_{Ax \in M - y} (f(x) - \langle p, x \rangle) \tag{22}$$

with its associated dual problem

$$e_*(p) := \inf_{q \in Y^*} (f^*(p - A^*q) + \sigma_M(q) - \langle q, y \rangle). \tag{23}$$

Corollary 5.1. *If we suppose that*

$$p \in \text{Int} \, (\text{Dom} \, f^* + A^* b(M)) \tag{24}$$

then there exists a solution $\bar{x}$ (satisfying $A\bar{x} \in M - y$) of the problem $h(y)$. If we suppose further that

$$y \in \text{Int} \, (M - A \, \text{Dom} \, f) \tag{25}$$

then the solutions $\bar{x}$ of the problem $h(y)$ are the solutions of the inclusion

$$p \in \partial f(\bar{x}) + A^* N_M(A\bar{x} + y) \tag{26}$$

and the set of solutions $\bar{x}$ of $h(y)$ is the subdifferential $\partial e_*(p)$ of the marginal function e_*.

The following solutions are then equivalent:

(i) $\bar{q} \in \partial h(y)$;

(ii) $\bar{q}$ *is a solution of the problem* $e_*(p)$;

(iii) $\bar{q}$ *is a solution of the inclusion* $y \in \partial \sigma_M(\bar{q}) - A \partial f^*(p - A^*\bar{q})$.   (27)

The optimal solutions $\bar{x}$ *and* $\bar{q}$ *of the problems* $h(y)$ *and* $e_*(p)$ *are related by*

$$p \in \partial f(\bar{x}) + A^*\bar{q} \quad \text{and} \quad \bar{q} \in N_M(A\bar{x} + y). \quad (28)$$

The minimisation problem

$$h(y) := \inf_{Ax+y=0} (f(x) - \langle p, x \rangle) \quad (29)$$

which is a minimisation problem with 'equality constraints' is obtained as the particular case in which $M = \{0\}$. Its dual problem is

$$e_*(p) := \inf_{q \in Y^*} (f^*(p - A^*q) - \langle q, y \rangle). \quad (30)$$

Corollary 5.2. *If we suppose that*

$$p \in \text{Int} \, (\text{Dom} \, f^* + \text{Im} \, A^*) \quad (31)$$

then there exists a solution $\bar{x}$ *of the problem* $h(y)$.

If we suppose further that

$$-y \in \text{Int} \, (A \, \text{Dom} \, f) \quad (32)$$

then the solutions $\bar{x}$ *of the problem* $h(y)$ *are the solutions of the inclusion*

$$p \in \partial f(\bar{x}) + \text{Im} \, A^*, \qquad A\bar{x} + y = 0 \quad (33)$$

and the set of solutions $\bar{x}$ *is the subdifferential* $\partial e_*(p)$.

The following conditions are equivalent

(i) $\bar{q} \in \partial h(y)$;

(ii) $\bar{q}$ *is a solution of the problem* $e_*(p)$;

(iii) $\bar{q}$ *is a solution of the inclusion* $y \in -A \partial f^*(p - A^*\bar{q})$.   (34)

The optimal solutions $\bar{x}$ *and* $\bar{q}$ *of the problems* $h(y)$ *and* $e_*(p)$ *are related by*

$$p \in \partial f(\bar{x}) + A^*\bar{q}. \quad (35)$$

Suppose that $P \subset Y$ is a convex closed cone and denote its negative polar cone by P^-.

The cone P defines an order relation $\geq$ by

$$y_1 \geq y_2 \text{ if and only if } y_1 - y_2 \in P \tag{36}$$

and the cone P^- defines the order relation

$$q_1 \leq q_2 \text{ if and only if } q_1 - q_2 \in P^-. \tag{37}$$

The minimisation problem

$$h(y) := \inf_{Ax+y \geq 0} (f(x) - \langle p, x \rangle) \tag{38}$$

which is a minimisation problem with 'inequality constraints' is obtained in the special case in which $M = P$. Its dual problem is

$$e_*(p) := \inf_{q \in P^-} (f^*(p - Aq) - \langle q, y \rangle). \tag{39}$$

Corollary 5.3. *If we suppose that*

$$p \in \text{Int}\,(\text{Dom}\, f^* + A^*P^-) \tag{40}$$

then there exists a solution $\bar{x}$ of the problem $h(y)$.
If we suppose further that

$$y \in \text{Int}\,(P - A\,\text{Dom}\, f) \tag{41}$$

then the solutions $\bar{x}$ of the problem $h(y)$ are the solutions of the inclusion

$$p \in \partial f(\bar{x}) + A^* N_P(A\bar{x} + y) \tag{42}$$

and the set of solutions $\bar{x}$ is equal to $\partial e_*(p)$.
The following conditions are equivalent

(i) $\bar{q} \in \partial h(y)$;
(ii) $\bar{q}$ *is a solution of* $e_*(p)$;
(iii) $\bar{q}$ *is a solution of the inclusion* $y \in N_P - (\bar{q}) - A\partial f^*(p - A^*\bar{q})$. (43)

The solutions $\bar{x}$ and $\bar{q}$ of the problems $h(y)$ and $e_*(p)$ are related by

(i) $$p \in \partial f(\bar{x}) - A^*\bar{q}$$
(ii) $$A\bar{x} + y \geq 0, \ \bar{q} \leq 0 \text{ and } \langle \bar{q}, A\bar{x} + y \rangle = 0. \tag{44}$$

5.4 Principle of Price Decentralisation

We consider

(i) n Hilbert spaces X_i $(i = 1, \ldots, n)$;
(ii) n nontrivial, convex, lower semi-continuous functions $f_i : X_i \to \mathbb{R} \cup \{+\infty\}$;
(iii) a Hilbert space Y;
(iv)   continuous linear operators $A_i \in L(X_i, Y)$;
(v)   a convex closed subset $M \subset Y$. (45)

We shall now solve the minimisation problem

$$h(y) := \inf_{\sum_{i=1}^{n} A_i x_i + y \in M} \left(\sum_{i=1}^{n} (f_i(x_i) - \langle p, x \rangle) \right). \tag{46}$$

This is a particular case of problem (22), in which

$$X := \sum_{i=1}^{n} X_i, \quad f(x) := \sum_{i=1}^{n} f_i(x_i) \quad \text{and} \quad Ax = \sum_{i=1}^{n} A_i x_i.$$

The dual problem is

$$e_*(p_1, \ldots, p_n) = \inf_{q \in Y^*} \left(\sum_{i=1}^{n} f_i^*(p_i - A_i^* q) + \sigma_M(q) - \langle q, y \rangle \right). \tag{47}$$

Corollary 5.4. *If we suppose that*

$$(p_1, \ldots, p_n) \in \text{Int} \left\{ \prod_{i=1}^{n} \text{Dom } f_i^* + A_i^* q \right\}_{q \in b(M)} \tag{48}$$

then there exists a solution $(\bar{x}_1, \ldots, \bar{x}_n)$ of the problem $h(y)$.
If we suppose further that

$$y \in \text{Int} \left(M - \sum_{i=1}^{n} A_i \text{ Dom } f_i \right) \tag{49}$$

then the solutions $(\bar{x}_1, \ldots, \bar{x}_n)$ and $\bar{q}$ of the problems $h(y)$ and $e_*(p_1, \ldots, p_n)$ are the solutions of the system

(i) $\qquad\qquad \bar{x}_i \in \partial f_i^*(p_i - A_i^* \bar{q}) \quad (i = 1, \ldots, n)$

(ii) $\qquad\qquad \bar{q} \in N_M \left(\sum_{i=1}^{n} A_i \bar{x}_i + y \right) \tag{50}$

where $\bar{q}$ is a solution of the dual problem $e_*(p_1, \ldots, p_n)$ and the set of solutions $(\bar{x}_1, \ldots, \bar{x}_n)$ is the subdifferential $\partial e_*(p_1, \ldots, p_n)$.
The following conditions are equivalent

(i) $\bar{q} \in \partial h(y)$;
(ii) $\bar{q}$ *is a solution of* $e_*(p_1, \ldots, p_n)$;
(iii) $\bar{q}$ *is a solution of the inclusion* $y \in \partial \sigma_M(\bar{q}) - \sum_{i=1}^{n} A_i \partial f_i^*(p_i - A_i^* \bar{q})$. (51)

When the conjugate functions f_i^* are differentiable, the solutions $(\bar{x}_1, \ldots, \bar{x}_n)$ are obtained using the Lagrange multipliers $\bar{q}$ (the solutions $\bar{q}$ of $e_*(p_1, \ldots, p_n)$) from the formulae

$$\bar{x}_i = \nabla f_i^*(p_i - A_i^* \bar{q}) \quad (i = 1, \ldots, n).$$

In other words, once we know a Lagrange multiplier $\bar{q}$, we may obtain an optimal solution $(\bar{x}_1, \ldots, \bar{x}_n)$ by solving n independent minimisation problems

$$\inf_{x_i \in X_i} (f_i(x_i) - \langle p_i, x_i \rangle + \langle \bar{q}, A_i x_i \rangle) \tag{52}$$

which are obtained by adding a 'cost of violating the constraints' $\langle \bar{q}, A_i x_i \rangle$ to the initial loss function $f_i(\cdot) - \langle p_i, \cdot \rangle$.

It is this result which justifies the role of prices $\bar{q}$ (subgradients of the marginal function $h(y)$) in economic models as a means of decentralising decisions; in other words, a means of solving the n problems (52) independently. We shall return to this fundamental problem of decentralisation in Chapter 10, in the context of the theory of economic equilibrium.

5.5 Regularisation and Penalisation

Consider a nontrivial, convex, lower semi-continuous function f from a Hilbert space X to $\mathbb{R} \cup \{+\infty\}$. With any $\lambda > 0$ we associate the function f_λ defined by

$$f_\lambda(x) := \inf_{y \in X} \left[f(y) + \frac{1}{2\lambda} \|y - x\|^2 \right]. \tag{53}$$

We shall show that the functions f_λ are convex differentiable functions which are simply convergent to the function f as y tends to 0. This provides us with a *regularisation* procedure which enables us to approximate f by a more regular function.

When y tends to infinity, we may interpret the minimisation problem (53) as a *penalised* version of the minimisation problem

$$-f^*(0) = \inf_{x \in X} f(x). \tag{54}$$

We recall that the minimisation problem $f_\lambda(x)$ has a unique solution denoted by $J_\lambda x$ (see Theorem 2.2 (Proximation Theorem)).

Theorem 5.2. *Suppose that $f : X \to \mathbb{R} \cup \{+\infty\}$ is a nontrivial, convex, lower semi-continuous function. Then the functions f_λ are convex and differentiable and*

$$\nabla f_\lambda(x) = \frac{1}{\lambda}(x - J_\lambda x) \in \partial f(J_\lambda x). \tag{55}$$

Moreover, when λ tends to 0,

$$\forall x \in \text{Dom} f, \quad f_\lambda(x) \to f(x) \quad \text{and} \quad J_\lambda x \to x \tag{56}$$

and when $\lambda \to \infty$,

$$f_\lambda(x) \quad \text{tends to} \quad -f^*(0) = \inf_{x \in X} f(x). \tag{57}$$

Proof. a) In the proof of Theorem 4.3, we showed that $\|J_\lambda x - x\|$ converges to 0. Moreover, $f_\lambda(x) \le f(x) + \frac{1}{2\lambda}\|x - x\|^2 = f(x)$. Since $f(x) \le \liminf_{\lambda \to 0} f(J_\lambda x)$ (because f is lower semi-continuous) and since

$$f(J_\lambda x) = f_\lambda(x) - \frac{1}{2\lambda}\|J_\lambda x - x\|^2 \le f_\lambda(x),$$

it follows that $f(x) \le \liminf_{\lambda \to 0} f_\lambda(x)$. Thus, $f(x) = \lim_{\lambda \to 0} f_\lambda(x)$.

b) Provisionally, we set $A_\lambda(x) := \frac{1}{\lambda}(x - J_\lambda x)$. In Theorem 4.3, we showed that $A_\lambda(x)$ belongs to $\partial f(J_\lambda x)$. Thus,

$$
\begin{aligned}
f_\lambda(x) - f_\lambda(y) &= f(J_\lambda x) - f(J_\lambda y) + \frac{\lambda}{2}\|A_\lambda(x)\|^2 - \frac{\lambda}{2}\|A_\lambda(y)\|^2 \\
&\le \langle A_\lambda(x), J_\lambda x - J_\lambda y \rangle + \frac{\lambda}{2}\|A_\lambda(x)\|^2 - \frac{\lambda}{2}\|A_\lambda(y)\|^2
\end{aligned}
$$

(because $A_\lambda(x) \in \partial f(J_\lambda x)$)

$$
\le \langle A_\lambda(x), x - y \rangle - \lambda\langle A_\lambda(x), A_\lambda(x) - A_\lambda(y) \rangle + \frac{\lambda}{2}\|A_\lambda(x)\|^2 - \frac{\lambda}{2}\|A_\lambda(y)\|^2
$$

(because $J_\lambda = 1 - \lambda A_\lambda$)

$$
\begin{aligned}
&= \langle A_\lambda(x), x - y \rangle - \lambda\left(\frac{1}{2}\|A_\lambda(x)\|^2 + \frac{1}{2}\|A_\lambda(y)\|^2 - \langle A_\lambda(x), A_\lambda(y) \rangle\right) \\
&= \langle A_\lambda(x), x - y \rangle - \frac{\lambda}{2}\|A_\lambda(x) - A_\lambda(y)\|^2 \\
&\le \langle A_\lambda(x), x - y \rangle.
\end{aligned}
$$

Thus, we have shown that

$$A_\lambda(x) \in \partial f_\lambda(x). \tag{58}$$

Moreover, since $A_\lambda(y)$ belongs to $\partial f_\lambda(y)$ for all y, we obtain the inequalities

$$
\begin{aligned}
f_\lambda(x) - f_\lambda(y) &\ge \langle A_\lambda(y), x - y \rangle \\
&= \langle A_\lambda(x), x - y \rangle + \langle A_\lambda(y) - A_\lambda(x), x - y \rangle \\
&\ge \langle A_\lambda(x), x - y \rangle - \|A_\lambda(y) - A_\lambda(x)\|\|x - y\| \\
&\ge \langle A_\lambda(x), x - y \rangle - \frac{1}{\lambda}\|x - y\|^2
\end{aligned}
$$

since $\|A_\lambda(x) - A_\lambda(y)\| \le \frac{1}{\lambda}\|x - y\|$ (see Proposition 2.7, since $A_\lambda = \frac{1}{\lambda}(1 - J_\lambda)$). Thus,

$$\left|\frac{f_\lambda(x) - f_\lambda(y) - \langle A_\lambda(x), x - y \rangle}{\|x - y\|}\right| \le \frac{1}{\lambda}\|x - y\|,$$

whence $A_\lambda(x) = \nabla f_\lambda(x)$.

c) From Fenchel's Theorem, we know that

$$f_\lambda(x) + \inf_{q \in X^*} \left(f^*(-q) + \frac{\lambda}{2}\|q\|^2 - \langle q, x \rangle \right) = 0.$$

In other words, we may write

$$f_\lambda(x) + (f^* - x)_{1/\lambda}(0) = 0. \tag{59}$$

Consequently, when $\lambda \to \infty$, $(f^* - x)_{1/\lambda}(0)$ tends to $(f^* - x)(0) = f^*(0) = -\inf_{y \in X} f(y)$ from the above.

d) From Theorem 5.1, we know that $J_\lambda x$ and the optimal solution $\bar{q}_{1/\lambda}$ of the problem $(f^* - x)_{1/\lambda}$ are related as follows:

(i) $\qquad\qquad\qquad 0 \in \partial f(J_\lambda x) - \bar{q}_{1/\lambda}$

(ii) $\qquad\qquad\qquad x = -J_\lambda x + \lambda\bar{q}_{1/\lambda}. \tag{60}$

This shows that $\nabla f_\lambda(x)$ is the unique solution of the problem $(f^* - x)_{1/\lambda}$. Thus, if 0 belongs to the domain of ∂f^* (in other words, if there exists a minimum of f), then $\nabla f_\lambda(x)$ converges to 0 as λ tends to infinity. Consequently, if the limit of $J_\lambda x$ as λ tends to infinity exists, it belongs to $\partial f^*(0)$; in other words, it generates the minimum of f. $\qquad\qquad\Box$

6. Generalised Gradients of Locally Lipschitz Functions

6.1 Introduction

Since both continuously differentiable functions and convex continuous functions are locally Lipschitz functions, it is natural to wonder if the latter are 'differentiable' in some weak sense. In 1975, Frank H. Clarke introduced the concept of the *generalised gradient* of a locally Lipschitz function which is a convex, closed, bounded subset. The generalised gradient reduces to the gradient of the function if the function is continuously differentiable and is equal to the subdifferential of the function if the function is convex and continuous. Since the upper envelope of Lipschitz functions is Lipschitz, this upper envelope also has a generalised gradient. Finally, Fermat's rule applies: if x is a *local minimum* of a locally Lipschitz function, 0 belongs to its generalised gradient at x.

Furthermore, the concept of generalised gradient will enable us to define the *normal cone at x for an arbitrary non-empty subset*; we shall show that this coincides with the normal cone at x for a convex closed subset.

6.2 Definitions

The concept of differentiation plays such an important role that it has been generalised and extended in many directions, according to specific applications. We shall only describe the concepts which one meets when trying to define directional derivatives for locally Lipschitz functions.

Definition 6.1. *Let f be a function from X to $\mathbb{R} \cup \{+\infty\}$ with a non-empty domain. We shall call the following limit (when it exists), the* **Clarke right directional derivative** *of f at x in the direction v:*

$$D_c f(x)(v) := \limsup_{\substack{h \to 0+ \\ y \to x}} \frac{f(y + hv) - f(y)}{h}. \tag{1}$$

We shall say that f is Clarke right differentiable at x if for all $v \in X$, the limit $D_c f(x)(v)$ exists and is finite.

We recall that (when it exists) the limit

$$Df(x)(v) := \lim_{h \to 0+} \frac{f(x + hv) - f(x)}{h} \qquad (2)$$

is called the *right derivative of* f *at* x *in the direction* v, and that f is *right differentiable at* x if $Df(x)(v)$ exists for all v.

We shall say that f is *Gâteaux differentiable* at x if f is right differentiable and $v \to Df(x)(v) = \langle \Delta f(x), v \rangle$ is linear and continuous.

We shall call $\nabla f(x) \in X^*$ the gradient of f at x. We shall say that f is *continuously differentiable* if for all $v \in U$, the function $y \to \langle \nabla f(y), v \rangle$ is continuous at x. We shall say that f is *Fréchet differentiable* at x if

$$\lim_{v \to 0} \left| \frac{f(x + v) - f(x) - \langle \nabla f(x), v \rangle}{\|v\|} \right| = 0 \qquad (3)$$

and that f is *strictly Fréchet differentiable* at x if

$$\lim_{\substack{y \to x \\ v \to 0}} \left| \frac{f(y + v) - f(y) - \langle \nabla f(x), v \rangle}{\|v\|} = 0 \right|.$$

We note that the function $v \to D_c f(x)(v)$ is *positively homogeneous* and that, when the limits below exist, we have

$$Df(x)(v) \leq D_c f(x)(v). \qquad (4)$$

We also note that a Gâteaux-differentiable function f is not necessarily Clarke differentiable. However, we do have the following result.

Proposition 6.1. *Suppose that* f *is continuously differentiable at* x. *Then* f *is Clarke differentiable and*

$$\langle \nabla f(x), v \rangle = D_c f(x)(v). \qquad (5)$$

Proof. Since f is continuously differentiable at x, then for every $\varepsilon > 0$, there exists $\eta > 0$ such that

$$|\langle \nabla f(z), v \rangle - \langle \nabla f(x), v \rangle| \leq \varepsilon \text{ when } \|z - x\| \leq \eta. \qquad (6)$$

If $\|y\| \leq \eta/2$ and if $0 < t < \eta/2\|v\|$, we set $g(t) = f(y + tv)$. Then g is differentiable and

$$g'(t) = \lim_{\theta \to 0} \frac{f(y + tv + \theta v) - f(y + tv)}{\theta} = \langle \nabla f(y + tv), v \rangle. \qquad (7)$$

Thus, if $\theta \leq \eta/2\|v\|$, we have

$$\frac{f(y + \theta v) - f(y)}{\theta} - \langle \nabla f(x), v \rangle = \frac{g(\theta) - g(0)}{\theta} - \langle \nabla f(x), v \rangle$$

$$= \frac{1}{\theta} \int_0^\theta (\langle \nabla f(y + tv), v \rangle - \langle \nabla f(x), v \rangle) dt$$

and consequently, since $\|y + tv - x\| \le \eta$,

$$\left| \frac{f(y + \theta v) - f(y)}{\theta} - \langle \nabla f(x), v \rangle \right| \le \varepsilon$$

when

$$\|y - x\| \le \eta/2 \quad \text{and} \quad \theta \le \eta/2\|v\|.$$

This implies that if $\alpha \le \eta/2$ and $\beta \le \eta/2\|v\|$,

$$\sup_{\|y-x\|\le\alpha} \sup_{\theta\le\beta} \frac{f(y + \theta v) - f(y)}{\theta} \le \langle \nabla f(x), v \rangle + \varepsilon. \tag{8}$$

Taking the infimum with respect to α and β, we have

$$D_c f(x)(v) \le \langle \nabla f(x), v \rangle + \varepsilon, \tag{9}$$

which, taken with (4), implies (5). $\qquad \square$

We recall that any convex function continuous at a point x is Clarke differentiable (Theorem 4.1); we restate this result.

Proposition 6.2. *Suppose that a function $f : X \to \mathbb{R} \cup \{+\infty\}$ is continuous at a point x in the interior of its domain. Then f is Clarke differentiable and*

$$\forall v \in X, \ \ Df(x)(v) = D_c f(x)(v). \tag{10}$$

We recall that continuously differentiable functions and convex continuous functions are locally Lipschitz (see Theorem 2.1).

We shall show that not only continuously differentiable functions and convex continuous functions but also, more generally, locally Lipschitz functions are Clarke differentiable.

Theorem 6.1. *Any locally Lipschitz function $f : X \to \mathbb{R} \cup \{+\infty\}$ is Clarke differentiable on the interior of its domain. For all $x \in \text{Int Dom} f$,*

$$v \to D_c f(x)(v) \text{ is positively homogeneous, convex and continuous.} \tag{11}$$

Moreover,

$$\{x, v\} \in \text{Int Dom} f \times U \to D_x f(x)(v) \text{ is upper semi-continuous.} \tag{12}$$

Remark. Propositions 6.1 and 6.2 show that the Clarke derivative of locally Lipschitz functions provides a natural generalisation of the concepts of Fréchet and right derivatives in convex analysis.

Proof of Theorem 6.1. Suppose $x \in \text{Int Dom} f$. Since f is locally Lipschitz, there exist $\eta > 0$ and $L > 0$, such that

$$\forall y, z \in x + \eta B, \qquad |f(y) - f(z)| \leq L\|y - z\|. \tag{13}$$

Then, for all $\alpha \leq \eta/2$ and $\beta \leq \eta/2\|v\|$,

$$-L\|v\| \leq \frac{f(y + \theta v) - f(y)}{\theta} \leq L\|v\|$$

when $y \in x + \alpha B$ and $\theta \leq \beta$. It follows that

$$-L\|v\| \leq D_c f(x)(v) = \inf_{\alpha, \beta > 0} \sup_{\substack{\|y-x\| \leq \alpha \\ 0 < \theta \leq \beta}} \frac{f(y + \theta v) - f(y)}{\theta} \leq L\|v\|,$$

whence, f is Clarke differentiable; in particular, we obtain the inequality

$$|D_c f(x)(v)| \leq L\|v\|. \tag{14}$$

We already know that $v \to D_c f(x)(v)$ is positively homogeneous. We shall show that this function is convex. Writing

$$\frac{f(y + \theta(\lambda v + (1 - \lambda w)) - f(y)}{\theta}$$
$$= (1 - \lambda)\frac{f(z + \alpha w) - f(z)}{\alpha} + \lambda\frac{f(y + \beta v) - f(y)}{\beta},$$

where $z = y + \theta \lambda v$ converges to x, and $\alpha = (1 - \lambda)\theta$ and $\beta = \lambda\theta$ converge to 0, and taking the upper limits of the two sides, we obtain

$$D_c f(x)(\lambda v + (1 - \lambda)w) \leq D_c f(x)(v) + (1 - \lambda)D_c f(x)(w). \tag{15}$$

It remains to show that $\{x, v\} \to D_c f(x)(v)$ is upper semi-continuous. From the definition of $D_c f(x)(v)$, given $\varepsilon > 0$ there exists α_0 such that

$$\sup_{\substack{\|z-x\| \leq 2\alpha_0 \\ \lambda \leq \alpha_0}} \frac{f(z + \lambda v) - f(z)}{\lambda} \leq D_c f(x)(v) + \varepsilon/2. \tag{16}$$

If $\|z - y\| \leq \alpha_0$ and $\|y - x\| \leq \alpha_0$, then since f is locally Lipschitz, we obtain

$$\frac{f(z + \lambda w) - f(z)}{\lambda} \leq \frac{f(z + \lambda v) - f(z)}{\lambda} + L\|v - w\|. \tag{17}$$

Consequently, if $y \in x + \alpha_0 B$, $\alpha \leq \alpha_0$ and $\beta \leq \beta_0$, then

$$\sup_{\substack{\|z-y\| \leq \alpha \\ \lambda \leq \beta}} \frac{f(z + \lambda w) - f(z)}{\lambda} \leq \sup_{\substack{\|z-x\| \leq 2\alpha_0 \\ \lambda \leq \alpha_0}} \frac{f(z + \lambda v) - f(z)}{\lambda} + L\|v - w\|$$
$$\leq D_c f(x)(v) + \varepsilon/2 + L\|v - w\|$$
$$\leq D_c f(x)(v) + \varepsilon$$

when $\|v - w\| \leq \frac{\varepsilon}{2L}$.

Letting α and β tend to 0, we deduce that

$$D_c f(y)(w) \leq D_c f(x)(v) + \varepsilon \text{ when } \|y - x\| \leq \alpha_0 \text{ and } \|v - w\| \leq \frac{\varepsilon}{2L};$$

whence, $D_c f(x)(v)$ is upper semi-continuous at $\{x, v\}$. $\qquad\square$

6.3 Elementary Properties

Next we shall establish certain elementary properties of Clarke derivatives.

Proposition 6.3. *Suppose that f and g are two locally Lipschitz functions from X to $\mathbb{R} \cup \{+\infty\}$ and that $x \in \operatorname{Int} \operatorname{Dom} f \cap \operatorname{Int} \operatorname{Dom} g$. Then*

$$D_c(\alpha f + \beta g)(x)(v) \leq \alpha D_c f(x)(v) + \beta D_c g(x)(v) \tag{18}$$

if $\alpha, \beta > 0$. If $x \in \operatorname{Int} \operatorname{Dom} f$, then

$$D_c(-f)(x)(v) = D_c f(x)(-v). \tag{19}$$

Proof. Formula (18) is self evident. To prove (19), we write

$$\frac{-f(y + \lambda v) - (-f(y))}{\lambda} = \frac{f(z + \lambda(-v)) - f(z)}{\lambda} \tag{20}$$

where $z = y + \lambda v$ converges to x when y converges to x and $\lambda > 0$ tends to 0. Taking the upper limits as y and z converge to x and λ converges to 0, the term on the left converges to $D_c(-f)(x)(v)$ and that on the right to $D_c f(x)(-v)$. $\square$

Proposition 6.4. *Let f be a locally Lipschitz function from X to $\mathbb{R} \cup \{+\infty\}$ with a non-empty domain. Let P be a closed convex cone in X. If f is increasing on P in the sense that*

$$f(x) \leq f(x + v), \qquad \forall v \in P, \tag{21}$$

then

$$\forall v \in X, \qquad D_c f(x)(v) \leq \sigma(P^+, v) \tag{22}$$

where $\sigma(P^+, v)$ is the support function of the positive polar cone P^+ of P.

Proof. For all $v \in P$, we have the inequality

$$\frac{f(y + \theta v + \theta(-v)) - f(y + \theta v)}{\theta} \leq 0.$$

Taking the limit of the supremum as y tends to x and θ tends to 0, we deduce that $D_c f(x)(-v) \leq 0$, $\forall v \in P$. Moreover, $\sigma(P^+, v) = 0$ if $v \in -P$ and $\sigma(P^+, v) = +\infty$ if $v \notin -P$. Whence inequality (22) holds. $\square$

Proposition 6.5. *Let $f : X \to \mathbb{R} \cup \{+\infty\}$ be a locally Lipschitz function. Suppose that $x \in \operatorname{Int} \operatorname{Dom} f$ is a local minimum of f. Then, for all $v \in X$, $D_c f(x)(v) \geq 0$. If x is a global minimum of f on a convex set X then*

$$\forall y \in K, \qquad D_c f(x)(y - x) \geq 0. \tag{23}$$

Proof. Suppose that x minimises f on the ball $x + \eta B$ with centre x and radius η. Then, if $\lambda \le \eta/2$ and $\beta \le \eta/2\|v\|$, we have, if $\theta \le \beta$

$$0 \le \frac{f(x + \theta v) - f(x)}{\theta} \le \sup_{\theta \le \beta} \frac{f(x + \theta v) - f(x)}{\theta}.$$

Taking the limit as β tends to 0, we obtain $D_c f(x)(v) \ge 0$. If K is convex, we may take $v = y - x$, since $x + \theta v = (1 - \theta)x + \theta y \in K$ if θ is sufficiently small. □

Proposition 6.6. *Suppose that a nontrivial function $f : X \to \mathbb{R} \cup \{+\infty\}$ is positively homogeneous and locally Lipschitz. Then, for all $x \in \mathrm{Int\ Dom}\ f$,*

$$D_c f(x)(x) = f(x). \tag{24}$$

Proof. We note firstly that

$$\begin{aligned} f(x) &= \frac{f(x + hx) - f(x)}{h} \\ &\le \limsup_{\substack{y \to x \\ h \to 0+}} \frac{f(y + hx) - f(y)}{h} \\ &= D_c f(x)(x). \tag{25} \end{aligned}$$

Suppose L is the Lipschitz constant of f at x. We may write

$$\begin{aligned} \frac{f(y + hx) - f(y)}{h} &= f(y) + \frac{f(y + hx) - f(y + hy)}{h} \tag{26} \\ &\le f(y) + L\|y - w\|. \end{aligned}$$

Whence, taking the upper limit as $h \to 0+$ and y tends to x, we obtain $D_c f(x)(x) \le f(x)$. □

Next we shall study the differentiability of the composition $g = f \circ G$ where G maps an open subset Ω of a Hilbert space Y into Dom f, the domain of f.

Definition 6.2. *We shall say that G is Gâteaux differentiable at $x \in \Omega$ if there exists $\nabla G(x) \in L(Y, X)$ such that*

$$\forall v \in Y, \quad \frac{G(x + \theta y) - G(x)}{\theta} \quad \textit{converges to } \nabla G(x) \cdot v \textit{ in } X \textit{ as } \theta \to 0. \tag{27}$$

We shall say that G is Fréchet differentiable if

$$\lim_{\|v\| \to 0} \frac{\|G(x + v) - G(x) - \nabla G(x) \cdot v\|}{\|v\|} = 0. \tag{28}$$

G is said to be strictly Fréchet differentiable if

$$\lim_{\substack{\|y - x\| \to 0 \\ \|v\| \to 0}} \frac{\|G(y + v) - G(y) - G(x) \cdot v\|}{\|v\|} = 0. \tag{29}$$

Proposition 6.7. *Suppose that f is locally Lipschitz. If G is strictly Fréchet differentiable at x then*

$$D_c g(x)(v) \leq (D_c f)(G(x))(\nabla G(x) \cdot v). \tag{30}$$

Proof. Since f is locally Lipschitz, given $\varepsilon > 0$ and $u \in X$, there exist numbers α, β such that

$$\frac{f(z + \theta w) - f(z)}{\theta} \leq D_c f(G(x))(u) + \varepsilon \tag{31}$$

when $\|z - G(x)\| \leq \alpha$, $\theta \leq \beta$ and $\|w - u\| \leq \frac{\varepsilon}{2L}$. We take $u = \nabla G(x) \cdot v$. There exists $\eta \leq \beta$ such that if $\|y - x\| \leq \eta$ and $\theta \leq \eta$, then $\|G(y) - G(x)\| \leq \alpha$ and

$$\left\| \nabla(x) \cdot v - \frac{G(y + \theta v) - G(y)}{\theta} \right\| \leq \frac{\varepsilon}{2L}$$

since G is strictly differentiable. Taking $z = G(y)$ and $w = \frac{G(y+\theta v) - G(y)}{\theta}$, it follows from (31) that

$$
\begin{aligned}
D_c g(x)(v) &\leq \sup_{\substack{\|y-x\| \leq \eta \\ \theta \leq \eta}} \frac{g(y + \theta v) - g(y)}{\theta} \\
&\leq D_c f(G(x))(\nabla G(x) \cdot v) + \varepsilon. \qquad \square
\end{aligned}
$$

Corollary 6.1. *Let $A \in L(Y, X)$ be a continuous linear operator. Then*

$$D_c(fA)(x)(v) \leq (D_c f)(Ax)(Av). \tag{32}$$

If $A \in L(Y, X)$ is surjective, then

$$D_c(fA)(x)(v) = (D_c f)(Ax)(Av). \tag{33}$$

Proof. In this case, Banach's theorem (see Theorem 4.3.1 of (Aubin 1979a)) implies that $A(x + \alpha B_Y)$ contains a ball $Ax + \gamma(\alpha)B_X$. Consequently,

$$\sup_{\substack{\|z - Ax\| \leq \gamma(\alpha) \\ \theta \leq \beta}} \frac{f(z + \theta Av) - f(z)}{\theta} \leq \sup_{\substack{\|y-x\| \leq \alpha \\ \theta \leq \beta}} \frac{f(Ay + \theta Av) - f(Ay)}{\theta}$$

which implies that $D_c f(Ax)(Av) \leq D_c(fA)(x)(v)$. $\qquad \square$

Suppose that f is a *Lipschitz* function from X to $\mathbb{R} \cup \{+\infty\}$ and that $B \in L(X, Y)$ is a continuous, linear, *surjective* operator from X to Y. We define

$$\alpha(y) = \inf_{Bx=y} f(x), \tag{34}$$

setting $\alpha(y) = +\infty$ if $y \notin B\mathrm{Dom}\, f$.

Proposition 6.8. *Suppose that f is Lipschitz and that $B \in L(X, Y)$ is surjective. Then α is Lipschitz if its domain is non-empty. If $\alpha(y) = f(\bar{x})$ where $B\bar{x} = y$, then*

$$\forall u \in X, \qquad 0 \le D_c\alpha(y)(Bu) + D_c f(\bar{x})(u). \tag{35}$$

Proof. a) We pick arbitrary y and z in Y and $\varepsilon > 0$. There exists $y_\varepsilon \in X$ such that $f(y_\varepsilon) \le \alpha(y) + \varepsilon$ and $By_\varepsilon = y$. Since B is surjective, Banach's theorem (see, for example, Theorem 4.3.1 of (Aubin 1979a)) implies that there exist a constant $c > 0$ and a solution z_ε of the equation $Bz_\varepsilon = z$ satisfying $\|y_\varepsilon - z_\varepsilon\| \le c\|y - z\|$. Whence,

$$\alpha(z) \le f(z_\varepsilon) \le f(y_\varepsilon) + L\|y_\varepsilon - z_\varepsilon\| \le \alpha(y) + \varepsilon + Lc\|y - z\|.$$

This implies that α is Lipschitz with constant Lc, where L is the Lipschitz constant of f.

b) Consider the inequality

$$\begin{aligned}
0 &\le \frac{\alpha(y + \theta Bu - \theta Bu) - \alpha(y + \theta Bu)}{\theta} + \frac{f(\bar{x} + \theta v) - f(\bar{x})}{\theta} \\
&\le \sup_{\substack{\|z - y\| \le \alpha \\ \theta \le \beta}} \frac{\alpha(z - \theta Bv) - \alpha(z)}{\theta} + \sup_{\theta \le \beta} \frac{f(\bar{x} + \theta v) - f(\bar{x})}{\theta}.
\end{aligned}$$

Passing to the limit as α and β tend to 0, we obtain the inequality (35). □

We consider m functions $f_i : X \to \mathbb{R} \cup \{+\infty\}$ and their upper envelope g defined by $g(x) = \max_{i \in I} f_i(x)$ where $I = \{1, \dots, n\}$. We shall denote $I(x) = \{i \in I | g(x) = f_i(x)\}$.

We note that *if the functions f_i are locally Lipschitz, the same is true of their upper envelope.* (If $|f_i(y) - f_i(z)| \le L_i\|y - x\|$ and $y, z \in x + \eta_i B$, then $|g(y) - g(z)| \le L\|y - z\|$ if $y, z \in x + \eta B$ where $\eta = \min_{i \in I} \eta_i > 0$ and $L = \max_{i \in I} L_i > 0$.) Whence, the functions f_i and g are Clarke differentiable.

Proposition 6.9. *Suppose that the m functions f_i are locally Lipschitz and that $x \in \cap_{i \in I} \text{Int Dom } f_i$. Then,*

$$D_c g(x)(v) \le \max_{i \in I(x)} D_c f_i(x)(v). \tag{36}$$

Proof. a) We first note that there exists $\alpha_1 > 0$ such that if $\|x - y\| \le \alpha_1$ then $I(y) \subset I(x)$. (Suppose $a = g(x) - \max_{j \notin I(x)} f_j(x) > 0$, $\varepsilon = \frac{a}{3}$ and $\alpha_1 > 0$ are such that for all $i \in I$ $|f_i(y) - f_i(x)| \le \varepsilon$ whenever $\|y - x\| \le \alpha_1$. Then if $j \in I(y)$

$$\begin{aligned}
f_j(x) &\ge f_j(y) - \varepsilon = g(y) - \varepsilon \ge g(x) - 2\varepsilon \\
&= a - 2\varepsilon + \max_{i \notin I(x)} f_i(x) > \max_{i \notin I(x)} f_i(x).
\end{aligned}$$

Thus $j \in I(x)$.)

b) If $\alpha \leq \alpha_1/2$ and $\beta \leq \alpha_1/2\|v\|$, we obtain the inequality

$$\frac{g(y + \theta v) - g(y)}{\theta} \leq \max_{i \in I(y+\theta v)} \frac{f_i(y + \theta v) - f_i(y)}{\theta} \leq \max_{i \in I(x)} \frac{f_i(y + \theta v) - f_i(v)}{\theta}.$$

Whence,

$$D_c g(x)(v) \leq \inf_{\alpha,\beta>0} \max_{i \in I(x)} \sup_{\substack{\|y-x\| \leq \alpha \\ \theta \leq \beta}} \frac{f_i(y + \theta v) - f_i(y)}{\theta}.$$

Moreover, for all $\varepsilon > 0$, and for all $i \in I$, there exist $\alpha_i > 0$ and $\beta_i > 0$ such that

$$\sup_{\substack{\|y-x\| \leq \alpha_i \\ \theta \leq \beta_i}} \frac{f_i(y + \theta v) - f_i(y)}{\theta} \leq \max_{i \in I(x)} D_c f_i(x)(v) + \varepsilon.$$

Taking $\alpha = \min_{i \in I(x)} \alpha_i > 0$ and $\beta = \min_{i \in I(x)} \beta_i > 0$, it then follows that:

$$
\begin{aligned}
D_c g(x)(v) &\leq \max_{i \in I(x)} \sup_{\substack{\|y-x\| \leq \alpha \\ \theta \leq \beta}} \frac{f_i(y + \theta v) - f_i(y)}{\theta} \\
&\leq \max_{i \in I(x)} \sup_{\substack{\|y-x\| \leq \alpha \\ \theta \leq \beta}} \frac{f_i(y + \theta v) - f_i(y)}{\theta} \leq \max_{i \in I(x)} D_c f_i(x)(v) + \varepsilon.
\end{aligned}
$$

Letting ε tend to 0 completes the proof of the proposition. $\square$

Remark. If the functions f_i are continuously differentiable at x, then their upper envelope g satisfies $Dg(x)(v) = \max_{i \in I(x)} \langle \nabla f_i(x), v \rangle$. If $I(x)$ consists of a single index, then g is Gâteaux differentiable and $\nabla g(x) = \nabla f_{i_0}(x)$ where $g(x) = f_{i_0}(x)$.

6.4 Generalised Gradients

Definition 6.3. *Suppose that the function* $f : X \rightarrow \mathbb{R} \cup \{+\infty\}$ *is Clarke differentiable at* x. *Then the subset* $\partial f(x)$ *of* X *defined by*

$$\partial f(x) = \{p \in X^* | \langle p, v \rangle \leq D_c f(x)(v), \ \forall v \in X\} \tag{37}$$

is called the **generalised gradient** *of* f *at* x.

Theorem 6.2. *Suppose the function* $f : X \rightarrow \mathbb{R} \cup \{+\infty\}$ *is locally Lipschitz. Then it has a non-empty generalised gradient* $\partial f(x)$ *at any point* x *in the interior of* Dom f, *which is convex, closed and bounded and has a support function* $\sigma(\partial f(x), v) := \sup\{\langle p, v \rangle | p \in \partial f(x)\}$ *which satisfies*

$$\sigma(\partial f(x), v) = D_c f(x)(v). \tag{38}$$

Moreover, the set-valued map $x \in \text{Int Dom } f \to \partial_c f(x) \in X^*$ satisfies

$$(x, v) \in \text{Int Dom } f \times X \to \sigma(\partial f(x), v) \text{ is upper semi-continuous.} \tag{39}$$

Proof. The theorem follows from Definition 6.1 and Theorem 6.1. Since $D_c f(x)(\cdot)$ is convex, positively homogeneous and continuous, it is the support function of the convex closed subset of elements $p \in X^*$ such that $\langle p, v \rangle \leq D_c f(x)(v)$ for all $v \in X$; in other words, of the generalised gradient $\partial f(x)$ of f at x. Thus, $\partial f(x)$ is non-empty, convex and closed and (38) applies. Since $\sigma(\partial f(x), v) \leq L\|v\| = L\sigma(B^*, v)$ (where B^* is the unit ball of X^*), it follows that

$$\partial f(x) \subset LB^* \tag{40}$$

This completes the proof of Theorem 6.2. □

Proposition 6.10. *If f is continuously differentiable then the generalised gradient $\partial f(x) = \{\nabla f(x)\}$ reduces to the usual gradient $\nabla f(x)$. If f is convex and continuous at x, then the generalised gradient is equal to the subdifferential of f at x.*

Proof. Proposition 6.1 shows that if f is continuously differentiable, then $\sigma(\partial f(x), v) = D_c f(x)(v) = \langle \nabla f(x), v \rangle$. Thus, $\partial f(x) = \{\nabla f(x)\}$. Proposition 6.2 shows that if f is convex and continuous, then $\sigma(\partial_c f(x), v) = D_c f(x)(v) = Df(x)(v)$. Now, the right derivative of a convex continuous function is the support function of the subdifferential $\partial f(x)$ (see Theorem 4.1). □

Propositions 6.3 to 6.7 translate as follows in terms of the generalised gradient.

Proposition 6.11. *Suppose f and g are two locally Lipschitz functions. If $x \in \text{Int Dom} f \cap \text{Int Dom } g$, then if α and $\beta > 0$,*

$$\partial(\alpha f + \beta g)(x) \subset \alpha \partial f(x) + \beta \partial g(x). \tag{41}$$

If $x \in \text{Int Dom} f$, then

$$\partial_c(-f)(x) = -\partial_c f(x). \tag{42}$$

If f is increasing on a convex closed cone P, then

$$\partial_c f(x) \subset P^+. \tag{43}$$

If x is a local minimum of f, then x is a solution of the inclusion

$$0 \in \partial f(x) \qquad \text{(Fermat's rule).} \tag{44}$$

If f is positively homogeneous, then

$$\forall p \in \partial f(x), \qquad \langle p, x \rangle = f(x). \tag{45}$$

If $A \in L(Y, X)$ and if $A^* \in L(X^*, Y^*)$ denotes its transpose, then

$$\partial_c(fA)(x) \subset A^* \partial_c f(Ax) \tag{46}$$

with equality if A is surjective.

Proof. It suffices to note that since the generalised gradients $\partial f(x)$ are convex, closed and bounded, they are weakly compact; whence $\partial f(x) + A$ is convex and closed when A is convex and closed (see (Schwartz 1970)).   □

Proposition 6.8 translates in the following way.

Proposition 6.12. *Suppose that f is a Lipschitz function from X to $\mathbb{R} \cup \{+\infty\}$ and that $B \in L(X, Y)$ is a surjective operator. If $y \in \mathrm{Int}(B \cdot \mathrm{Dom} f)$ and if $\bar{x}$ is the solution of the minimisation problem*

$$\alpha(y) = \inf_{Bx=y} f(x) = f(\bar{x}) \quad \text{where } B\bar{x} = y, \tag{47}$$

then there exists $\bar{p} \in V^*$ satisfying

$$\bar{p} \in \partial \alpha(y) \quad \text{with } B^* \bar{p} \in \partial f(\bar{x}). \tag{48}$$

Proof. Following Proposition 6.8,

$$0 \le D_c \alpha(y)(-Bv) + D_c f(\bar{x})(v) = \sigma(-B^* \partial \alpha(y) + \partial f(\bar{x}), v).$$

Whence $0 \in \partial f(\bar{x}) - B^* \partial \alpha(y)$.   □

Remark. An element $\bar{p} \in Y^*$ satisfying (48) is called a *Lagrange multiplier* for the problem of minimisation of a Lipschitz function f under the linear equality constraints $Bx = y$.

Proposition 6.9 may be restated as follows:

Proposition 6.13. *Suppose that the m functions f_i are locally Lipschitz and that $x \in \cap_{i \in I} \mathrm{Int} \, \mathrm{Dom} f_i$. Then*

$$\partial g(x) \subset \overline{\mathrm{co}} \bigcup_{i \in I(x)} \partial_c f_i(x). \tag{49}$$

6.5 Normal and Tangent Cones to a Subset

Suppose that K is an arbitrary *non-empty* subset of X. We let d_K denote the distance function measuring the distance from K, defined by

$$d_K(y) = \inf_{y \in K} \|x - y\|. \tag{50}$$

This is clearly Lipschitz with constant 1:

$$|d_K(Y) - d_K(z)| \le \|y - z\|$$

Consequently, it is Clarke differentiable.

Definition 6.4. *Suppose $x \in K$. We shall say that the set*

$$T_K(x) := \{v \in X | D_c d_K(x)(v) \le 0\} \tag{51}$$

is the **tangent cone** *to K at x and that*

$$N_K(x) = T_K(x)^- = \{p \in X^* | \langle p, v \rangle \le 0, \ \forall v \in T_K(x)\} \tag{52}$$

is the **normal cone** *to K at x.*

Since $v \to D_c d_K(x)(v)$ is convex, positively homogeneous and continuous, $T_K(x)$ is a *convex closed cone*. Since the normal cone is the negative polar cone of $T_K(x)$, $N_K(x)$ is a *convex closed cone*.
Whence

$$T_K(x) = N_K(x)^-. \tag{53}$$

It is useful to define the normal cone in terms of the generalised gradient of d_K.

Proposition 6.14. *The normal cone $N_K(x)$ is the closed convex cone generated by the generalised gradient of d_K at x*

$$N_K(x) = (\partial d_K(x))^{--}. \tag{54}$$

Proof. We show that $\partial d_K(x)$ is contained in the normal cone $N_K(x)$. If $p \in \partial d_K(x)$ and if $v \in T_K(x)$, then $\langle p, v \rangle \le D_c d_K(x)(v) \le 0$. Thus, the convex, closed cone generated by $\partial d_K(x)$ is contained in the normal cone $N_K(x)$. To prove the inverse inclusion, it is sufficient to show that $(\partial d_K(x))^- \subset T_K(x)$. Suppose then that $v_0 \in (\partial d_K(x))^-$. Then $D_c d_K(x)(v_0) = \sigma(\partial d_K(x), v_0) = \sup\{\langle p, v_0 \rangle | p \in \partial_c d_K(x)\} \le 0$. Thus v_0 belongs to $T_K(x)$. $\square$

Next we state certain elementary properties of tangent and normal cones. First we mention the following fact

If $\mathrm{Int}\ K \ne \emptyset$ and if $x \in \mathrm{Int}\ K$, then $T_K(x) = X$ and $N_K(x) = 0$. $\tag{55}$

Indeed, if $x + \eta B \subset K$, then, for all $v \in X$, $y + \theta v$ belongs to K if $\|y - x\| \le \alpha$ and $\theta \le \beta$ whenever $\alpha \le \eta/2$ and $\beta \le \eta/2\|v\|$. Thus,

$$D_c d_K(x)(v) = \inf_{\alpha, \beta > 0} \sup_{\substack{\|y - x\| \le \alpha \\ \theta \le \beta}} \frac{d_K(y + \theta v) - d_K(y)}{\theta} \le 0.$$

6.6 Fermat's Rule for Minimisation Problems with Constraints

Proposition 6.15. *Let f be a locally Lipschitz function. Suppose that $x \in K$ minimises f over K. Then, there exists $L > 0$ such that*

$$\forall y \in X, \qquad f(x) + Ld_K(x) \le f(y) + Ld_K(y) \tag{56}$$

and consequently, $0 \in \partial f(x) + N_K(x)$ (or $-\partial f(x) \cap N_K(x) \ne \emptyset$).

Proof. Since f is locally Lipschitz, there exists a neighbourhood $x + \eta B$ on which f is Lipschitz with constant L. We take $\alpha < \eta/2$ and $\varepsilon < \frac{\eta - 2\alpha}{2}$ which is destined to tend to 0. Let $y \in x + \alpha B$. Then, we may associate y with some $y_\varepsilon \in K$ such that $\|y - y_\varepsilon\| \le d_K(y)(1 + \varepsilon)$. Moreover, $\|x - y\| \le \alpha \le \eta$, $d_K(y) \le \|x - y\| \le \alpha$ and $\|x - x_\varepsilon\| \le \|x - y\| + \|y - y_\varepsilon\| \le \alpha + \alpha(1 + \varepsilon) = \alpha(2 + \varepsilon) \le \eta$. Consequently, $f(y_\varepsilon) \le f(y) + L\|y - y_\varepsilon\| \le f(y) + L(1 + \varepsilon)d_K(y)$. Since $d_K(x) = 0$ (because $x \in K$) and since $f(x) \le f(y_\varepsilon)$ (because $y_\varepsilon \in K$) we obtain $f(x) + Ld_K(x) \le f(y) + L(1 + \varepsilon)d_K(y)$ for all $y \in x + \alpha B$. Letting ε tend to 0, we deduce that x is a *local minimum of the function* $y \to f(y) + Ld_K(y)$. Whence, by virtue of Propositions 6.3, 6.5 and 6.14, we obtain the inequality

$$
\begin{aligned}
0 &\le D_c(f + Ld_K)(x)(v) \le D_c f(x)(v) + LD_c d_K(x)(v) \\
&\le \sigma(\partial_c f(x) + L\partial d_K(x), v)
\end{aligned}
$$

for all $v \in X$.

This implies that $0 \in \partial f(x) + L\partial d_K(x) \subset \partial f(x) + N_K(x)$, following Proposition 6.15. $\qquad\square$

Remark. The first assertion is *very important* in the sense that it allows us to replace a minimisation problem with constraints by a minimisation problem without constraints.

Remark. If K is convex, $N_K(x)$ is the normal cone of convex analysis (see Definition 4.3). We note that the function d_K is convex. Take $p \in N_K$. If $y \in K$, we have

$$\langle p, y - x \rangle \le D_c d_K(x)(y - x) = Dd_K(x)(y - x) \le d_K(y) - d_K(x) = 0.$$

Conversely, if x maximises $y \to \langle p, y \rangle$ over K, Proposition 6.15 implies that $0 \in -p + N_K(x)$. Thus, the two concepts of normal cones coincide.

7. Two-person Games. Fundamental Concepts and Examples

7.1 Introduction

Let us consider two subsets E and F. Our aim is to choose pairs $(x, y) \in E \times F$ using various optimisation techniques motivated by so-called decision theory. This means that we shall provide mechanisms for selecting elements (called decisions or strategies) of sets (decision sets, strategy sets) which should reflect real decision-taking techniques as closely as possible.

The history of science shows us that parlour games have presented mathematicians with numerous problems. Thus, the chevalier de Méré consulted Blaise Pascal about the problems of dice games. This led to a correspondence between Blaise Pascal and Pierre de Fermat; the six letters they exchanged served as a departure point for modern probability theory (which proves that mathematicians may also profit from immoral company!) [1].

The terms players and strategies have been used since the start and tradition (conservatism) has led to their retention. The current status of the theory of games as a mathematical theory is due to John von Neumann who, between 1928 and 1941, proposed a general framework, with a view to applications in the social sciences, within which conflicts and cooperation of players may be taken into account. This fundamental work formed the skeletal structure of the

---

[1] '... My best teacher of this worldly science was Antoine Gombaud, chevalier de Méré.... He was a strong little man, very elegant and scented, who voluntarily established himself as a judge of etiquette and graces. After several sea campaigns, he limited his gallantry to the conquest of the salons and gave up the sword for the pen. He was very friendly with Pascal, Balzac, Ménage, Clérambault and other men of letters of his time and himself perpetrated a number of treatises on 'true honesty', 'eloquence', 'the delicacy of expression' and 'worldly intercourse'.... However, I was taken by the idea of passions and the feelings which engender them stole into my mind with no specific object in view. It is true that they could have settled on the chevalier himself and that it was not because of him they did not rest there. In fact, Monsieur de Méré was enamoured of his fourteen year old school girl; he told me as much in poems in which, because of my journeys to the islands, he referred to me as the 'beautiful Indian'.... For my part, Monsieur de Méré was not to my liking.... However, I was flattered that he took a liking to me: the first and the last conquests are those for which one is most grateful.' FRANÇOISE CHANDERNAGOR. *L'Allée du Roi*. Recollections of Françoise d'Aubigné, marquess of Maintenon, wife of the King of France.

book *Theory of Games and Economic Behaviour*, which he published in 1944 in collaboration with the economist O. Morgenstern.

In fact, this change of direction is due to Léon Walras, who introduced the description of a consumer as *an automaton seeking to maximise a utility function subject to various types of constraints imposed by its environment*. In this case, the strategies are commodities and prices and the players are consumers, manufacturers and other economic agents. The individuals who adopt this behaviour of an automaton are said to be "rational". This should not be taken as the definition of the adjective 'rational' in the philosophical context (in the sense of natural knowledge, as opposed to that which comes from myths or from faith). Anyway, the concept of reason is the subject of cognitive psychology and little is known about this topic except that most of the time a rational individual cannot maximise his utility function, assuming that he has one. Some ten years ago, H. Simon and others questioned the universality of this behaviour and proposed replacing the notion of optimality by a less nontrivial notion of satisfactoriness.

Whilst we await the psychologists' findings about knowledge, one way of resolving the dilemma is to realise that the first point of view is static, whilst the second is dynamic. Taking into account evolutionary phenomena, we need no longer assume that an individual is looking for a permanent optimum but may suppose that he seeks to increase his utility as he goes along. The second point of view is less unrealistic in this sense.

We shall restrict ourselves here to the static case (anyway, investigations of the dynamic framework are now under course). Even with these limitations, game theory has provided economists with useful tools for clarifying concepts. In order to avoid being distracted, one must always remember that these are only imperfect and perfectible tools and that one should beware of all dogmatism when using them.

Curiously enough, the mathematical problems which have been motivated by game theory have led to major contributions to nonlinear analysis which have ultimately been useful in very many areas; this is another example of the universality of mathematical results which we mentioned when talking about Banach at the end of Chapter 2.

7.2 Decision Rules and Consistent Pairs of Strategies

Let us christen our two players Emil and Frances. Emil's task is to choose a strategy x in the set E and that of Frances is to choose a strategy y in the set F. The pair $(x, y) \in E \times F$ is also called a *bistrategy*.

One elementary mechanism which allows Emil and Frances to select their respective strategies involves providing them with *decision rules*.

Definition 7.1. *A* **decision rule** *for Emil is a set-valued map C_E from F to E which associates each strategy $y \in F$ of Frances with the strategies $x \in C_E(y)$ which may be played by Emil when he knows that Frances is playing y.*

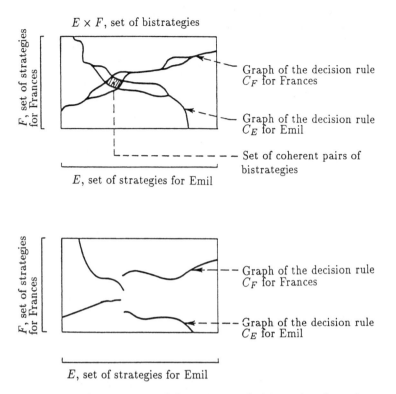

Fig. 7.1. Examples of one-to-one and discontinuous decision rules where there are no consistent strategy pairs.

Similarly, a decision rule for Frances is a set-valued map C_F from E to F which associates each strategy $x \in E$ played by Emil and known to Frances with the strategies $y \in C_F(x)$ which Frances may implement.

Once the players Emil and Frances have been described in terms of their decision rules C_E and C_F, we may distinguish pairs of strategies (x, y) which are in *static equilibrium,* in the sense that

$$\bar{x} \in C_E(\bar{y}) \quad \text{and} \quad \bar{y} \in C_F(\bar{x}) \tag{1}$$

Definition 7.2. *A pair of strategies $(\bar{x}, \bar{y})$ which satisfies condition (1) for the decision rules C_E of Emil and C_F of Frances is called a* **consistent pair** *of strategies or a* **consistent bistrategy**.

The interest of this concept of consistent bistrategies naturally depends on the choice of decision rules.

The most simple example of a decision rule is that of a *constant decision rule.* A strategy $x \in E$ of Emil may be identified with the constant decision rule

$y \in F \to x \in E$, which describes obstinate behaviour by Emil, who plays the strategy x irregardless of the strategy chosen by Frances.

Consequently, when Emil and Frances play the strategies x and y, respectively, the pair (x, y) forms a consistent pair.

The set of consistent pairs may be empty or very large or it may reduce to a small number of bistrategies. A mechanism will only be of interest to a game theorist if, firstly, the set of consistent pairs is non-empty and, secondly, this set is small (in the best case consisting of a single pair).

We note that the problem of finding consistent strategy pairs is a so-called *fixed-point problem*. We use $\mathbf{C}$ to denote the set-valued map of $E \times F$ into itself defined by

$$\forall (x, y) \in E \times F, \qquad \mathbf{C}(x, y) := C_E(y) \times C_F(x). \tag{2}$$

The inclusions (1) which define the consistent pairs may clearly be written in the form

$$(\bar{x}, \bar{y}) \in \mathbf{C}(\bar{x}, \bar{y}). \tag{3}$$

This is a primary motivation behind the derivation of fixed-point theorems. We shall quote (and admit without proof the prototype of these theorems, the Brouwer Theorem) the most famous of these theorems.

7.3 Brouwer's Fixed-point Theorem (1910)

Theorem 7.1. *Let K be a convex compact subset of a finite-dimensional space. Any continuous mapping of K into itself has a fixed point.*

We shall discuss the consequences of this theorem, which turn out to be convenient and easy to handle, above all in applications to game theory.

The Dutch mathematician Brouwer (1881–1966) is famous for his contributions to mathematical logic and was one of the founders of combinatorial topology, where he innovatively introduced the important notion of a simplex and the triangulation technique which he used to prove this famous theorem which is at the root of nonlinear analysis.

Thus, we obtain the following corollary.

Corollary 7.1. *Suppose that the behaviours of Emil and Frances are described by one-to-one continuous decision rules and that the strategy sets E and F are* **convex compact** *subsets of finite-dimensional vector spaces, then there is at least one* **consistent strategy pair**.

Proof. We take $K := E \times F$ which is convex and compact. Then the set-valued map $\mathbf{C}$ defined by (2) is a continuous mapping and thus has a fixed point. □

We shall generalise this theorem to the case of multi-valued decision rules.

7.4 The Need to Convexify: Mixed Strategies

Brouwer's Theorem which, in practice, is the fundamental tool for finding consistent strategies together with the Separation Theorem which, as we have already seen, is at the root of optimisation theory, both assume that the strategy sets are *convex*. This is an exorbitant assumption which excludes, for example, the use of *finite strategy sets*.

What then should we do? As so often in mathematics, starting from a situation which appears hopeless because of the absence of desirable properties, one boldly invents another situation in which the validity of these properties is re-established.

We shall follow this course, beginning with a finite strategy set. Suppose that $E = \{1, \ldots, n\}$ is a set of n elements.

We associate E with the subset

$$M^n := \{\lambda \in \mathbb{R}^n_+ \mid \sum_{i=1}^n \lambda_i = 1\} \tag{4}$$

called the $(n-1)$ simplex of $\mathbb{R}^n$.

This is clearly a convex compact subset of $\mathbb{R}^n$. We can embed E in M^n by the mapping δ which associates the ith element of E with the ith element e^i of the canonical basis of $\mathbb{R}^n$:

$$\delta : i \in \{1, \ldots, n\} \to \delta(i) := e^i := (0, \ldots, 1, \ldots, 0) \tag{5}$$

We also note that

$$M^n = \mathrm{co}\{e^1, \ldots, e^n\} \tag{6}$$

Interpretation

J. von Neumann proposed interpreting the elements $\lambda \in M^n$ as *mixed strategies*. In this framework, a player does not choose a single strategy as before but plays all the strategies and chooses only the probabilities with which he plays them.

One important justification for a player's use of mixed strategies is the protection which he obtains by disguising his intentions from his partners. By playing the different strategies randomly, in such a way that only their probabilities are determined, he prevents his partners from discovering the strategy which he is going to play, since he does not know it himself.

We must not hide the fact that, in 'convexifying' strategy sets, we are moving away from our original static framework, since random play assumes that the game will be repeated!

However, one might reason that there is a 'game' if there is *uncertainty* in the choices of the players and, thus, taking this uncertainty into account we may rejoin the static framework.

Psychologists and sociologists suggest that this uncertainty which enables the players (actors) to take a detached view of the decisions with which they are faced should be considered as a *component* of the notion of *power*.

We shall also have recourse to this *convexification process* later in the context of cooperative games, where we convexify the set of *coalitions of players* (see Chapter 13).

In fact, this is a mathematical necessity which provides a palliative improvement of the static case bringing it halfway towards the dynamic framework which has just been invented.

Any set-valued map C from $E = \{1, \ldots, n\}$ to a vector space X may be extended to a set-valued map $\widehat{C}$ from $\mathbb{R}^n$ to X as follows

$$\forall \lambda \in \mathbb{R}^n, \quad \widehat{C}(\lambda) := \sum_{i=1}^{n} \lambda_i C(i) \tag{7}$$

If δ is the mapping from $\{1, \ldots, n\}$ to M^n defined by (5), we have the following scheme

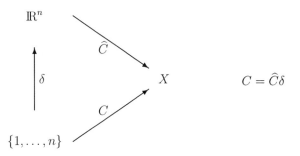

If C is a one-to-one mapping, it is clear that its extension $\widehat{C}$ is a linear mapping from $\mathbb{R}^n$ to X.

The process which associates the mapping $C : \{1, \ldots, n\} \to X$ with the linear mapping $\widehat{C} : M^n \to X$ may be thought of as a *linearisation process* associated with the *convexification* process which associates the convex compact set M^n with the finite set $\{1, \ldots, n\}$.

7.5 Games in Normal (Strategic) Form

The traditional way of modelling game theory is to assume that each player classifies the bistrategies using an *evaluation function* f. This function has several names, for example, *criterion function*, *utility function*, *gain function*, *loss function*, *cost function*, etc. The terminology is a matter of taste. Whatever terminology is used, such a function may be associated with a partial order $\geq$ (called the partial order of preferences) as follows

$$(x_1, y_1) \in E \times F \text{ is preferred to } (x_2, y_2) \in E \times F$$
$$\text{if and only if } f(x_1, y_1) \leq f(x_2, y_2) \tag{8}$$

(for loss functions or cost functions; for utility functions and gain functions, the direction of the inequality is inverted).

A player behaves so as to *minimise his losses* as far as possible.

Remark. We have associated a partial order with a loss function f. We note that the partial order remains unchanged (in fact this is the only thing that matters) if we replace the function f by any function $\varphi \circ f$ where φ is a strictly increasing bijection from $\mathbb{R}$ to $\mathbb{R}$.

In particular, if $a > 0$ and $b \in \mathbb{R}$ are given, the function $af + b$ defines the same partial order as f.

The inverse question then arises: can we represent any partial order by an evaluation function? Sadly, the most common partial order on $\mathbb{R}^n$, the lexicographic partial order cannot be represented by a continuous utility function. This led to a debate lasting several decades between supporters and adversaries of utility functions, until Gérard Debreu derived a theorem showing that a large class of partial orders may be represented by continuous functions (for a simple version of this theorem, see Theorem 5.4.1 of (Aubin 1977)).

We shall not become embroiled in this debate, especially since considerations of cognitive psychology seem to indicate that the mechanisms of choice do not obey (globally) rules for classification according to a partial order. Moreover, this notion is of little meaning in a dynamic framework.

Nonetheless, this is still a source of intrinsically interesting mathematical problems. The relevance of the assumption that the behaviour of the players is based on evaluation functions is a concern of economics, which is not an exact science.

None of this is very serious since, whilst the use of utility functions may legitimately be rejected, it is more difficult to take issue over the use of decision rules at this level of generality.

Let us return to our problem. We now suppose that the players Emil and Frances choose (separately) their strategies using their loss functions f_E and f_F from $E \times F$ to $\mathbb{R}$.

We set

$$\mathbf{f}(x, y) := (f_E(x, y), f_F(x, y)) \in \mathbb{R}^2. \tag{9}$$

Definition 7.3. *A two-person game in* **normal (strategic) form** *is defined by a mapping* $\mathbf{f}$ *from* $E \times F$ *into* $\mathbb{R}^2$ *called a* **biloss mapping**.

We have described a natural way of associating decision rules with the players of a game in strategic form. Let us now consider Emil's loss function f_E. If he happens to know the strategy $y \in F$ played by Frances, he may be tempted to choose the strategy $\bar{x} \in E$ which minimises his loss $x \to f(x, y)$, assuming Frances's strategy is fixed. In other words, he may choose a strategy in the set $\overline{C}_E(y)$ defined by

$$\overline{C}_E(y) := \{\bar{x} \in E | f_E(\bar{x}, y) = \inf_{\{x \in E\}} f_E(x, y)\}. \tag{10}$$

This enables us to define a decision rule $\overline{C}_E : F \to E$ for Emil. Similarly, we define the decision rule $\overline{C}_F : E \to F$ for Frances by the formula

$$\overline{C}_F(x) := \{\bar{y} \in |f_F(x, \bar{y}) = \inf_{\{y \in F\}} f_F(x, y)\}. \tag{11}$$

Definition 7.4. *The decision rules $\overline{C}_E$ and $\overline{C}_F$ associated with the loss functions by formulae (10) and (11) are called the* **canonical decision rules**.

A consistent pair of strategies $(\bar{x}, \bar{y})$ based on the canonical decision rules is called a **non-cooperative equilibrium (or a Nash equilibrium) of the game.**

Thus, a pair $(\bar{x}, \bar{y})$ is a non-cooperative equilibrium if and only if

(i) $$f_E(\bar{x}, \bar{y}) = \inf_{x \in E} f_E(x, \bar{y})$$
(ii) $$f_F(\bar{x}, \bar{y}) = \inf_{y \in F} f_F(\bar{x}, y). \tag{12}$$

Consequently, a non-cooperative equilibrium is a situation in which each player optimises his own criterion, assuming that his partner's choice is fixed. This is called a situation with *individual stability*.

One convenient way of finding non-cooperative equilibria is to introduce the functions

(i) $$f_E^b(y) := \inf_{x \in E} f_E(x, y) \qquad f_E \text{ flat}$$
(ii) $$f_F^b(y) := \inf_{y \in F} f_F(x, y) \qquad f_F \text{ flat}. \tag{13}$$

Thus, we note that a pair $(\bar{x}, \bar{y}) \in E \times F$ is a non-cooperative equilibrium if and only if

(i) $$f_E(\bar{x}, \bar{y}) = f_E^b(y)$$
(ii) $$f_F(\bar{x}, \bar{y}) = f_F^b(\bar{x}) \tag{14}$$

7.6 Pareto Optima

Does the concept of non-cooperative equilibrium provide the only reasonable scheme for solution of a game in strategic form? This is not necessarily the case, particularly if we assume that the players communicate, exchange information and cooperate. In this case, they may notice that there exist strategy pairs (x, y) satisfying

$$f_E(x, y) < f_E(\bar{x}, \bar{y}) \text{ and } f_F(x, y) < f_F(\bar{x}, \bar{y}) \tag{15}$$

where the two players Emil and Frances have losses *strictly less* than in the case of non-cooperative equilibrium $(\bar{x}, \bar{y})$. When this situation occurs, it betrays a lack of *collective stability* , since the two players can each find 'better' strategies for themselves.

Definition 7.5. *A strategy pair* $(x_*, y_*) \in E \times F$ *is said to be* **Pareto optimal**[2] *if there are no other strategy pairs* $(x, y) \in E \times F$ *such that* $f_E(x, y) < f_E(x_*, y_*)$ *and* $f_F(x, y) < f_F(x_*, y_*)$.

The idea is that there exist non-cooperative equilibria which are Pareto optimal. Regrettably, there are few examples of such equilibria and no general theorem is known.

We denote the set of losses of each player by $\mathbb{R}_E = \mathbb{R}$ and $\mathbb{R}_F := \mathbb{R}$ (respectively) and the set of the players' bilosses by $\mathbb{R}^2 := \mathbb{R}_E \times \mathbb{R}_F$.

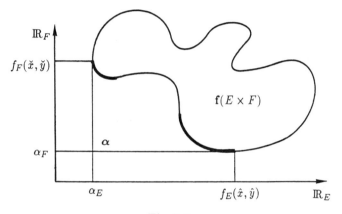

Fig. 7.2.

Figure 7.2 shows the set $f(E \times F) \subset \mathbb{R}^2$ of bilosses $\mathbf{f}(x, y) = (f_E(x, y), f_F(x, y))$ suffered by the two players. The bilosses corresponding to the Pareto optima are shown by thick lines. We note that the selection of the Pareto optima is not a very precise mechanism.

Suppose, for example, that there exists a pair $(\breve{x}, \breve{y})$ which minimises Emil's loss function f_E on $E \times F$:

$$f_E(\breve{x}, \breve{y}) = \inf_{\substack{x \in E \\ y \in F}} f_E(x, y) =: \alpha_E. \tag{16}$$

Clearly, such a pair is Pareto optimal. For Frances to accept this situation, we must assume that her only goal in life is to please Emil. Similar comments apply to any Pareto-optimal strategy pair $(\hat{x}, \hat{y})$ which minimises f_F on $E \times F$

$$f_F(\hat{x}, \hat{y}) = \inf_{\substack{x \in E \\ y \in F}} f_F(x, y) =: \alpha_F. \tag{17}$$

---

[2]In fact, to be exact, we should use the term 'weakly Pareto optimal'. We commit this abuse of terminology consciously.

We observe that if a strategy pair $(\tilde{x}, \tilde{y})$ minimises both f_E and f_F simultaneously on $E \times F$, then it is the best candidate for a solution scheme. In this case, we would have

$$\alpha_E = f_E(\tilde{x}, \tilde{y}) \qquad \text{and} \qquad \alpha_F = f_F(\tilde{x}, \tilde{y})$$

which only happens in exceptional cases. This is why the vector

$$\boldsymbol{\alpha} := (\alpha_E, \alpha_F) \in \mathbb{R}^2 \tag{18}$$

is called the *virtual or 'shadow' minimum* of the game.

We also note that the bistrategy $(\check{x}, \check{y})$ of (16) (or the bistrategy $(\hat{x}, \hat{y})$ of (17)) is not a propitious choice if one takes into account sensible psychological considerations. It is reasonable to think (or rather to expect) that the players will agree to replace a given strategy pair by another strategy pair which will result in lower losses *for each of them*. One cannot sensibly imagine that one player would let the other player be the sole beneficiary of this operation.

In fact, one of the objectives of the theory of cooperative games is to provide mechanisms for selecting Pareto optima. One example of this selection process is the case in which Frances's behaviour consists of pleasing Emil without taking her own interest into account (devoted behaviour); this leads to the strategy pairs $(\check{x}, \check{y})$ of (16).

7.7 Conservative Strategies

If behaviour of this type exists, it is not universal. There is also the contrary behaviour, in which Frances's only goal is to annoy Emil and where Emil is aware of this (we assume that Emil is convinced of Frances's dark designs, or that he is paranoic, etc.). In this case, Emil evaluates the loss associated with a strategy x using the function $f_E^\sharp$ (f_E sharp) defined by

$$f_E^\sharp(x) := \sup_{y \in F} f_E(x, y). \tag{19}$$

$f_E^\sharp$ is said to be Emil's *worst-loss function*. In this case, Emil's behaviour consists of finding strategies $x^\sharp \in E$ which minimise the worst loss, namely solutions of

$$f_E^\sharp(x^\sharp) = \inf_{x \in E} f_E^\sharp(x). \tag{20}$$

We shall say that Emil's strategy x is *conservative*. We set

$$v_E^\sharp = \inf_{x \in E} \sup_{y \in F} f_E(x, y) =: \inf_{x \in E} f_E^\sharp(x) \tag{21}$$

and call $v_E^\sharp$ Emil's *conservative value*.

This conservative value may be used as a *threat*. Emil may always reject a strategy pair $(x, y) \in E \times F$ satisfying

$$f_E(x, y) > v_E^\sharp \tag{22}$$

since, by playing a conservative strategy $x^\sharp \in E$, Emil ensures that his loss $f_E(x^\sharp, y)$ is strictly less than $f_E(x, y)$ since $f_E^\sharp(x^\sharp, y) \leq f_E(x^\sharp) =: v_E^\sharp < f_E(x, y)$.

Consequently, if he cannot reach agreement with his partner, he can always threaten to play a conservative strategy $x^\sharp$ so as to limit his loss to $v_E^\sharp$.

In a symmetric fashion, Frances's *conservative value* is defined by

$$v_F^\sharp := \inf_{y \in F} \sup_{x \in E} f_F(x, y) =: \inf_{y \in F} f_F^\sharp(y) \tag{23}$$

where we have set

$$f_F^\sharp(y) := \sup_{x \in E} f_F(x, y) \qquad (f_F \text{ sharp}). \tag{24}$$

We shall call the vector

$$\mathbf{v}^\sharp := (v_E^\sharp, v_F^\sharp) \tag{25}$$

the *conservative vector for the game*.

Thus, the only strategy pairs of any interest are those which satisfy

$$\mathbf{f}(x, y) \leq \mathbf{v}^\sharp. \tag{26}$$

Thus, the set of bilosses of the strategies of interest is contained in the rectangle $[\alpha_E, v_E^\sharp] \times [\alpha_F, v_F^\sharp]$ (see Fig. 7.3). Here we have a first selection process.

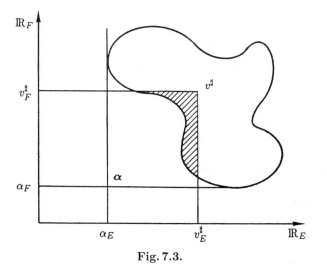

Fig. 7.3.

The idea is to find non-cooperative equilibria which are Pareto optimal or pairs of conservative strategies which are Pareto optimal. *Games in which the conservative vector is Pareto optimal* are usually called *inessential games*. We shall show that certain zero-sum games are inessential.

7.8 Some Finite Games

We shall give examples of games to illustrate the concepts described above.

In all these games, Emil and Frances have strategy sets consisting of two elements $E := \{I, II\}$, $F = \{1, 2\}$. The biloss mapping is represented by bimatrices

Emil / Frances	1	2
I	(a, b)	(c, d)
II	(e, f)	(g, h)

For example, if the strategies $\{I, 2\}$ are played, Emil's loss is equal; to c and that of Frances is equal to d.

We begin with the well-known game of the prisoner's dilemma.

Prisoner's dilemma

Emil and Frances are accomplices to a crime which leads to their imprisonment. Each has to choose between the strategies of confession (strategies I and 1, respectively) or accusation (strategies II and 2, respectively).

If neither confesses, moderate sentences (a years in prison) are handed out. If Emil confesses and Frances accuses him, Frances is freed (0 years in prison) and Emil is sentenced to $c > a$ years in prison. If both confess, they will each have to serve b years in prison, where $a < b < c$.

Emil / Frances	1 (peaceable)	2 (aggressive)
I (peaceable)	(a, a)	$(c, 0)$
II (aggressive)	$(0, c)$	(b, b)

Many authors have embroidered on this game. The original interpretation may also be modified in favour of diplomatic or military illustrations.

For example, strategies I and 1 may be interpreted as being *peaceable* whilst strategies II and 2 are *aggressive* .

Figure 7.4 shows the losses incurred in each case. We illustrate this game in the space of bilosses. We have

$$f_E^\sharp(I) = c, \quad f_E^\sharp(II) = b, \quad f_F^\sharp(1) = c, \quad f_F^\sharp(2) = b$$

whence

$$v_E^\sharp = f_E^\sharp(II) = b, \quad v_F^\sharp = f_F^\sharp(2) = b$$

and the strategy pair $(II, 2)$ is *conservative*. It follows that the pairs $(I, 2)$ and $(II, 1)$ are useless, since, for example, by playing I, Emil risks a loss of c and by playing II, Emil limits his loss to $b < c$.

In addition, we have

$$f_E^{\flat}(1) = 0, f_E^{\flat}(2) = b, \quad f_F^{\flat}(I) = 0, \quad f_F^{\flat}(II) = b$$

whence the strategy pair $(II, 2)$ is also a *non-cooperative equilibrium*, since

(i) $\qquad\qquad\qquad f_E(II, 2) = b < f_E(I, 2) = c$

(ii) $\qquad\qquad\qquad f_F(II, 2) = b < f_F(II, 1) = c.$

The strategy pairs $(I, 1)$, $(II, 1)$ and $(1, 2)$ are Pareto optimal.

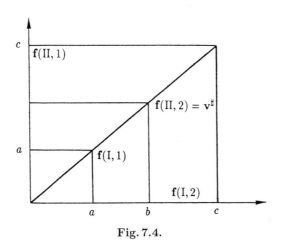

Fig. 7.4.

If there is no cooperation or communication between the players, aggressive strategies will be chosen, whilst an examination of the situation and a minimum of cooperation will enable the players to choose peaceable strategies. However, when playing a peaceable strategy, a player runs a large risk if he allows his partner to play an aggressive strategy.

The paradoxes which arise *en masse* stem from the elementary and simplistic nature of this game, which it is improper of us to have interpreted in terms of war and peace.

The fact that it is impossible to propose a strategy pair as a candidate for a 'solution to the game' is due, amongst other things, to the static nature of the game and the obligation to choose once and for all between polarised strategies with no room for compromise, etc. But this game does provide a direct illustration of some of the difficulties one meets.

Game of Chicken

If $a < b < c$, this game is represented by the matrix

Frances Emil	1	2
I	(a, a)	$(b, 0)$
II	$(0, b)$	(c, c)

The bilosses are shown in Fig. 7.5:

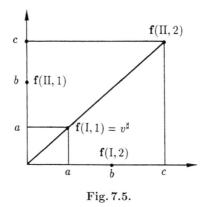

Fig. 7.5.

This may be interpreted as follows. Emil and Frances are driving and reach a crossroads with no signals and no rules of priority. The strategies in each case are to stop (strategies I and 1) or to cross (strategies II and 2). If both cross, the cost to each player of the subsequent accident is c. If both stop, they are only penalised by a slight delay, represented by a loss $a < c$. If one crosses and the other stops, the one who crosses loses nothing, while the one who stops incurs a delay and a loss costing $b \in]a, c[$.

We have

$$f_E^\sharp(\mathrm{I}) = b, \quad f_E^\sharp(\mathrm{II}) = c, \quad f_F^\sharp(1) = b, \quad f_F^\sharp(2) = c$$

whence the strategies I and 1 are conservative and $\mathbf{v}^\sharp = (\mathrm{I}, 1)$. The game is inessential since $(\mathrm{I}, 1)$ is Pareto optimal as are the pairs $(\mathrm{I}, 2)$ and $(\mathrm{II}, 1)$.

Since

$$f_E^\flat(1) = 0, \quad f_E^\flat(2) = b, \quad f_F^\flat(\mathrm{I}) = 0 \text{ and } f_F^\flat(\mathrm{II}) = b$$

we note that the pairs $\{\mathrm{I}, 2\}$ and $\{\mathrm{II}, 1\}$ are non-cooperative equilibria which are Pareto optimal. However, they are not interchangeable.

Battle of the Sexes

The strategies of Emil and Frances consist of going to a football match or going shopping. Emil prefers the match, Frances prefers window-shopping; however, they both prefer to be together. This game is represented by

Emil \ Frances	1	2
I	$(0, a)$	(b, b)
II	(b, b)	$(a, 0)$

where $0 < a < b$.

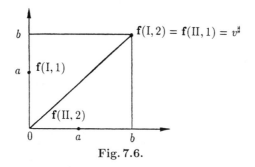

Fig. 7.6.

We note that the pairs $(I, 1)$ and $(II, 2)$ are Pareto optimal, that

$$f_E^\sharp(I) = f_E^\sharp(II) = f_F^\sharp(1) = f_F^\sharp(2) = b$$

and that

$$\mathbf{v}^\sharp = (b, b).$$

Whence, the four strategy pairs are conservative. We also note that the pairs $(I, 1)$ and $(II, 2)$ are the non-cooperative equilibria for the game.

Coordination Game

Emil and Frances have to open a door to escape from a fire. The strategies of the players are, respectively, to go through the doorway (strategies I, 1) or to push the door open (strategies II, 2). If no one opens the door, they stay in the fire, incurring a loss of c. If Emil pushes the door open (strategy II) and Frances passes through (strategy 1), she escapes

first (zero loss) and Emil escapes second with a loss $a < c$.

If they both push the door open at the same time, it takes longer and both come out with slight burns (loss $b \in]a, c[$). This game is represented by the matrix

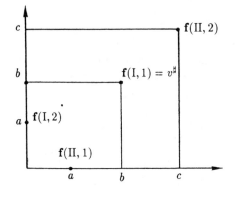

Fig. 7.7.

Emil \ Frances	1 (go through)	2 (push)
I (go through)	(b, b)	$(0, a)$
II (push)	$(a, 0)$	(c, c)

and the bilosses are as shown in Fig. 7.7:

The Pareto-optimal pairs are the strategies $(I, 2)$ and $(II, 1)$, where one of the players pushes the door open and the other passes through the doorway.

Since

$$f_E^\sharp(I) = b, \quad f_E^\sharp(II) = c, \quad f_F^\sharp(1) = b, \quad f_F^\sharp(2) = c,$$

we deduce that

$$\mathbf{v}^\sharp = (b, b)$$

and that the conservative strategies involve both players pushing the door open.

Since

$$f_E^b(1) = a, \quad f_E^b(2) = 0, \quad f_F^b(I) = a, \quad f_F^b(II) = 0$$

we deduce that the pairs $(I, 2)$ and $(II, 1)$ are the non-cooperative equilibria of the game, which are Pareto optimal.

7.9 Cournot's Duopoly

We next describe the fundamental example of the duopoly, where the two players are each manufacturers of the same single commodity. In this case, the loss functions are cost functions which depend on the production of the two players. This game and the concept of non-cooperative equilibrium were introduced by Antoine Cournot in 1838. He was the first to propose the concept of non-cooperative equilibrium, which he introduced in the framework of an economic model. This model has played an important historical role in explaining the behaviour of competing manufacturers in the same market.

Description of the Model

We suppose that Emil and Frances manufacture the same single commodity. We denote the quantities of this commodity produced by our players by $x \in \mathbb{R}_+$ and $y \in \mathbb{R}_+$.

We assume that the price $p(x, y)$ is an affine function of the total production $x + y$

$$p(x, y) := \alpha - \beta(x + y) \qquad (\alpha \geq 0, \beta > 0) \tag{27}$$

and that the cost functions C_E and C_F of each manufacturer are affine functions of the production

$$C_E(x) := \gamma x + \delta, \qquad C_F(y) := \gamma y + \delta, \quad (\gamma > 0, \delta \geq 0). \tag{28}$$

Emil's net cost is equal to

$$f_E(x, y) := \gamma x + \delta - p(x, y)x = \beta x \left(x + y + \frac{\gamma - \alpha}{\beta} \right) + \delta$$

and that of Frances is

$$f_F(x, y) := \gamma x + \delta - p(x, y)y = \beta y \left(x + y + \frac{\gamma - \alpha}{\beta} \right) + \delta.$$

Taking $\beta = 1$ and $\delta = 0$ does not modify the game. Then, setting $u = \alpha - \gamma$, the duopoly may be viewed as a two-person game, where

$$E := [0, u], \qquad F := [0, u], \tag{29}$$

with loss functions defined by

$$f_E(x, y) := x(x + y - u), \qquad f_F(x, y) := y(x + y - u). \tag{30}$$

The biloss mapping is then defined by

$$\mathbf{f}(x, y) := (x(x + y - u), y(x + y - u)). \tag{31}$$

This maps the upper triangle

$$T_+ := \{(x, y) \in [0, u]^2 | x + y \geq u\} \tag{32}$$

into the rectangle $S_+ = [0, u^2]^2$, the diagonal

$$T_0 := \{(x, y) \in [0, u]^2 | x + y = u\} \tag{33}$$

onto $\{0\}$ and the lower triangle

$$T_- := \{(x, y) \in [0, u]^2 | x + y \leq u\} \tag{34}$$

onto the triangle

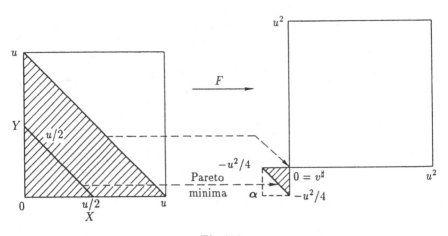

Fig. 7.8.

$$S_- := \left\{ (f,g) \in \left[-\frac{u^2}{4}, 0 \right]^2 \bigg| f + g \geq -\frac{u^2}{4} \right\}. \tag{35}$$

We observe that the set

$$P := \left\{ (x,y) \in [0,u]^2 \bigg| x + y = \frac{u}{2} \right\} \tag{36}$$

is mapped into the subset

$$\mathbf{f}(P) := \left\{ (f,g) \in \left[-\frac{u^2}{4}, 0 \right]^2 \bigg| f + g = -\frac{u^2}{4} \right\} \tag{37}$$

(see Fig. 7.8).

We then note that *the subset P of (36) is the set of Pareto strategies.*

The strategy pair:

$$x_P := \frac{u}{4}, \qquad y_P := \frac{u}{4} \tag{38}$$

results in a loss to each player of $-\frac{u^2}{8}$. Thus, if the manufacturers agree to cooperate, this Pareto-optimal strategy pair is a reasonable compromise.

It is clear that Emil's worst-loss function $f_E^\sharp$

$$f_E^\sharp(x) = \sup_{0 \leq y \leq u} x(x + y - u) = x^2 \tag{39}$$

attains its minimum at $x^\sharp = 0$. In a symmetric fashion, $f_F^\sharp$, defined by $f_F^\sharp(y) = y^2$ attains its minimum at $y^\sharp = 0$. Consequently the *conservative strategies of Emil and Frances are equal to 0,* whence the production of each player is zero.

The conservative vector $v^\sharp$ for the game is equal to $(0,0)$.

We note that the virtual minimum is equal to $\boldsymbol{\alpha} = \left[-\frac{u^2}{4}, -\frac{u^2}{4} \right]$.

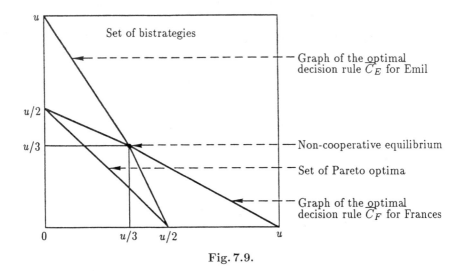

Fig. 7.9.

Non-cooperative Equilibria

Suppose that Frances produces y units. In this case, Emil will produce $\bar{x}$ units to minimise his cost function $x \to x(x + y - u)$ on $[0, u]$, where

$$\bar{x} = \overline{C}_E(y) = \frac{1}{2}(u - y). \tag{40}$$

Thus, $\overline{C}_E : y \to \frac{1}{2}(u - y)$ is Emil's *canonical decision rule*. Similarly, Frances's canonical decision rule is given by $\overline{C}_F : x \to \frac{1}{2}(u - x)$. Consequently, the non-cooperative equilibrium of the duopoly (also called Cournot's equilibrium) is the fixed point of the mapping $(x, y) \to (\overline{C}_E(y), \overline{C}_F(x))$; in other words, this is the strategy pair

$$x_C := \frac{u}{3}, \qquad y_C := \frac{u}{3} \tag{41}$$

which results in a cost of $-\frac{u^2}{9}$ to each player. We note that, in this game, the non-cooperative equilibrium is not Pareto optimal.

We also note that the non-cooperative equilibrium may be attained algorithmically. Consider the following scenario. We suppose that the players play alternately, Emil in the even periods and Frances in the odd periods. When Frances produces y_{2n-1} in the period $2n - 1$, Emil produces $x_{2n} := \overline{C}_E(y_{2n-1})$ in the period $2n$. Frances then changes her production rate and produces $y_{2n+1} := \overline{C}_F(x_{2n})$, and so on.

The sequences of elements x_{2n} and y_{2n-1} are subsequences (indexed by the even and odd indices, respectively) extracted from a sequence of elements z_k which satisfies the recurrence relation

$$2z_{k+1} + z_k = u.$$

Multiplying each of these equalities by $(-1)^{k+1}2^k$ and adding them, we obtain

$$z_n = \frac{u}{2}\frac{1+2^{-n}}{1+2-1} + (-1)^{n+1}2^{-e-1}z_1.$$

Whence, the sequence z_n, and thus also the subsequences x_{2n} and y_{2n+1}, converge to $\frac{u}{3}$.

Unfortunately, this algorithm does not converge in general games.

Associated Game Relating to the Decision Rules

A duopoly may be associated with another game, which involves choosing not the strategies but the decision rules.

Let us consider Emil's point of view. He may decide to play an affine decision rule C^a of the form

$$C^a_E(y) := a(u - y) \quad \text{where} \quad a \in]0, 1[. \tag{42}$$

This means that he does not produce anything when Frances produces the maximum u and that he decides to produce u if Frances produces nothing.

When, in turn, Frances decides to behave according to an affine decision rule C^b_F defined by

$$C^b_F(x) := b(u - x) \quad \text{where} \quad b \in]0, 1[, \tag{43}$$

the subsequent consistent strategy pair is equal to

$$\left(\frac{a(1-b)u}{1-ab}, \frac{b(1-a)u}{1-ab} \right). \tag{44}$$

This subjects the players to the following costs:

(i) $$g_E(a, b) := -\frac{a(1-a)(1-b)^2u^2}{(1-ab)^2}$$

(ii) $$g_F(a, b) := -\frac{b(1-b)(1-a)^2u^2}{(1-ab)^2} \tag{45}$$

Thus, we have constructed a new game, the strategies of which are the slopes of the affine decision rules. In this new game, if Frances plays a slope b, Emil will play the slope $\bar{a} = \sigma_E(b)$ which minimises the function $a \to g_E(a, b)$. We obtain

$$\bar{a} = \sigma_E(b) = \frac{1}{2-b}. \tag{46}$$

Similarly, Frances's canonical decision rule σ_F in this new game is given by

$$\sigma_F(a) = \frac{1}{2-a}. \tag{47}$$

The non-cooperative equilibrium of this game is formed by the pair of slopes

$$(\bar{a} = 1, \bar{b} = 1) \tag{48}$$

Adopting this concept, Emil and Frances implement the decision rules

$$C_E^A(y) = u - y \ \text{ and } \ C_F^A(x) = u - x. \tag{49}$$

The set of consistent strategy pairs associated with these decision rules is equal to:

$$A := \{(x, y) \in [0, u]^2 | x + y = u\}. \tag{50}$$

These consistent strategy pairs result in a *zero cost* to the players.

Stackelberg Equilibrium

As in the initial game, the non-cooperative equilibrium $(\bar{a}, \bar{b}) = (1, 1)$ may be obtained algorithmically, as follows. Frances, who starts, plays the slope $\frac{1}{2}$, which is just her canonical decision rule $\overline{C}_F$. If Emil knows (or guesses) that Frances will play $\overline{C}_F = C_F^{1/2}$, in the new game, he will play the slope $\sigma_E \left(\frac{1}{2}\right) = \frac{2}{3}$. The associated decision rule $C_E^{2/3}$ is called Emil's *Stackelberg decision rule*. The consistent strategy pair associated with the decision rules $C_E^{2/3}$ and $C_F^{1/2}$ is equal to

$$x_S = \frac{u}{2}, \qquad y_S := \frac{u}{4}. \tag{51}$$

This pair is called Emil's *Stackelberg equilibrium* after the economist H. von Stackelberg who described this behaviour in 1933 in a review of price theory.

The associated costs are given by the formulae

$$f_E(x_S, y_S) = -\frac{u^2}{8}, \qquad f_F(x_S, y_S) = -\frac{u^2}{16}. \tag{52}$$

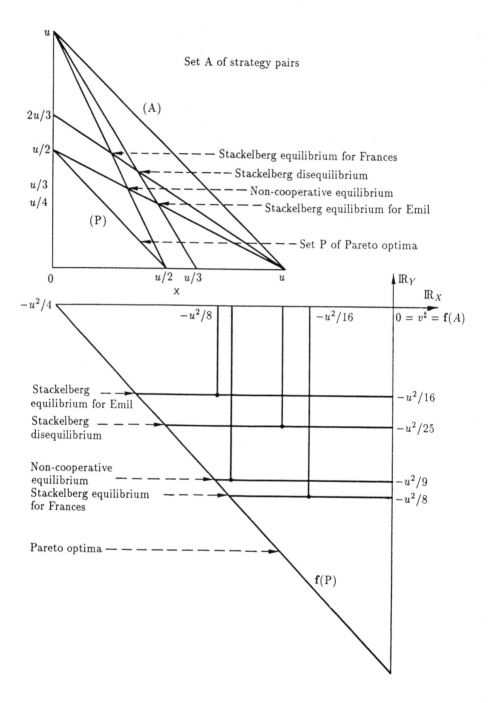

Fig. 7.10.

By playing his *Stackelberg equilibrium*, Emil achieves a cost better than that provided by the non-cooperative equilibrium, $-\frac{u^2}{8}$ as against $-\frac{u^2}{9}$, whilst Frances loses.

Suppose now that Frances follows the same reasoning as Emil. In this case, both will play their Stackelberg decision rules $C_E^{2/3}$ and $C_F^{2/3}$. The consistent strategy pair associated with these two decision rules is

$$x_D := \frac{2u}{5}, \qquad y_D := \frac{2u}{5}. \tag{53}$$

This is called the *Stackelberg disequilibrium*, since the costs incurred

$$f_E(x_D, y_D) = -\frac{2u^2}{25}, \qquad f_F(x_D, y_D) = \frac{2u^2}{25} \tag{54}$$

are greater than those in the case of the non-cooperative equilibrium!

We are in the paradoxical situation where unilateral use of the Stackelberg decision rule is advantageous for the player who uses it, whilst simultaneous use of the Stackelberg decision rule is unfavourable to both players. We have rediscovered the prisoner's dilemma:

Emil \ Frances	canonical (1/2)	Stackelberg (2/3)
canonical (1/2)	$\left(-\frac{u^2}{9}, -\frac{u^2}{9}\right)$	$\left(-\frac{u^2}{16}, -\frac{u^2}{8}\right)$
Stackelberg (2/3)	$\left(-\frac{u^2}{8}, -\frac{u^2}{16}\right)$	$\left(-\frac{2u^2}{25}, -\frac{2u^2}{25}\right)$

The algorithm associated with the new game yields the series of slopes $\frac{1}{2}, \frac{2}{3} = \sigma_E\left(\frac{1}{2}\right), \frac{3}{4} = \sigma_F\left(\frac{2}{3}\right), \ldots, 1 - \frac{1}{n}, \ldots$, since

$$\sigma\left(1 - \frac{1}{n}\right) = \frac{1}{2 - 1 + \frac{1}{n}} = 1 - \frac{1}{n+1}$$

These clearly converge to the slope 1. Thus, Emil will successively play the slopes $1 - \frac{1}{3}, 1 - \frac{1}{5}, \ldots, 1 - \frac{1}{2n+1}$ in the even periods, whilst Frances will play the slopes $\frac{1}{2}, 1 - \frac{1}{4}, 1 - \frac{1}{6}, \ldots, 1 - \frac{1}{2n}$ in the odd periods.

In the even periods, the consistent pairs are

$$x_{2n} := \frac{u}{2}, \qquad y_{2n} := \frac{(2n-2)n}{2(2n-1)}$$

and in the odd periods they are

$$x_{2n+1} := \frac{(2n-1)u}{4n}, \qquad y_{2n+1} := \frac{u}{2}.$$

They converge to the pair $\left(\frac{u}{2}, \frac{u}{2}\right) \in A$.

8. Two-person Zero-sum Games:
Theorems of Von Neumann and Ky Fan

8.1 Introduction

It is in the context of two-person zero-sum games (called duels) that we shall prove the two fundamental theorems of this book, theorems which have applications in many other domains outside game theory. The first statement and proof of the minimax theorem are due to John von Neumann in 1928. Since then, many different proofs and variations on this theorem have been given. The proof we describe here is in our opinion the most elementary.

In 1972, Ky Fan proved another minimax inequality, which is stronger since it has been shown to be equivalent to Brouwer's fixed-point theorem.

This inequality also plays a crucial role, not only in game theory, but also as a useful tool for proving many theorems of nonlinear analysis. Experience shows that it is better to use Ky Fan's Inequality than the fixed-point theorems due to Brouwer or Kakutani, although all these results are equivalent (see Chapter 9).

8.2 Value and Saddle Points of a Game

We now consider the important class of two-person zero-sum games, which by definition satisfy

$$\forall x \in E, \quad \forall y \in F, \quad f_E(x, y) + f_F(x, y) = 0. \tag{1}$$

In other words, Frances's loss is Emil's gain and vice-versa. Since $f(E \times F)$ is contained in the second bisectrix of $\mathbb{R}^2$, any strategy pair is Pareto optimal, so that this concept is not of interest here. Condition (1) enables us to set

$$f(x, y) := f_E(x, y), \qquad -f(x, y) := f_F(x, y) \tag{2}$$

and consequently

$$f^\sharp(x) \quad := \quad \sup_{y \in F} f(x, y), \qquad v^\sharp := \inf_{x \in E} \sup_{y \in F} f(x, y) \tag{3}$$

$$f^\flat(y) \quad := \quad \inf_{x \in E} f(x, y), \qquad v^\flat := \sup_{y \in F} \inf_{x \in E} f(x, y). \tag{4}$$

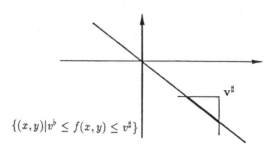

$\{(x,y)|v^\flat \le f(x,y) \le v^\sharp\}$

Fig. 8.1.

We set

$$E^\sharp := \{x \in E | f^\sharp(x) = v^\sharp\}$$
$$F^\flat = \{y^\flat \in F | f^\flat(y^\flat) = v^\flat\}. \tag{5}$$

Then we have

$$f_E^\sharp(x) = f^\sharp(x), \quad f_F^\sharp(y) = -f^\flat(y) \tag{6}$$

and

$$v_E^\sharp = v^\sharp, \quad v_F^\sharp = -v^\flat, \quad \mathbf{v}^\sharp = (v^\sharp, -v^\flat). \tag{7}$$

The subsets $E^\sharp$ and $F^\flat$ consist of the conservatives strategies of Emil and Frances, respectively.

Since

$$\forall x \in E, \quad \forall y \in F, \quad f^\flat(y) \le f^\sharp(x)$$

we deduce that

$$v^\flat \le v^\sharp \tag{8}$$

in other words that

$$\mathbf{v}^\sharp = (v^\sharp, -v^\flat) \tag{9}$$

lies above the second bisectrix.

We shall call $[v^\flat, v^\sharp]$ the *duality interval*. The set K of strategy pairs

$$K := \{(x,y) \in E \times F | v^\flat \le f(x,y) \le v^\sharp\}, \tag{10}$$

which is equal to the set of strategy pairs $(x,y) \in E \times F$ such that $\mathbf{f}(x,y) \le \mathbf{v}^\sharp$, contains $E^\sharp \times F^\flat$. There are situations in which $v^\flat$ is strictly less that $v^\sharp$.

Example. In November 1713, in a letter to Nicoli Bernoulli, Rémond de Montmort proposed the following game of 'pure reason':

A father wishes to give his son a Christmas present and says to him: I shall take an odd or an even number of tokens in my hand, as I think fit.

- If you guess that the number is even and the number in my hand is odd, I shall give you two écus.

- If you guess that the number is odd and the number in my hand is even, you will give me one écu.

- If you guess that the number is odd and the number in my hand is odd, you shall have one écu.

- If you guess that the number is even and the number in my hand is odd, you will give me one écu.

I wonder

1. What rule should one prescribe for the father, so that he saves as much money as possible?

2. What rule should one prescribe for the son, so that he turns the situation to his advantage?

3. Determine the advantage that the father gives his son, and calculate the value of the gift, assuming that each behaves in the way which is most advantageous to himself.

R. de Montmort's intuition is that 'it would be absolutely impossible to prescribe any rule for such a game between equally astute and perceptive players'.

Let us call the father Emil and the son Francis. This moving family scene translates into the finite game

Emil \ Francis	even (1)	odd (2)
even (I)	2	−1
odd (II)	−1	1

where Francis plays the columns and Emil plays the rows.

The coefficients of this matrix represent Emil's losses or Francis's gains.

Let us calculate the conservative values.

Since Emil's worst losses are 2 and 1, respectively, his conservative value is given by $v^\sharp = 1$, which he obtains by taking an odd number of tokens in his hand.

Francis's worst gain is -1 in both cases and, consequently, his conservative value is given by $v^\flat = -1 < v^\sharp$. The strategy pairs (odd, even) and (odd, odd) belong to the set K.

Let us now analyse the different ways of playing. Suppose that Francis plays the odd strategy (2), which is conservative. Anticipating this choice and using his canonical decision rule, Emil would be well advised to play the strategy even (I), which gives him a loss of -1. But, at that moment, Francis, guessing this ruse, actually announces an even strategy (1), which causes Emil to lose two écus. He should have been satisfied with his conservative strategy (odd) which would have limited his loss to one.

This situation illustrates the consequences of the absence of non-cooperative equilibria. In fact, Emil's canonical decision rule $\overline{C}_E$ is given by

$$\overline{C}_E(1) = \{\text{II}\} \quad \text{and} \quad \overline{C}_E(2) = \{\text{I}\}$$

and that of Francis is given by

$$\overline{C}_F(\text{I}) = \{1\} \quad \text{and} \quad \overline{C}_F(\text{II}) = \{2\}.$$

We note that the mapping $\mathbf{C} := (\overline{C}_F \times \overline{C}_E)$ is defined by

$$\begin{aligned}
\mathbf{C}(\text{I}, 1) &= (\text{II}, 1), & \mathbf{C}(\text{I}, 2) = (\text{I}, 1) \\
\mathbf{C}(\text{II}, 1) &= (\text{II}, 2), & \mathbf{C}(\text{II}, 2) = (\text{I}, 2)
\end{aligned}$$

which we represent by the following scheme

Emil \ Francis	even (1)	odd (2)
even (I)	↓	←
odd(II)	→	↑

graph of $\mathbf{C}$

This scheme illustrates not only the absence of fixed points, but also the circular nature of the evolution of the 'natural' algorithm. If Emil plays I, Francis plays $\overline{C}_F(\text{I}) = \{1\}$, Emil plays $\overline{C}_E(1) = \{\text{II}\}$, Francis plays $\overline{C}_F(\text{II}) = \{2\}$, Emil plays $\overline{C}_E(2) = \{\text{I}\}$ and so on.

We note that we have only left the static framework to illustrate R. de Montmort's intuition.

It was two centuries before Émile Borel suggested the notion of mixed strategies and von Neumann proved the theorem mentioned above and the deterministic framework was left behind.

Example. Let us consider the finite game where $E := \{1, 2\}$, $F := \{1, 2, 3\}$ and f is described by the matrix

Emil \ Frances	1	2	3
1	−6	2	−3
2	4	−5	−4

where Emil plays the rows and Frances plays the columns.

The coefficients of this matrix represent Emil's losses and Frances' gains.

Let us calculate the conservative values. Emil's worst losses are 2 and 4, respectively, his conservative value is given by $v^\sharp = 2$ and his conservative strategy is 1. Frances's worst gains are −6, −5 and −4, respectively, it follows that $v^\flat = -4$ and that Frances's conservative strategy is strategy 3. The strategy pairs (1,2), (1,3) and (2,3) belong to the set K.

Here again, there are no non-cooperative equilibria. Emil's canonical decision rule $\overline{C}_E$ is given by

$$\overline{C}_E(1) = \{1\}, \quad \overline{C}_E(2) = \{2\}, \quad \overline{C}_E(3) = \{2\}$$

and Frances's canonical decision rule $\overline{C}_F$ is given by

$$\overline{C}_F(1) = \{2\}, \quad \overline{C}_F(2) = \{1\}.$$

We note that the mapping $\mathbf{C} := (\overline{C}_F \times \overline{C}_E)$ has no fixed points.

The absence of non-cooperative equilibria when $v^\flat$ is strictly less than $v^\sharp$ is a general fact.

Proposition 8.1. *The following conditions are equivalent:*

(a) $(\bar{x}, \bar{y})$ *is a non-cooperative equilibrium;*
(b) $\forall (x, y) \in E \times F, \ f(\bar{x}, y) \le f(\bar{x}, \bar{y}) \le f(x, \bar{y});$
(c) $v^\sharp = v^\flat$ *and* $\bar{x} \in E^\sharp$, $\bar{y} \in F^\flat$ *are conservative strategies.* (11)

Proof. The equivalence of properties (11)(a) and (11)(b) clearly follows from (2) and (11)(b) evidently implies (11)(c). The converse is also easy. We let v denote the common value $v^\sharp = v^\flat$. If $\bar{x} \in E^\sharp$ and $\bar{y} \in F^\flat$ are conservative strategies, then $v = f^\flat(\bar{y}) \le f(\bar{x}, \bar{y}) \le f^\sharp(\bar{x}) = v$, which implies the inequalities (11)(b). □

Definition 8.1. *When* $v^\flat = v^\sharp$, *the common value* $v = v^\sharp = v^\flat$ *is called the* **value of the game** *and the non-cooperative equilibria are called* **saddle points**.

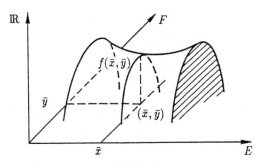

Fig. 8.2.

Example of a saddle point. We take $E := \{1, 2\}$, $F := \{1, 2, 3\}$ and f described by the matrix

Frances Emil	1	2	3
1	-2	-1	-4
2	1	0	-6

We note that $v = -1$ and that the pair $(1, 2)$ is a non-cooperative equilibrium or saddle point of the game.

8.3 Existence of Conservative Strategies

To find non-cooperative equilibria, we must first find the conditions which imply the equality of $v^{\sharp}$ and $v^{\flat}$. To this end, we introduce an intermediate value $v^{\natural}$ (v natural) and prove successively that $v^{\natural} = v^{\sharp}$ (under topological assumptions) and that $v^{\natural} = v^{\flat}$ (under convexity assumptions).

We denote the family of finite subsets K of F by $\mathcal{K}$. We set

$$v^{\sharp}_K := \inf_{x \in E} \sup_{y \in K} f(x, y) \tag{12}$$

and

$$v^{\natural} := \sup_{K \in \mathcal{K}} v^{\sharp}_K = \sup_{K \in \mathcal{K}} \inf_{x \in E} \sup_{y \in K} f(x, y). \tag{13}$$

Since every point y of F may be identified with the finite subset $\{y\} \in \mathcal{K}$, we note that $v^{\sharp}_{\{y\}} = f^{\flat}(y)$ and consequently, that $v^{\flat} = \sup_{y \in F} v^{\sharp}_{\{y\}} \leq \sup_{K \in \mathcal{K}} v^{\sharp}_K =: v^{\natural}$. Since $\sup_{y \in K} f(x, y) \leq \sup_{y \in F} f(x, y)$, we deduce that $v^{\sharp}_K \leq v^{\sharp}$, whence, that $v^{\natural} \leq v^{\sharp}$. In summary, we have shown that

$$v^{\flat} \leq v^{\natural} \leq v^{\sharp}. \tag{14}$$

We shall now prove that reasonable topological assumptions imply that $v^{\natural} = v^{\sharp}$.

Proposition 8.2. *We assume that*

$$E \text{ is compact} \tag{15}$$

and that

$$\forall y \in F, \quad x \to f(x, y) \text{ is lower semi-continuous.} \tag{16}$$

Then, there exists $\bar{x} \in E$ such that

$$\sup_{y \in F} f(\bar{x}, y) = v^{\sharp} \tag{17}$$

and

$$v^{\natural} = v^{\sharp}. \tag{18}$$

Remark. Since the functions $x \to f(x, y)$ are lower semi-continuous, the same is true of the function $f^{\sharp}$ (see Proposition 1.5).

Since E is compact, Weierstrass's theorem implies the existence of $\bar{x} \in E$ which minimises $f^{\sharp}$. Following (3), this may be written as

$$\sup_{y \in F} f(\bar{x}, y) = f^{\sharp}(\bar{x}) = \inf_{x \in E} f^{\sharp}(x) = \inf_{x \in E} \sup_{y \in F} f(x, y) = v^{\sharp}. \tag{19}$$

Proposition 8.2 gives a stronger result with the same assumptions, which are the reasonable assumptions for obtaining conservative strategies (solutions of (17)).

Proof. It suffices to show that there exists $\bar{x} \in E$ such that

$$\sup_{y \in F} f(\bar{x}, y) \leq v^{\natural}. \tag{20}$$

Since $v^{\sharp} \leq \sup_{y \in F} f(\bar{x}, y)$ and $v^{\natural} \leq v^{\sharp}$, we shall deduce that $v^{\natural} = v^{\sharp}$.

We set

$$S_y := \{x \in E | f(\bar{x}, y) \leq v^{\natural}\}.$$

The inequality (20) is equivalent to the inclusion

$$\bar{x} \in \bigcap_{y \in F} S_y. \tag{21}$$

Thus, we must show that this intersection is non-empty.

For this, we shall prove that the S_y are closed sets with the finite-intersection property.

The set S_y is closed since S_y is a lower section of the lower semi-continuous function $x \to f(x, y)$.

We show that for any finite sequence $K := \{y_1, \ldots, y_n\} \in \mathcal{K}$ of F, the finite intersection

$$\bigcap_{i=1,\ldots,n} S_{y_i} \neq \emptyset$$

is non-empty. In fact, since E is compact, and since

$$x \to \max_{i=1,\ldots,n} f(x, y_i) = \max_{y \in K} f(x, y)$$

is lower semi-continuous, it follows that there exists $\hat{x} \in E$ which minimises this function. Such an $\hat{x} \in E$ satisfies

$$\max_{y \in K} f(\hat{x}, y) = \inf_{x \in E} \max_{y \in K} f(x, y) \leq \sup_{K \in \mathcal{K}} \inf_{x \in E} \max_{y \in K} f(x, y) = v^{\natural}.$$

Since E is compact, the intersection of the closed sets S_y is non-empty and there exists $\bar{x}$ satisfying (21) and thus (20). $\qquad \square$

We shall now show that convexity assumptions imply the equality $v^{\flat} = v^{\natural}$.

Proposition 8.3. *Suppose that E and F are convex sets and that*

(i) $\forall y \in F$, $x \to f(x, y)$ *is convex, and*
(ii) $\forall x \in E$, $y \to f(x, y)$ *is concave*   (22)

then $v^\flat = v^\natural$.

Proof. We set $M^n := \{\lambda \in \mathbb{R}_+^n \mid \sum_{i=1}^n \lambda_i = 1\}$. With any finite subset $K := \{y_1, \ldots, y_n\}$ we associate the mapping ϕ_K from E to $\mathbb{R}^n$ defined by

$$\phi_K(x) := (f(x, y_1), \ldots, f(x, y_n)).$$ (23)

We also set

$$w_K := \sup_{\lambda \in M^n} \inf_{x \in E} \langle \lambda, \phi_K(x) \rangle.$$ (24)

We shall prove successively that

a) $\phi_K(E) + \mathbb{R}_+^n$ is a convex subset (Lemma 8.1);

b) $\forall K \in \mathcal{K}$, $v_K^\natural \le w_K$ (Lemma 8.2);

c) $\forall K \in \mathcal{K}$, $w_K \le v^\flat$ (Lemma 8.3).

Whence, the inequalities

$$v^\natural := \sup_{K \in \mathcal{K}} v_K^\natural \le \sup_{K \in \mathcal{K}} w_K \le v^\flat \le v^\natural$$ (25)

will imply the desired equality $v^\flat = v^\natural$.

Lemma 8.1. *If E is convex and if the functions $x \to f(x, y)$ are convex, then the set $\phi_K(E) + \mathbb{R}_+^n$ is convex.*

Proof. Any convex combination $\alpha_1(\phi_K(x_1) + u_1) + \alpha_2(\phi_K(x_2) + u_2)$ where $\alpha_1, \alpha_2 \ge 0$, $\alpha_1 + \alpha_2 = 1$ $(x_1, x_2 \in E, u_1, u_2 \in \mathbb{R}_+^n)$ may be written in the form $\phi_K(x) + u$ where $x := \alpha_1 x_1 + \alpha_2 x_2$ belongs to E (since E is convex) and $u := \alpha_1 u_1 + \alpha_2 u_2 + \alpha_1 \phi_K(x_1) + \alpha_2 \phi_K(x_2) - \phi_K(x)$. Since the functions $x \to f(x, y)$ are convex, $\alpha_1 \phi_K(x_1) + \alpha_2 \phi_K(x_2) - \phi_K(\alpha_1 x_1 + \alpha_2 x_2)$ is a vector in $\mathbb{R}_+^n$. Thus, u belongs to $\mathbb{R}_+^n$ and consequently, $\alpha_1(\phi_K(x_1) + u_1) + \alpha_2(\phi_K(x_2) + u_2) = \phi_K(x) + u$ belongs to $\phi_K(E) + \mathbb{R}_+^n$. $\square$

We recall that $M^n := \{\lambda \in \mathbb{R}_+^n \mid \sum_{i=1}^n \lambda_i = 1\}$ is convex and compact and that we set

$$w_K := \sup_{\lambda \in M^n} \inf_{x \in E} \langle \lambda, \phi_K(x) \rangle.$$

Lemma 8.2. *If E is convex and if the functions $x \to f(x, y)$ are convex, then for any finite set K, we have the inequality*

$$v_K^\natural \le w_K.$$ (26)

Proof. Let $\varepsilon > 0$ and denote $\mathbf{1} := (1, \ldots, 1)$. We shall show that

$$(w_K + \varepsilon)\mathbf{1} \in \phi_K(E) + \mathbb{R}^n_+. \tag{27}$$

Suppose that this is not the case. Since $\phi_K(E) + \mathbb{R}^n_+$ is convex, following Lemma 8.1, we may use Theorem 2.5 (Large Separation Theorem). There exists $\lambda \in \mathbb{R}^n$, $\lambda \neq 0$, such that

$$\sum_{i=1}^n \lambda_i(w_K + \varepsilon) = \langle \lambda, (w_K + \varepsilon)\mathbf{1} \rangle \leq \inf_{v \in \phi_K(E) + \mathbb{R}^n_+} \langle \lambda, v \rangle$$
$$= \inf_{x \in E} \langle \lambda, \phi_K(x) \rangle + \inf_{u \in \mathbb{R}^n_+} \langle \lambda, u \rangle.$$

Then $\inf_{u \in \mathbb{R}^n_+} \langle \lambda, u \rangle$ is bounded below and consequently, λ belongs to $\mathbb{R}^n_+$ and $\inf_{u \in \mathbb{R}^n_+} \langle \lambda, u \rangle$ is equal to 0. Since λ is non-zero, then $\sum_{i=1}^n \lambda_i$ is strictly positive. We set $\bar{\lambda} = \lambda / \sum_{i=1}^n \lambda_i \in M^n$ and deduce that

$$w_K + \varepsilon \leq \inf_{x \in E} \langle \lambda, \phi_K(x) \rangle \leq \sup_{\lambda \in M^n} \inf_{x \in E} \langle \lambda, \phi_K(x) \rangle = w_K.$$

This is impossible and thus we have established the property (27).

This implies that there exist $x_\varepsilon \in E$ and $u_\varepsilon \in \mathbb{R}^n_+$ such that $(w_K + \varepsilon)\mathbf{1} = \phi_K(x_\varepsilon) + u_\varepsilon$.

From the definition of ϕ_K, we deduce that

$$\forall i = 1, \ldots, n, \quad f(x_\varepsilon, y_i) \leq w_K + \varepsilon.$$

Whence,

$$v_K^\sharp \leq \max_{i=1,\ldots,n} f(x_\varepsilon, y_i) \leq w_K + \varepsilon.$$

We complete the proof of the lemma by letting ε tend to 0. $\square$

Lemma 8.3. *Suppose that F is convex and that the functions $y \to f(x, y)$ are concave. Then, for any finite subset K of F, we have $w_K \leq v^\flat$.*

Proof. With each $\lambda \in M^n$, we associate the point $y_\lambda := \sum_{i=1}^n \lambda_i y_i$ which belongs to F, since the latter is convex. The concavity of the functions $y \to f(x, y)$ implies that

$$\forall x \in E, \quad \sum_{i=1}^n \lambda_i f(x, y_i) \leq f(x, y_\lambda).$$

Consequently,

$$\inf_{x \in E} \sum_{i=1}^n \lambda_i f(x, y_i) \leq \inf_{x \in E} f(x, y_\lambda) \leq \sup_{y \in F} \inf_{x \in E} f(x, y) := v^\flat.$$

The proof of Lemma 8.3 is completed by taking the supremum over M^n. $\square$

Lemmas 8.1 to 8.3 may now be applied, as indicated, to complete the proof of Proposition 8.3. $\square$

Propositions 8.2 and 8.3 imply the existence of a value.

Theorem 8.1. *Suppose that*

(i)   *E is convex and compact*
(ii) $\forall y \in F, \ x \to f(x,y)$ *is convex and lower semi-continuous*   (28)

and that

(i)   *F is convex*
(ii) $\forall x \in E, \ y \to f(x,y)$ *is concave.*   (29)

   Then f has a value:

$$v := v^{\sharp} = v^{\flat}$$ (30)

and there exists $\bar{x} \in E$ satisfying:

$$\sup_{y \in F} f(\bar{x}, y) = v.$$ (31)

Applying Theorem 8.1 to f and to $-f$, we obtain the minimax theorem.

Theorem 8.2 (von Neumann). *Suppose that*

(i)   *E is convex and compact*
(ii) $\forall y \in F, \ x \to f(x,y)$ *is convex and lower semi-continuous*   (32)

and that

(i)   *F is convex and compact*
(ii) $\forall x \in E, \ y \to f(x,y)$ *is concave and upper semi-continuous.*   (33)

Then there exists a saddle point $(\bar{x}, \bar{y}) \in E \times F$.

Corollary 8.1. *Consider a zero-sum game defined on finite strategy sets $\{1, \ldots, n\}$ and $\{1, \ldots, p\}$ by a matrix $\{a_{ij}\}_{\substack{1 \le i \le n \\ 1 \le j \le p}}$ (a_{ij} is Emil's loss and Frances's gain).*

   We associate this with the game defined on the mixed strategy sets M^n and M^p by

$$\tilde{f}(\lambda, \mu) = \sum_{i=1}^{n} \sum_{j=1}^{p} \lambda_i \mu_j a_{ij}.$$ (34)

Then there exists a saddle point formed from mixed strategies.

This provided an answer to R. de Montmort's question. In this case, identifying λ with $(\lambda, 1 - \lambda)$ and μ with $(\mu, 1 - \mu)$, the function $\tilde{f}(\lambda, \mu)$ may be written as

$$\begin{aligned}\tilde{f}(\lambda, \mu) &:= 2\lambda\mu - \lambda(1 - \mu) - \mu(1 - \lambda) + (1 - \lambda)(1 - \mu) \\ &= 5\lambda\mu - 2\lambda - 2\mu + 1.\end{aligned}$$

Thus, we see that the value of the game is equal to $v = \frac{1}{5}$ and that the saddle point is formed by the pair $\left(\frac{2}{5}, \frac{2}{5}\right)$, which involves playing the even strategies with probability $\frac{2}{5}$ and the odd strategies with probability $\frac{3}{5}$.

8.4 Continuous Partitions of Unity

In the following paragraphs, we shall use convex combinations $f(x) \in M^n$ which depend *continuously* on a parameter $x \in E$. This will enable us to cover the convex hull of n points in a continuous fashion. Even better than this, we shall construct these functions $f(x)$ so that the components f_i are zero outside open sets A_i covering the space E. Such functions are called continuous partitions of unity.

Definition 8.2. *Let f be a real-valued function defined on a metric space E. The smallest closed set S such that $f(x) = 0, x \notin S$ is called the **support** of f and is denoted by* $\mathrm{supp}(f)$.

In other words, the support of f is the closure in E of the set of elements $x \in E$ such that $f(x) \neq 0$. It is also the set of elements $x \in E$ such that *any* neighbourhood V of x contains a point y with $f(y) \neq 0$.

Definition 8.3. *Consider an open covering $\{A_i\}_{i=1,\ldots,n}$ of E. A family $\{f_i\}_{i \in I}$ of continuous functions from E to $[0, 1]$ such that:*

(i) $\forall x \in E, \ \sum_{i=1}^{n} f_i(x) = 1$

(ii) $\forall i = 1, \ldots, n, \ \mathrm{supp} f_i \subset A_i$ (35)

*is called a **partition of unity subordinate to this covering**.*

Before we prove the existence of a partition of unity subordinate to a finite open covering, we shall need the following propositions.

Proposition 8.4 (separation of two closed sets by a continuous function). *Let M and N be two non-empty, disjoint, closed subsets of a metric space E. Then there exists a **continuous** function g from E to $[0, 1]$ such that*

$$g(x) = 0 \ \forall x \in M, \qquad g(x) = 1 \ \forall x \in N. \tag{36}$$

Proof. Since M and N are disjoint, $d(x, M) + d(x, N) > 0$ for all $x \in E$. Thus, the function g defined by

$$g(x) = \frac{d(x, M)}{d(x, M) + d(x, N)} \tag{37}$$

is a continuous function from E to $[0, 1]$ which takes the value 0 on M and is equal to 1 on N. □

Proposition 8.5. *Suppose that $E = A \cup B$ is the union of two open sets. Then there exists an open set W such that*

$$\overline{W} \subset A \ \text{ and } \ E = W \cup B \tag{38}$$

Proof. If $A = E$, we take $W = E$, and if $B = E$ we take $W = \emptyset$. Suppose now that $A \neq E$ and that $B \neq E$. The non-empty closed sets $\complement A$ and $\complement B$ are disjoint. Thus, we consider a continuous function f which takes the value 0 on $\complement A$ and 1 on $\complement B$ and we take $W := \{x \in E | f(x) > \frac{1}{2}\}$. This is an open set. Since $\overline{W}$ is contained in $\{x \in E | f(x) \geq \frac{1}{2}\}$, and since $\complement A$ is contained in $\complement \overline{W}$, it follows that $\overline{W}$ is contained in A. If x does not belong to B, $f(x) = 1$ and thus x belongs to W. Thus, E is covered by W and B. $\qquad\square$

Proposition 8.6. *Let $\{A_i\}_{i=1,\ldots,n}$ be a finite open covering of E. Then there exists an open covering $\{W_i\}_{i=1,\ldots,n}$ such that*

$$\forall i = 1, \ldots, n, \quad \overline{W}_i \subset A_i. \tag{39}$$

Proof. We construct the covering W_i recursively, using Proposition 8.5.

Setting $B_1 = \cup_{i=2}^n A_i$, we obtain $E = A_1 \cup B_1$.

Proposition 8.5 implies that there exists an open set $W_1 \subset A_1$ such that

$$\overline{W}_1 \subset A_1 \text{ and } E = W_1 \cup B_1 = W_1 \cup \bigcup_{j=1}^n A_j.$$

Suppose that we have constructed the open sets W_j $(1 \leq j \leq k-1)$ such that

$$\overline{W}_i \subset A_j \text{ if } 1 \leq j \leq k-1; \qquad E = \bigcup_{i=1}^{k-1} W_i \cup \bigcup_{j=k}^n A_j. \tag{40}$$

We introduce the open set

$$B_k = \bigcup_{i=1}^{k-1} W_i \cup \bigcup_{j=k+1}^n A_j$$

such that $E = A_k \cup B_k$ following (40). Proposition 8.5 implies that there exists an open subset $W_k \subset A_k$ such that $\overline{W}_k \subset A_k$ and $E = W_k \cup B_k$. Thus, we have constructed k open subsets W_i such that

$$\overline{W}_i \subset A_i \text{ if } 1 \leq i \leq k; \qquad E = \bigcup_{i=1}^k W_i \cup \bigcup_{j=k+1}^n A_j.$$

Thus, the recurrence may be continued and the proof of the proposition is complete. $\qquad\square$

Theorem 8.3. *Given any finite open covering of a metric space E, there exists a continuous partition of unity which is subordinate to it.*

Proof. Suppose that $E = \cup_{i=1}^n A_i$ for some open sets A_i.

Following Proposition 8.3, there exist n open sets $W_i \subset A_i$ such that $\overline{W}_i \subset A_i$ and $E = \cup_{i=1}^n W_i$.

Since $\cap_{i=1}^{n}[W_i = \emptyset$, the function $\sum_{i=1}^{n} d(x, [W_i])$ is strictly positive. The functions f_i defined by

$$f_i(x) := \frac{d(x, [W_i])}{\sum_{i=1}^{n} d(x, [W_i])} \qquad (41)$$

form a continuous partition of unity. If $f_i(x) > 0$, it follows that $x \notin [W_i$ (which is closed), whence $x \in W_i$. Thus, the support of f_i is contained in $\overline{W}_i$, which is itself contained in A_i. $\qquad \square$

8.5 Optimal Decision Rules

What should we do when the convexity assumptions are missing? As in Corollary 8.1, we could embed the strategy sets in other sets of mixed strategies (we did this for finite sets). There is another approach which involves considering a strategy as a constant decision rule.

Let us consider, for example, the case of Frances. What value could she attribute to a decision rule $C_F : E \to F$. Since she plays $C_F(x)$ whenever Emil plays x, the worst gain than she may incur is

$$f^\flat(C_F) := \inf_{x \in E} f(x, C_F(x)) \qquad (42)$$

when she has no means of knowing Emil's choice in advance.

We note that this definition is consistent with the definition of the worst gain incurred by a strategy y_0 considered as a constant decision rule $x \to y_0$, since

$$f^\flat(y_0) := \inf_{x \in E} f(x, y_0) = \inf_{x \in E} f(x, y_0(x)).$$

Consequently, if $\mathcal{C}_F$ is a set of decision rules which contains the set F (of constant decision rules) we have:

$$v^\flat := \sup_{y \in F} \inf_{x \in E} f(x, y) \leq \sup_{C_F \in \mathcal{C}_F} f^\flat(C_F) \leq \inf_{x \in E} \sup_{y \in F} f(x, y) =: v^\sharp. \qquad (43)$$

Proposition 8.7. *We denote the set of all the decision rules of E in F by F^E. Then*

$$\sup_{C_F \in F^E} f^\flat(C_F) = v^\sharp. \qquad (44)$$

Proof. By definition, we may associate any $\varepsilon > 0$ and any $x \in E$ with a strategy $D_\varepsilon(x) \in F$ such that

$$\sup_{y \in F} f(x, y) \leq f(x, D_\varepsilon(x)) + \varepsilon. \qquad (45)$$

It then follows that

$$v^{\sharp} = \inf_{x \in E} \sup_{y \in F} f(x, y) \le f^{\flat}(D_{\varepsilon}) + \varepsilon \le \sup_{C_F \in F^E} \inf_{x \in E} f(x, C_F(x)) + \varepsilon.$$

Since this inequality holds for all $\varepsilon > 0$, we obtain the inequality $v^{\sharp} \le$ $\sup_{C_F \in F^E} \inf_{x \in E} f(x, C_F(x))$ which, taken with the inequalities (43), implies the desired equality (44).   □

We shall show that, under additional assumptions, the equation (44) remains true if Frances is forced to use *only continuous decision rules* (this enables us to model a regular behaviour for Frances).

Theorem 8.4. *Suppose that*

(i)   *E is compact*
(ii) $\forall y \in F, \quad x \to f(x, y)$ *is lower semi-continuous*   (46)

and that

(i)   *F is a convex subset*
(ii) $\forall x \in E, \quad y \to f(x, y)$ *is concave.*   (47)

Then if $C(E, F)$ denotes the set of continuous mappings from E to F, we have

$$\sup_{D \in C(E,F)} \inf_{x \in E} f(x, D(x)) = \inf_{x \in E} \sup_{y \in F} f(x, y).$$ (48)

Proof. We already know that $\sup_{D \in C(E,F)} \inf_{x \in E} f(x, D(x)) \le v^{\sharp}$, from (43). Thus, it remains to prove the opposite inequality. Firstly, we may associate any $\varepsilon > 0$ with a mapping (not necessarily continuous) D_{ε} from E to F which satisfies (45).

In addition, since the functions $x \to f(x, y)$ are lower semi-continuous, there exist neighbourhoods $B(x, \eta(x))$ of x such that

$$\forall z \in B(x, \eta(x)), \qquad f(x, D_{\varepsilon}(x)) \le f(z, D_{\varepsilon}(x)) + \varepsilon.$$ (49)

Since E is compact, it can be covered by n balls $B(x_i, \eta(x_i))$. Let $\{g_i\}_{i=1,\dots,n}$ be a continuous partition of unity subordinate to this covering. We introduce the function D defined by

$$D(x) = \sum_{i=1}^{n} g_i(x) D_{\varepsilon}(x_i),$$

which is continuous since the functions g_i are continuous. Finally, since the functions $y \to f(x, y)$ are concave, since $g_i(x) \ge 0$ for all i and $\sum_{i=1}^{n} g_i(x) = 1$, we have:

$$f(x, D(x)) \ge \sum_{i \in I(x)} g_i(x) f(x, D_{\varepsilon}(x_i))$$ (50)

where $I(x)$ is the set of integers $i = 1, \ldots, n$ such that $g_i(x) > 0$. This set is non-empty since $\sum_{i=1}^{n} g_i(x) = 1$.

Moreover, if $g_i(x) > 0$, x belongs to the support of g_i, which is continuous on the ball $B(x_i, \eta(x_i))$. It follows from (49) that $f(x, D_\varepsilon(x_i)) \geq f(x_i, D_\varepsilon(x_i)) - \varepsilon$ and from (45) that $f(x_i, D_\varepsilon(x_i)) \geq \sup_{y \in F} f(x_i, y) - \varepsilon \geq v^\sharp - \varepsilon$.

Thus, if $i \in I(x)$, we have $f(x, D_\varepsilon(x_i)) \geq v^\sharp - 2\varepsilon$. Then (50) implies that

$$f(x, D(x)) \geq \sum_{i \in I(x)} g_i(x)(v^\sharp - 2\varepsilon) = v^\sharp - 2\varepsilon.$$

It follows that $\inf_{x \in E} f(x, D(x)) \geq v^\sharp - 2\varepsilon$, whence that $\sup_{D \in \mathcal{C}(E,F)} \inf_{x \in E} f(x, D(x)) \geq v^\sharp - 2\varepsilon$.

We obtain the desired inequality by letting ε tend to 0. $\square$

We shall now establish another expression for $v^\sharp$.

In the game-theory context, we now suppose that Emil has information about Frances's choice of strategy and that he has the right to choose continuous decision rules $C \in \mathcal{C}(F, E)$. Thus, he may continuously associate any strategy $y \in F$ played by Frances with a strategy $C(y) \in E$.

Theorem 8.5. *We retain the assumptions of (46) and (47) of Theorem 8.4. If $\mathcal{C}(F, E)$ denotes the set of continuous decision rules of F in E, then*

$$\inf_{C \in \mathcal{C}(F,E)} \sup_{y \in F} f(C(y), y) = \inf_{x \in E} \sup_{y \in F} f(x, y). \tag{51}$$

Proof. We shall use the convex compact set

$$M^n := \{\lambda \in \mathbb{R}_+^n \mid \sum_{i=1}^{n} \lambda_i = 1\}.$$

The inequality $\inf_{C \in \mathcal{C}(F,E)} \sup_{y \in F} f(C(y), y) \leq v^\sharp$ is clearly always true.

Since E is compact and the functions $x \to f(x, y)$ are lower semi-continuous, Proposition 8.2 implies that there exists $\bar{x} \in E$ such that

$$\sup_{y \in F} f(\bar{x}, y) = v^\sharp = \sup_{K = \{y_1, \ldots, y_n\} \in \mathcal{K}} \inf_{x \in E} \max_{i=1, \ldots, n} f(x, y_i) \tag{52}$$

Thus, it is sufficient to prove that for any finite set $K = \{y_1, \ldots, y_n\}$ and any continuous mapping $C \in \mathcal{C}(F, E)$, we have

$$\inf_{x \in E} \max_{i=1, \ldots, n} f(x, y_i) \leq \sup_{y \in F} f(C(y), y). \tag{53}$$

Firstly, we note that

$$\inf_{x \in E} \max_{i=1, \ldots, n} f(x, y_i) = \inf_{x \in E} \sup_{\lambda \in M^n} \sum_{i=1}^{n} \lambda_i f(x, y_i) \tag{54}$$

$$\leq \inf_{\mu \in M^n} \sup_{\lambda \in M^n} \sum_{i=1}^{n} \lambda_i f(C(\sum_{j=1}^{n} \mu_j y_j), y_i)$$

$$= \inf_{\mu \in M^n} \sup_{\lambda \in M^n} \phi(\mu, \lambda)$$

where the function ϕ is defined on $M^n \times M^n$ by

$$\phi(\mu, \lambda) = \sum_{i=1}^{n} \lambda_i f(C(\sum_{j=1}^{n} \mu_j y_j), y_i). \tag{55}$$

Since C is continuous and since the functions $x \to f(x, y_i)$ are lower semi-continuous, it follows that the functions $\mu \to \phi(\mu, \lambda)$ are lower semi-continuous. The functions $\lambda \to \phi(\mu, \lambda)$ are linear, whence concave. The set M^n is convex and compact.

Thus, Theorem 8.4 implies that

$$\inf_{\mu \in M^n} \sup_{\lambda \in M^n} \phi(\mu, \lambda) = \sup_{D \in \mathcal{C}(M^n, M^n)} \inf_{\mu \in M^n} \phi(\mu, D(\mu)). \tag{56}$$

But, Brouwer's Theorem implies that any mapping $D \in \mathcal{C}(M^n, M^n)$ has a fixed point $\mu_D \in M^n$. Thus,

$$\inf_{\mu \in M^n} \phi(\mu, D(\mu)) \leq \phi(\mu_D, D(\mu_D)) = \phi(\mu_D, \mu_D)) \leq \sup_{\mu \in M^n} \phi(\mu, \mu). \tag{57}$$

This then implies that

$$\sup_{D \in \mathcal{C}(M^n, M^n)} \inf_{\mu \in M^n} \phi(\mu, D(\mu)) \leq \sup_{\mu \in M^n} \phi(\mu, \mu). \tag{58}$$

Moreover, since the functions $y \to f(x, y)$ are assumed to be concave we have

$$\begin{aligned}
\phi(\mu, \mu) &= \sum_{i=1}^{m} \mu_i f(C(\sum_{j=1}^{n} \mu_j y_j), y_i) \\
&\leq f(C(\sum_{j=1}^{n} \mu_j y_j), \sum_{j=1}^{n} \mu_j y_j) \\
&\leq \sup_{y \in F} f(C(y), y).
\end{aligned} \tag{59}$$

Thus, the inequalities (54), (57), (58) and (59) imply the desired inequality (53). □

In particular, we deduce the following important inequality.

Theorem 8.6 (Ky Fan's Inequality). *Suppose that E is a convex compact subset of a Hilbert space and that f is a function from $E \times E$ to $\mathbb{R}$ satisfying*

(i) $\forall y \in E, \quad x \to f(x, y)$ *is lower semi-continuous*
(ii) $\forall x \in E, \quad y \to f(x, y)$ *is concave.* (60)

Then there exists $\bar{x} \in E$ such that

$$\sup_{y \in E} f(\bar{x}, y) \leq \sup_{y \in E} f(y, y). \tag{61}$$

Proof. Theorem 8.5 implies that there exists $\bar{x} \in E$ such that

$$\sup_{y \in E} f(\bar{x}, y) \leq v^{\sharp} \leq \inf_{C \in \mathcal{C}(E,E)} \sup_{y \in E} f(C(y), y) \leq \sup_{y \in E} f(y, y)$$

since the identity mapping is continuous from E to E.   □

Remark. We have deduced Ky Fan's Inequality from Brouwer's Fixed-point Theorem. In fact, these two results are equivalent and we can deduce Brouwer's Fixed-point Theorem from Ky Fan's Inequality.

Let D be a continuous mapping of a convex compact subset K of a finite-dimensional vector space $\mathbb{R}^n$ into itself. Set

$$f(x, y) := \langle x - D(x), x - y \rangle \tag{62}$$

where $\langle \cdot, \cdot \rangle$ is the Euclidean scalar product on $\mathbb{R}^n$.

This function clearly satisfies the assumptions of Theorem 8.6 (Ky Fan's Inequality); thus, there exists an element $\bar{x} \in K$ such that

$$\langle \bar{x} - D(\bar{x}), \bar{x} - y \rangle \leq 0 \tag{63}$$

for all $y \in K$. Taking $y = D(\bar{x})$, we have $\|\bar{x} - D\bar{x}\|^2 \leq 0$, whence $\bar{x} = D(\bar{x})$.

Remark. We can provide a direct proof of the Ky Fan inequality based on the Brouwer Fixed-Point Theorem by deriving a contradiction from the negation of the conclusion:

$$\forall\, x \in K, \ \exists\, y \in K \ \text{ such that } \ f(x, y) > 0$$

Hence K can be covered by the subsets

$$\mathcal{V}_y := \{ x \in K \mid f(x, y) > 0 \}$$

which are open since f is lower semicontinuous with respect to x. Since K is compact, it can be covered by n such open subsets $\mathcal{V}_{y_i}$. Let us consider a continuous partition of unity $(\alpha_i)_{i=1,\dots,n}$ subordinate to this open covering of K and define the map $c : K \mapsto X$ by

$$\forall\, x \in K, \ c(x) := \sum_{i=1}^{n} \alpha_i(x) y_i$$

It maps K to itself because K is convex and the elements y_i belong to K. It is also continuous, so that Brouwer's Fixed Point Theorem implies the existence of a fixed point $\bar{y} = c(\bar{y}) \in K$ of f. Since f is concave with respect to y, we deduce that

$$f(\bar{y}, \bar{y}) = f\left(\bar{y}, \sum_{i=1}^{n} \alpha_i(\bar{y}) y_i\right) \geq \sum_{i=1}^{n} \alpha_i(\bar{y}) f(\bar{y}, y_i)$$

Let us introduce

$$I(\overline{y}) := \{i = 1, \ldots, n \mid \alpha_i(\overline{y}) > 0\}$$

It is not empty because $\sum_{i=1}^{n} \alpha_i(\overline{y}) = 1$. Furthermore

$$\sum_{i=1}^{n} \alpha_i(\overline{y}) f(\overline{y}, y_i) = \sum_{i \in I(\overline{y})} \alpha_i(\overline{y}) f(\overline{y}, y_i) > 0$$

because, whenever i belongs to $I(\overline{y})$, $\alpha_i(\overline{y}) > 0$, so that $\overline{y}$ belongs to $\mathcal{V}_{y_i}$, and thus, by the very definition of this subset, $f(\overline{y}, y_i) > 0$. Hence, we have proved that $f(\overline{y}, \overline{y})$ is strictly positive, a contradiction of the assumption that $f(\overline{y}, \overline{y}) \leq 0$.

9. Solution of Nonlinear Equations and Inclusions

9.1 Introduction

Ky Fan's Inequality (which is equivalent to Brouwer's Fixed-point Theorem) implies a whole series of *existence theorems* for the solutions of nonlinear equations or inclusions. Such theorems are very useful in many applications, particularly in mathematical economics and game theory, as we shall see in the following chapters.

We shall begin by indicating how to adapt the concepts of continuity to the case of set-valued maps; we shall consider only *upper semi-continuous set-valued maps with convex closed values*.

Then we shall describe sufficient conditions for the existence of a solution

$$\bar{x} \in M^n = \{x \in \mathbb{R}_+^n \,|\, \sum_{i=1}^n x_i = 1\}$$

of the problem

$$C(\bar{x}) \cap \mathbb{R}_+^n = \emptyset \ (\text{where } 0 \in C(\bar{x}) - \mathbb{R}_+^n)$$

when C is a set-valued map from M^n to $\mathbb{R}^n$. In addition to certain technical assumptions, we shall assume that the condition

$$\forall x \in M^n, \quad \sup_{v \in C(x)} \langle v, x \rangle \geq 0$$

is satisfied.

Then we shall study the existence of zeros $\bar{x}$ of the inclusion

$$0 \in C(\bar{x})$$

when C is a set-valued map from a convex compact subset $K \subset X$ to X.

In addition to technical conditions, we assume that the tangential condition

$$\forall x \in K, \ C(x) \cap T_K(x) \neq \emptyset$$

is satisfied, where (we recall) $T_K(x)$ denotes the tangent cone to K at x which we studied in detail in Chapter 4.

This result has a number of consequences. Firstly, there is the famous fixed-point theorem due to Kakutani, which generalises Brouwer's Fixed-point Theorem to the case of set-valued maps. This says that any upper semi-continuous set-valued map from a convex compact subset into itself, with convex closed values, has a fixed point.

We shall then describe another consequence which will be very useful in economic models (viability theorem).

We assume we have convex closed subsets $L \subset X$, $M \subset Y$ and $P \subset Y^*$, a continuous linear operator $A \in L(X, Y)$ and a continuous mapping c from $L \times P$ to Y which is affine in its second argument.

We also suppose that

$$\forall x \in L, \quad \forall p \in P, \quad \langle p, Ac(x, p) \rangle \leq 0$$

together with certain technical assumptions. We shall then prove that there exists $(\bar{x}, \bar{p})$ satisfying

(i) $\bar{x} \in L, \quad A\bar{x} \in M, \quad \bar{p} \in P$
(ii) $c(\bar{x}, \bar{p}) = 0.$

We shall also prove other theorems which will be useful in game theory. The implications of these results are summarised in a diagram at the end of the chapter.

9.2 Upper Hemi-continuous Set-valued Maps

We shall study a whole class of nonlinear problems which reduce to an inclusion of the following form

$$\text{find } \bar{x} \in K \text{ such that } 0 \in C(\bar{x}) \tag{1}$$

where C is a set-valued map from K to a Hilbert space Y which associates $x \in K$ with a *subset $C(x)$ of Y which is always non-empty, convex and closed*. If C is an ordinary pointwise mapping, problem (1) may be written in the more familiar form of the solution of an equation:

$$\text{find } \bar{x} \in K \text{ such that } C(\bar{x}) = 0. \tag{2}$$

A solution $\bar{x}$ of (1) is called a zero of C or an equilibrium or stationary point.

The use of set-valued maps is mainly motivated by problems in optimisation theory, game theory and mathematical economics.

In fact, we only use a few elements of the general theory of set-valued maps. We use the fact that the images $C(x)$ are *convex closed sets* to represent them by their support functions

$$\forall p \in Y^*, \text{ we set } \sigma(C(x), p) = \sup_{y \in C(x)} \langle p, y \rangle \tag{3}$$

(since $y \in C(x)$ if and only if $\langle p, y \rangle \leq \sigma(C(x), p)$ for all $p \in Y^*$).

Definition 9.1. *We shall say that a set-valued map C is* **upper hemi-continuous** *at $x_0 \in K$ if and only if for all $p \in Y^*$, the function $x \rightarrow \sigma(C(x), p)$ is upper semi-continuous at x_0. It is upper hemi-continuous if it is upper hemi-continuous at all points $x_0 \in K$.*

Any continuous mapping C from K to Y clearly defines an upper hemi-continuous set-valued map (it is even sufficient if the functions $x \rightarrow \langle p, C(x) \rangle$ are continuous for any $p \in Y^*$).

Let B be the unit ball of Y.

Definition 9.2. *We shall say that a set-valued map C from K to Y is* **upper semi-continuous** *at x_0 if, for all $\varepsilon > 0$, there exists a neighbourhood $N(x_0)$ of x_0 such that $C(x) \subset C(x_0) + \varepsilon B$ for all $x \in N(x_0)$. It is upper semi-continuous if it is upper semi-continuous at all points $x_0 \in K$.*

Thus, we see that upper semi-continuity is a generalisation of the notion of continuity to set-valued maps.

First we indicate the link between these two notions.

Proposition 9.1. *Any upper semi-continuous mapping is upper hemi-continuous.*

Proof. For fixed $\varepsilon > 0$ and $p \in Y^*$, there exists a neighbourhood $N(x_0)$ such that

$$\forall x \in N(x_0), \quad C(x) \subset C(x_0) + \varepsilon B \tag{4}$$

whence also

$$\forall x \in N(x_0), \quad \sigma(C(x), p) \leq \sigma(C(x_0), p) + \varepsilon \|p\|_* \tag{5}$$

since $\sigma(\varepsilon B, p) = \varepsilon \|p\|_*$. Thus, $x \rightarrow \sigma(C(x), p)$ is upper semi-continuous at x_0. □

Theorems 4.2 and 6.2 state that the subdifferentials of convex continuous functions and, more generally, the generalised gradients of locally Lipschitz functions, are upper hemi-continuous.

Theorem 9.1. *Consider a nontrivial function $f : X \rightarrow \mathbb{R} \cup \{+\infty\}$ which is locally Lipschitz on the interior of its domain (in particular, a convex continuous function on $\mathrm{Int}\,\mathrm{Dom}f$). Then the set-valued map $x \in \mathrm{Int}\,\mathrm{Dom}f \rightarrow \partial f(x)$ is upper hemi-continuous.*

We now note a useful property of upper hemi-continuous set-valued maps.

Definition 9.3. *The* **graph** *of a set-valued map C from K to X is the subset*

$$\mathrm{Graph}\,(C) := \{(x, y) \in K \times Y | y \in C(x)\} \tag{6}$$

and the **inverse** *C^{-1} of the set-valued map C is the set-valued map from Y to K defined by*

$$x \in C^{-1}(y) \quad \text{if and only if} \quad y \in C(x). \tag{7}$$

We note that the graph of a set-valued map characterises the set-valued map C and its inverse. We also note that by inverting one-to-one mappings that are not injective, we obtain examples of set-valued maps.

Lastly, we recall that if $f : X \to \mathbb{R} \cup \{+\infty\}$ is a nontrivial, convex, lower semi-continuous function, then the inverse of the set-valued map $x \to \partial f(x)$ is the set-valued map $p \to \partial f^*(p)$ where f^* is the conjugate function of f (see Corollary 4.1).

Proposition 9.2. *The graph of an upper hemi-continuous set-valued map with convex closed values is closed.*

Proof. Consider a sequence of elements $(x_n, y_n) \in \text{Graph}\,(C)$ converging to the pair (x, y). Since the functions $x \to \sigma(C(x), p)$ are upper semi-continuous, the inequalities

$$\langle p, y_n \rangle \leq \sigma(C(x_n), p)$$

imply, by passing to the limit, that

$$\langle p, y \rangle = \lim_{n \to \infty} \langle p, y_n \rangle \leq \limsup_{n \to \infty} \sigma(C(x_n), p) \leq \sigma(C(x), p).$$

These inequalities imply that

$$y \in \overline{\text{co}}(C(x)) = C(x). \qquad \square$$

Remark. We recall that if C is a one-to-one mapping from K to Y then the following conditions are equivalent

(a) $\forall \varepsilon,\ \exists N(x_0)$ such that $\forall x \in N(x_0),\ C(x) \in C(x_0) + \varepsilon B$;
(b)   whenever a sequence x_n converges to x_0, $C(x_n)$ converges to $C(x_0)$.   (8)

If C is a set-valued map from K to Y, the notion of upper semi-continuity is the natural generalisation of condition (8)(a).

Generalisation of (8)(b) leads to the following definition.

Definition 9.4. *We shall say that a set-valued map C from K to Y is* **lower semi-continuous** *at $x_0 \in K$ if for any sequence x_n converging to $x_0 \in K$, for all $y_0 \in C(x_0)$, there exists a sequence of elements $y_n \in C(x_n)$ converging to y_0.*

In the case of set-valued maps, the concepts of upper and lower semi-continuity are no longer equivalent, as the examples of Figs. 9.1 and 9.2 show.

Definition 9.5. *We shall say that a set-valued map C is continuous (at x_0) if it is both lower and upper semi-continuous (at x_0).*

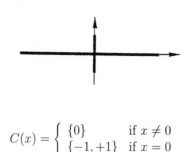

$$C(x) = \begin{cases} \{0\} & \text{if } x \neq 0 \\ \{-1, +1\} & \text{if } x = 0 \end{cases}$$

Fig. 9.1. Example of a set-valued map which is upper semi-continuous at 0, but not lower semi-continuous.

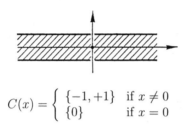

$$C(x) = \begin{cases} \{-1, +1\} & \text{if } x \neq 0 \\ \{0\} & \text{if } x = 0 \end{cases}$$

Fig. 9.2. Example of a set-valued map which is lower semi-continuous at 0, but not upper semi-continuous.

Remark. R.T. Rockafellar and R. Wets suggest to say that F is outer semi-continuous at x if $\text{Limsup}_{x' \to x} F(x') \subset F(x)$ and **inner semicontinuous** at x if $F(x) \subset \text{Liminf}_{x' \to x} F(x')$. The above proposition led several authors to call upper semicontinuous maps the ones which are outer semicontinuous in the Rockafellar-Wets terminology. Naturally, these two concepts coincide for compact-valued maps.

We shall need the following property of lower semi-continuous set-valued maps.

Proposition 9.3. *Suppose that*

(i) $f : X \times Y \to \mathbb{R}$ *is lower semi-continuous;*
(ii) *the set-valued map C from X to Y is lower semi-continuous.*   (9)

Then the function $\alpha : x \to \alpha(x) := \sup_{y \in C(x)} f(x, y)$ *is itself lower semi-continuous.*

Proof. We must show that if a sequence of elements $x_n \in X$ converges to x_0, then $\alpha(x_0) \leq \liminf_{n \to \infty} \alpha(x_n)$. We fix $\varepsilon > 0$.

From the definition of $\alpha(x_0)$, there exists $y \in C(x_0)$ such that, $\alpha(x_0) \leq f(x_0, y) + \varepsilon/2$.

Since C is lower semi-continuous at x_0, there exists a sequence of elements $y_n \in C(x_n)$ converging to y.

Since f is lower semi-continuous, it follows that there exists $N(\varepsilon)$ such that, for all $n \geq N(\varepsilon)$,

$$f(x_0, y) \leq f(x_n, y_n) + \varepsilon/2.$$

Since $y_n \in C(x_n)$, we have $f(x_n, y_n) \leq \alpha(x_n)$. Thus, the above inequalities imply that $\alpha(x_0) \leq \alpha(x_n) + \varepsilon$ whenever $n \geq N(\varepsilon)$.   □

9.3 The Debreu–Gale–Nikaïdo Theorem

We shall begin with a theorem which is used to prove the existence of solutions of many problems in mathematical economics.

Consider the simplex

$$M^n := \left\{ x \in \mathbb{R}^n_+ \,\middle|\, \sum_{i=1}^n x_i = 1 \right\}.$$

Theorem 9.2 (Debreu–Gale–Nikaïdo). *Let C be a set-valued map from M^n to $\mathbb{R}^n$ with non-empty values. If*

(i)   *C is upper hemi-continuous*
(ii) $\forall x \in M^n, \ C(x) - \mathbb{R}^n_+$ *is convex closed*
(iii) $\forall x \in M^n, \ \sigma(C(x), x) \geq 0$ *(Walras's law)*   (10)

then there exists $\bar{x} \in M^n$ such that $C(\bar{x}) \cap \mathbb{R}^n_+ \neq \emptyset$.

Proof. We introduce the function ϕ defined on $M^n \times M^n$ by

$$\phi(x, y) = -\sigma(C(x), y).$$

This function is concave in y (since $y \to \sigma(C(x), y)$ is convex) and lower semi-continuous in x (since, as C is upper hemi-continuous, $x \to \sigma(C(x), y)$ is upper semi-continuous). Since M^n is convex and compact, Ky Fan's Theorem, implies that there exists $\bar{x} \in M^n$ such that $\sup_{y \in M^n} \phi(\bar{x}, y) \leq \sup_{y \in M^n} \phi(y, y) \leq 0$ (following (10)(iii)), in other words that

$$0 \leq \sigma(C(\bar{x}), y) \text{ for all } y \in M^n.$$

This condition is equivalent to

$$0 \le \sigma(C(\bar{x}) - \mathbb{R}^n_+, y) \text{ for all } y \in \mathbb{R}^n \tag{11}$$

since $\sigma(-\mathbb{R}^n_+, y) = 0$ if $y \in \mathbb{R}^n$ and $\sigma(-\mathbb{R}^n_+, y) = +\infty$ if $y \notin \mathbb{R}^n_+$. Since $C(\bar{x}) - \mathbb{R}^n_+$ is convex and closed, (11) implies that $0 \in C(\bar{x}) - \mathbb{R}^n_+$, whence that $C(\bar{x}) \cap \mathbb{R}^n_+ \ne \emptyset$. $\qquad\square$

9.4 The Tangential Condition

Let us suppose once and for all that we have:

(i)   two Hilbert spaces X and Y;
(ii)   a continuous linear operator $A \in L(X, Y)$;
(iii) a convex compact subset $K \subset X$;
(iv)   an upper hemi-continuous set-valued map $C : K \to Y$ with
       non-empty, convex, closed values. $\tag{12}$

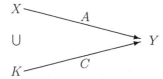

In order to solve the inclusions

$$0 \in C(\bar{x}) \text{ where } \bar{x} \in K \tag{13}$$

and

$$y \in A\hat{x} - C(\hat{x}) \text{ where } \hat{x} \in K \tag{14}$$

we shall impose a condition which interrelates the objects given in (12). We recall that the tangent cone to K at x is defined by:

$$T_K(x) := \text{closure} \left(\bigcup_{h>0} \frac{1}{h}(K - x) \right). \tag{15}$$

Definition 9.6. *We shall say the set-valued map C satisfies the* **tangential condition** *with respect to A if*

$$\forall x \in K, \ C(x) \cap \text{closure}(AT_K(x)) \ne \emptyset. \tag{16}$$

We note also the dual version of the tangential condition.

Proposition 9.4. *The tangential condition (16) implies the dual tangential condition*

$$\forall x \in K, \ \forall p \in A^{*-1} N_K(x), \ \sigma(C(x), -p) \geq 0. \tag{17}$$

The converse is true if the images $C(x)$ of the set-valued map C are compact (and convex).

Proof. a) Suppose $x \in K$ and $v \in C(x) \cap \text{closure}(AT_K(x))$ are fixed. Then $v = \lim_{n \to \infty} Au_n$ where u_n belongs to $T_K(x)$. Let us take p such that A^*p belongs to $N_K(x)$. Then

$$\sigma(C(x), -p) \geq \langle -p, v \rangle = \lim_{n \to \infty} \langle -p, Au_n \rangle = \lim_{n \to \infty} \langle -A^*p, u_n \rangle \geq 0$$

since $\langle A^*p, u_n \rangle \leq 0$ for all $u_n \in T_K(x) = N_K(x)^-$.

b) Let us now suppose that $C(x)$ is convex and compact and that the tangential condition is false.

$$0 \notin C(x) - \text{closure}(AT_K(x)) \tag{18}$$

(since this is equivalent to $C(x) \cap \text{closure}(AT_K(x)) = \emptyset$).

The Separation Theorem (Theorem 2.4) implies that there exist $p \in Y^*$ and $\varepsilon > 0$ such that

$$\sigma(C(x), -p) \leq \inf_{v \in T_K(x)} \langle -p, Av \rangle - \varepsilon.$$

Since $T_K(x)$ is a cone, this inequality implies that A^*p belongs to $T_K(x)^- = N_K(x)$ and that $\inf_{v \in T_K(x)} \langle -p, Av \rangle = 0$. Consequently, $\sigma(C(x), p) \leq -\varepsilon < 0$, which contradicts the dual tangential condition. $\qquad\square$

The properties of tangent cones to convex closed sets which we described in Chapter 4, in many cases enable us to check whether the tangential condition is satisfied. The following self-evident proposition is very useful.

Proposition 9.5. *If two set-valued maps C_1 and C_2 satisfy the tangential condition (or the dual tangential condition, respectively), so do the set-valued maps $\alpha_1 C_1 + \alpha_2 C_2$ where α_1 and α_2 are positive.*

We shall use this property as follows:

Corollary 9.1. *If a set-valued map C from K to Y satisfies the (dual) tangential condition and if y belongs to $A(K)$, then the set-valued map $x \to C(x) - A(x) + y$ also satisfies the (dual) tangential condition.*

9.5 The Fundamental Theorem for the Existence of Zeros of a Set-valued Map

Theorem 9.3. *We suppose that the assumptions (12) are in force (X and Y are Hilbert spaces, A belongs to $L(X, Y)$, $K \subset X$ is convex and compact and $C : K \to Y$ is upper hemi-continuous with non-empty, convex, closed values). If the tangential condition (16)*

$$\forall x \in K, \ C(x) \cap \text{closure}(AT_K(x)) \neq \emptyset$$

is satisfied, then

(a) $\exists \bar{x} \in K$, a solution of the inclusion $0 \in C(\bar{x})$

(b) $\forall y \in A(K)$, $\exists \hat{x} \in K$, a solution of the inclusion $y \in A\hat{x} - C(\hat{x})$.   (19)

Proof. (a) We shall prove a slightly stronger result, assuming that the dual tangential condition (17) is satisfied (instead of the tangential condition (16)).

(b) Corollary 9.1 implies that the second conclusion of the theorem follows from the first conclusion applied to the set-valued map $x \rightarrow C(x) - Ax + y$.

(c) To prove the existence of a zero of C, we shall argue by reduction to the absurd. Suppose therefore that for all $x \in K$, 0 does not belong to $C(x)$. Since the sets $C(x)$ are convex and closed, the Separation Theorem (Theorem 2.4) implies that

$$\forall x \in K, \quad \exists p \in Y^* \text{ such that } \sigma(C(x), -p) < 0. \tag{20}$$

We set

$$\Delta_p := \{x \in K | \sigma(C(x), -p) < 0\}. \tag{21}$$

The non-existence of zeros of C thus translates into the following:

$$K \subset \bigcup_{p \in Y^*} \Delta_p. \tag{22}$$

d) Since C is upper hemi-continuous, the sets Δ_p are open. Since the set K is compact, it can be covered by n open subsets Δ_{p_i}. Let $\{g_i\}_{i=1,\ldots,n}$ be a continuous partition of unity subordinate to this covering. We define the function $\phi : K \times K \rightarrow \mathbb{R}$ as follows

$$\phi(x, y) := - \sum_{i=1}^{n} g_i(x) \langle A^* p, x - y \rangle. \tag{23}$$

ϕ is continuous in x, affine in y and satisfies

$$\phi(y, y) = 0 \text{ for all } y \in K. \tag{24}$$

The assumptions of Ky Fan's Theorem (Theorem 8.6) are satisfied; whence, there exists $\bar{x} \in K$ such that

$$\forall y \in K, \ \phi(\bar{x}, y) = \langle -A^* \bar{p}, \bar{x} - y \rangle \leq 0 \tag{25}$$

where we have set $\bar{p} := \sum_{i=1}^{n} g_i(\bar{x}) p_i$. In other words, $A^* \bar{p}$ belongs to the normal cone $N_K(\bar{x})$.

The dual tangential condition implies that

$$\sigma(C(\bar{x}), -\bar{p}) \geq 0. \tag{26}$$

But this inequality is false. To see this, we let I be the subset of the indices i such that $g_i(\bar{x}) > 0$. I is non-empty since $\sum_{i=1}^{n} g_i(\bar{x}) = 1$. If i belongs to I, then $\bar{x}$ belongs to Δ_{p_i} and consequently

$$\sigma(C(\bar{x}), -\bar{p}) = \sigma\left(C(\bar{x}), -\sum_{i=1}^{n} g_i(\bar{x})p_i\right) \leq \sum_{i \in I} g_i(\bar{x})\sigma(C(\bar{x}), -p_i) < 0.$$

Thus, we have obtained a contradiction and proved our theorem.   □

This theorem has many corollaries and we shall state those which we shall use directly in our theorems.

Remark. By taking $C(x) := \{c(x)\}$ where c is a differentiable map and $A(x) = -c'(x)$ in Theorem 9.3, we derive the existence of a solution to the equation $c(\bar{x}) = 0$ where the solution $\bar{x}$ must belong to a compact convex subset K: *Let X and Y be Hilbert spaces, $K \subset X$ be a compact convex subset, $\Omega \supset K$ be an open neighborhood of K and $c : \Omega \mapsto Y$ be a continuously differentiable single-valued map. Assume that*

$$\forall\, x \in K, \quad -c(x) \in c'(x)T_K(x)$$

Then there exists a solution $\bar{x} \in K$ to the equation $c(\bar{x}) = 0$. In particular, when $x_0 \in K$ is given, there exists a sequence of elements $x_n \in K$ satisfying

$$\forall\, n \geq 0, \quad c'(x_n)(x_n - x_{n-1}) = -c(x_n)$$

i.e., the implicit version of the Newton algorithm.

The most important particular case is that in which X and Y are equal and A is the identity.

Theorem 9.4. *Suppose we have a Hilbert space X, a convex compact subset $K \subset X$ and an upper hemi-continuous set-valued map $C : K \to X$ with non-empty, convex, closed values. If the tangential condition*

$$\forall x \in K, \quad C(x) \cap T_K(x) \neq \emptyset \tag{27}$$

is satisfied, then

(a) $\exists \bar{x} \in K$ *such that* $0 \in C(\bar{x})$
(b) $\forall y \in K, \ \exists \hat{x} \in K$ *such that* $y \in \hat{x} - C(\hat{x})$.   (28)

9.6 The Viability Theorem

Since the velocity of a constant function $t \to \bar{x}$ is equal to zero, we can regard a zero $\bar{x} \in K$ of the set-valued map $C : K \to X$ as an equilibrium $\bar{x}$ (or a rest point) of the *differential inclusion*

$$x'(t) \in C(x(t))$$

governing the evolution of a time dependent function $t \to x(t)$ starting from an initial state $x(0) = x_0$ at the initial time $t = 0$.

Hence an equilibrium is a particular solution to this differential inclusion, so that it requires stronger assumptions than the mere existence of a solution.

The viability theorem states that *when K is (only) compact (but not necessarily convex) and C is upper hemicontinuous with non-empty, convex, compact values, the tangential condition (27) is necessary and sufficient for K to be viable under C*, in the sense that for any initial state x_0, there exists at least one solution to the differential inclusion $x' \in C(x)$ starting from x_0 and viable in K:

$$\forall\, t \geq 0, \quad x(t) \in K$$

We emphasize now this basic and curious link between the existence of the general equilibrium theorem 9.4 and the 'viability theorem': the General Equilibrium Theorem — which is an equivalent version of the 1910 Brouwer Fixed Point Theorem, the cornerstone of nonlinear analysis — states the existence of an equilibrium $\bar{x}$ of the set-valued map $C : K \to X$ when the dynamics of the uncertain dynamical system described by the set-valued map C confronted to the 'viability constraints' described by K are related by the tangential condition (27) and when K is furthermore assumed to be convex and compact.

Both the general equilibrium theorem 9.4 and the 'viability theorem' find here a particularly relevant formulation: viability implies stationarity.

Viability implies also stationarity not only when the convexity of K is traded with the convexity of the image $C(K)$: *If C is upper hemicontinuous with non-empty, convex, compact values, if $K \subset X$ is a compact subset such that $C(K)$ is convex and if there exists at least one viable solution to the differential inclusion $x' \in C(x)$, then there exists a viable equilibrium of C in K.*

Indeed, assume that there is no equilibrium. Hence, this means that 0 does not belong to the closed convex subset $C(K)$, so that the Separation Theorem implies the existence of some $p \in X^*$ and $\varepsilon > 0$ such that

$$\sup_{x \in K, v \in C(x)} \langle v, -p \rangle \;=\; \sigma(C(K), -p) \;<\; -\varepsilon$$

Hence, let us take any viable solution $x(\cdot)$ to differential inclusion $x' \in C(x)$ which exists by assumption. We deduce that

$$\forall\, t \geq 0, \quad \langle -p, x'(t) \rangle \;\leq\; -\varepsilon$$

so that, integrating from 0 to t, we infer that

$$\varepsilon\, t \;\leq\; \langle p, x(t) - x(0) \rangle$$

But K being bounded, we thus derive a contradiction.   □

We can even relax the assumption of the convexity of $C(K)$: *If C is upper hemicontinuous with non-empty, convex, compact values, if $K \subset X$ is a compact subset and if there exists a solution $x(\cdot)$ to the differential inclusion $x' \in C(x)$ viable in K such that*

$$\inf_{t>0} \frac{1}{t} \int_0^t \|x'(\tau)\| d\tau \ = \ 0$$

then there exists a viable equilibrium $\bar{x}$, i.e., a state $\bar{x} \in K$ solution to the inclusion $0 \in C(\bar{x})$.

The proof starts as in the proof of Theorem 9.3: We assume that there is no viable equilibrium, i.e., that for any $x \in K$, 0 does not belong to $C(x)$. Since the images of C are closed and convex, the Separation Theorem implies that there exists $p \in \Sigma$, the unit sphere, and $\varepsilon_p > 0$ such that $\sigma(C(x), -p) < -\varepsilon_p$. In other words, we can cover the compact subset K by the subsets

$$\Delta_p \ := \ \{ \, x \in K \mid \sigma(C(x), -p) < -\varepsilon_p \, \}$$

when p ranges over Σ. They are open thanks to the upper hemicontinuity of C, so that the compact subset K can be covered by q open subsets Δ_{p_j}. Set $\varepsilon := \min_{i=1,\ldots,q} \varepsilon_{p_i} > 0$.

Consider now a viable solution to the differential inclusion $x' \in C(x)$, which exists by assumption. Hence, for any $t \geq 0$, $x(t)$ belongs to some Δ_{p_j}, so that

$$-\|x'(t)\| \ \leq \ \langle -p_j, x'(t) \rangle \ \leq \ \sigma(C(x(t)), -p_j) \ < \ -\varepsilon$$

and thus, by integrating from 0 to t, we have proved that there exists $\varepsilon > 0$ such that, for all $t > 0$,

$$\varepsilon \ < \ \frac{1}{t} \int_0^t \|x'(\tau)\| d\tau$$

a contradiction of the assumption of the theorem. □

9.7 Fixed-point Theorems

The above in turn implies the famous fixed-point theorem due to Kakutani.

Theorem 9.5 (Kakutani). *Suppose that $K \subset X$ is a convex compact subset and that $D : K \to K$ is an upper hemi-continuous set-valued map with non-empty, convex, closed values. Then there exists a fixed point $x_* \in K$ of the set-valued map D.*

Proof. Since $D(x) - x \subset K - x \subset T_K(x)$, we note that the set-valued map $x \to D(x) - x$ satisfies the assumptions of Theorem 9.4 (above). Thus, it has a zero $x_* \in K$, which is a fixed point of D. □

In fact, the above proof implies a more general result.

Definition 9.7. *We shall say that a set-valued map $D : K \to X$ is **re-entrant** if*

$$\forall x \in K, \quad D(x) \cap (x + T_K(x)) \neq \emptyset \tag{29}$$

*and that it is **salient** if*

$$\forall x \in K, \quad D(x) \cap (x - T_K(x)) \neq \emptyset. \tag{30}$$

Theorem 9.6 (Kakutani–Fan). *Suppose that $K \subset X$ is a convex compact subset and that $D : K \to X$ is a* **re-entrant**, *upper hemi-continuous, set-valued map with non-empty, convex, closed values. Then the set-valued map D has a fixed point $x_* \in K$.*

Theorem 9.7. *Suppose that $K \subset X$ is a convex compact subset and that $D : K \to X$ is a* **salient**, *upper hemi-continuous, set-valued map with non-empty, convex, closed values. Then*

(a) *there exists a fixed point $x_* \in K$*
(b) $\forall y \in K, \exists x \in K$ *such that $y \in D(x)$ (whence $K \subset D(K)$)* $\qquad$ (31)

Proof. We apply Theorem 9.4 to the set-valued map $x \to x - D(x)$, which satisfies the tangential condition since D is salient. The zeros of this set-valued map are the fixed points of D and the solutions of $y \in \hat{x} - (\hat{x} - D(\hat{x}))$ are the elements of $D^{-1}(y)$. $\qquad\qquad\qquad\qquad\qquad\qquad\qquad\qquad\qquad\qquad\qquad$ □

A fixed-point $\bar{x}$ of a set-valued map D can be regarded as an equilibrium (or a rest-point) of the discrete dynamical system $x_{n+1} \in D(x_n)$ because, starting from $\bar{x}$, we may remain (or rest) at $\bar{x}$ forever.

9.8 Equilibrium of a Dynamical Economy

We can describe a dynamical economy (P, c) governing the evolution of an abstract **commodity** and an abstract **price**. The commodities evolve according to the laws

$$\begin{cases} i) & x'(t) = c(x(t), p(t)) \\[2mm] ii) & p(t) \in P \end{cases}$$

where the commodity $x(\cdot)$ ranges over a finite dimensional vector-space X, the price $p(\cdot)$ ranges over Y^*, $c : X \times Y^* \mapsto X$ describes the dynamics and where $P \subset Y^*$ is the set of feasible prices.

Here, the first equation describes how the price — regarded as a **message, or regulation control (in short, regulee)**, or again an **input** to the system — yields the commodity of the dynamical economy (once the initial commodity is fixed) — regarded as an **output**.

A solution to this system is a function $t \to x(t)$ satisfying this system for some time dependent price $t \to p(t)$ and an equilibrium $(\bar{x}, \bar{p})$ is a zero of c. Next we shall prove a theorem which is very useful for proving the existence of an equilibrium of a function $c(\cdot, \cdot)$, which is constrained to satisfy additional conditions of the form

$$Ax \in M$$

known as *viability conditions*. The choice of such a parameter p (which may be interpreted as an adaptive control) constitutes the so-called viability problem.

In other words, each p is associated with the set $Z(p)$ of zeros of $x \to c(x, p)$. Does there exist a parameter $\bar{p}$ such that $AZ(\bar{p})$ belongs to M?

To be more precise, we introduce:

(i)   two *Hilbert* spaces X and Y;
(ii)   two *convex closed* subsets $L \subset X$ and $M \subset Y$;
(iii)   a *continuous linear* operator $A \in L(X, Y)$;
(iv)   a *convex compact* subset $P \subset Y^*$;
(v)   a *continuous* mapping $c : L \times P \to X$.   (32)

We shall solve the following problem: find $\bar{x} \in L$ and $\bar{p} \in P$ such that

$$
\begin{aligned}
\text{(i)} && c(\bar{x}, \bar{p}) &= 0 \\
\text{(ii)} && A\bar{x} &\in M && \text{(viability condition)}
\end{aligned}
$$
(33)

Theorem 9.8. *We suppose that conditions (32) are in force, together with the following assumptions:*

$$
\begin{aligned}
\text{(i)} && \forall x \in L, \ p \to c(x, p) \text{ is affine} \\
\text{(ii)} && \forall x \in L, \ \forall p \in P, \ c(x, p) \in T_L(x)
\end{aligned}
$$
(34)

$$
\begin{aligned}
\text{(i)} && L \cap A^{-1}(M) \text{ is compact} \\
\text{(ii)} && 0 \in \text{Int}\,(A(L) - M) \\
\text{(iii)} && \forall y \in M, \ N_M(y) \subset \bigcup_{\lambda \geq 0} \lambda P
\end{aligned}
$$
(35)

and

$$
\forall x \in L, \ \forall p \in P, \ \langle p, Ac(x, p) \rangle \leq 0.
$$
(36)

Then there exists a solution $(\bar{x}, \bar{p}) \in L \times P$ of the problem (33).

Proof. This is again a consequence of Theorem 9.4. We introduce the convex compact subset $K := L \cap A^{-1}(M)$ and the set-valued map C from K to X defined by

$$
\forall x \in K, \ C(x) := \{c(x, p)\}_{p \in P}.
$$
(37)

Since P is convex and compact and $p \to c(x, p)$ is affine, the images $C(x)$ are convex and compact. Since $c : L \times P \to Y$ is continuous and P is compact, the set-valued map C is upper semi-continuous. In fact, if $x_0 \in L$ and $\varepsilon > 0$ are fixed, we may associate any $p \in P$ with neighbourhoods $N_p(x_0)$ and $N(p)$ of x_0 and p (respectively), such that

$$
\forall x \in N_p(x_0), \ \forall q \in N(p), \ c(x, q) \in c(x_0, p) + \varepsilon B \subset C(x_0) + \varepsilon B.
$$
(38)

Since P is compact, it can be covered by n neighbourhoods $N(p_i)$ $(i = 1, \ldots, n)$. Thus, $N(x_0) := \cap_{i=1}^n N(p_i)$ is a neighbourhood of x_0.

The properties (38) imply that

$$
\forall x \in N(x_0), \ C(x) \subset C(x_0) + \varepsilon B.
$$
(39)

Whence, C is upper hemi-continuous.

We shall prove that the dual tangential condition

$$\forall x \in K, \ \forall p \in N_K(x), \ \sup_{v \in C(x)} \langle -p, v \rangle \geq 0 \tag{40}$$

is satisfied. In fact, since 0 belongs to Int $(A(L) - M)$, we know from formula (49) of Chapter 4 that

$$N_K(x) = N_L(x) + A^* N_M(Ax).$$

Thus, any element p of $N_K(x)$ may be written as $p = p_0 + A^*q$ where $p_0 \in N_L(x)$ and $q \in N_M(Ax)$. There exists $p_1 \in P$ such that $q = \lambda p_1$ where $\lambda \geq 0$. Then we introduce $v := c(x, p_1) \in C(x)$. Since $c(x, p_1)$ belongs to $T_L(x) = N_L(x)^-$, by assumption, we have $\langle -p_0, c(x, p_1) \rangle \geq 0$. Moreover,

$$\langle -A^*q, c(x, p_1) \rangle = -\lambda \langle p_1, Ac(x, p_1) \rangle \geq 0.$$

Whence,

$$\sigma((Cx), -p) \geq \langle -p_0 - A^*q, c(x, p_1) \rangle \geq 0.$$

Thus, we may apply Theorem 9.4. There exists $\bar{x} \in K$, in other words $\bar{x} \in L$, satisfying $A\bar{x} \in M$, such that 0 belongs to $C(\bar{x})$; whence, there exists $\bar{p} \in P$ such that $0 = c(\bar{x}, \bar{p})$. □

9.9 Variational Inequalities

We shall consider

(i) a convex compact subset $K \subset X$;
(ii) an upper semi-continuous set-valued map C from K to X
     with convex compact values. $\hspace{3cm}$ (41)

which does not necessarily satisfy the tangential condition. The problem now is how to modify C in such a way that the new set-valued map satisfies this condition.

We note that this modification need only be carried out on the boundary ∂K of K, since for all $x \in \text{Int}(K)$, $T_K(x)$ is equal to the whole space.

For this, it is sufficient to subtract the set-valued map $x \rightarrow N_K(x)$ (which associates each x with the normal cone to K at x) from the set-valued map C and to find the zeros of the set-valued map $C - N_K$:

$$\bar{x} \in K \ \text{ such that } \ 0 \in C(\bar{x}) - N_K(\bar{x}). \tag{42}$$

By definition of the normal cone to K at $\bar{x}$, $N_K(\bar{x})$, the inclusion (42) is equivalent to

(i) $\bar{x} \in K$

(ii) $\exists \bar{v} \in C(\bar{x})$ such that $\langle \bar{v}, \bar{x} - y \rangle \geq 0$ $\forall y \in K$.   (43)

Definition 9.8. *The equivalent problems (42) and (43) are called* **variational inequalities**.

Remark. We saw that the solutions $\bar{x} \in K$ which minimise a nontrivial, convex, lower semi-continuous function $f : X \to \mathbb{R} \cup \{+\infty\}$ such that $0 \in \mathrm{Int}(K - \mathrm{Dom}f)$ over K, are the solutions of the inclusion $0 \in \partial f(\bar{x}) + N_K(\bar{x})$, whence solutions of the variational inequality (43) with $C(x) := -\partial f(x)$.

Remark. Since $T_K(x)$ is the negative polar cone of $N_K(x)$, Theorem 5.1 of (Aubin 1979a) implies that any element $v \in C(x)$ decomposes into the form $v = t + n$ where $t \in T_K(x)$, $n \in N_K(x)$ and $\langle t, n \rangle = 0$. Thus, for any $v \in C(x)$, the element $v - n = t$ belongs to $(C(x) - N_K(x)) \cap T_K(x)$, which shows that the set-valued map $(C - N_K)$ satisfies the tangential condition.

We also note that any zero $\bar{x}$ of $C - N_K$ which belongs to the interior of K is a zero of C and that if

$$\forall x \in K, \quad C(x) \subset T_K(x) \qquad (44)$$

then any zero $\bar{x}$ of $C - N_K$ is a zero of C, since in this case, there exists $\bar{v} \in C(\bar{x})$ which belongs to the intersection of $T_K(\bar{x})$ and $N_K(\bar{x})$, which is zero.

We could use this remark to apply Theorem 9.4 to deduce the existence of solutions of variational inequalities. But we can give a direct proof based on Ky Fan's Inequality.

Theorem 9.9. *Suppose that K is convex and compact and that C is an upper semi-continuous set-valued map from K to X with non-empty, convex, compact values. Then there exists a solution $\bar{x} \in K$ of the variational inequality (43).*

Proof. We set

$$\phi(x, y) = -\sigma(C(x), x - y). \qquad (45)$$

The function ϕ is concave in y and clearly satisfies $\phi(y, y) = 0$. Since C is upper semi-continuous with *compact* values, a variant of *Proposition 9.1* shows that $x \to \sigma(C(x), x - y)$ is upper semi-continuous. In fact, since $C(x_0)$ is bounded, $\|C(x_0)\| := \sup_{v \in C(x_0)} \|v\|$ is finite and the inclusion

$$\forall x \in N(x_0), \quad C(x) \subset C(x_0) + \eta B$$

implies that

$$\forall x \in N(x_0):$$
$$\begin{aligned}
\sigma(C(x), x - y) &\leq \sigma(C(x_0), x - y) + \eta \|x - y\| \\
&\leq \sigma(C(x_0), x_0 - y) + \sigma(C(x_0), x - x_0) + \eta \|x - y\| \\
&\leq \sigma(C(x_0), x_0 - y) + \|C(x_0)\| \|x - x_0\| + \eta \|x - x_0\| \\
&\quad + \eta \|x_0 - y\|.
\end{aligned}$$

Thus, taking $\eta \leq \frac{\varepsilon}{2\|x_0 - y\|}$ and replacing $N(x_0)$ by its intersection with the ball of radius $\frac{\varepsilon}{2(\eta + \|C(x_0)\|)}$, it follows that $x \to \sigma(C(x), x - y)$ is upper semi-continuous, whence $x \to \phi(x, y)$ is lower semi-continuous. Ky Fan's Inequality may then be applied (Theorem 8.6). Thus, there exists $\bar{x} \in K$ such that

$$\forall y \in K, \quad \sigma(C(\bar{x}), \bar{x} - y) \geq 0$$

in other words, such that

$$\inf_{y \in K} \sup_{v \in C(\bar{x})} \langle v, \bar{x} - y \rangle \geq 0.$$

Since K and $C(\bar{x})$ are convex compact sets, it follows that there exists $\bar{v} \in C(\bar{x})$ such that (following Theorem 9.8)

$$\inf_{y \in K} \langle \bar{v}, \bar{x} - y \rangle \geq 0.$$

Thus, this element $\bar{x} \in K$ is a solution of the variational inequality (43). $\square$

9.10 The Leray–Schauder Theorem

From Theorem 9.4, we may derive other theorems for the existence of zeros using the continuation technique due to Poincaré.

Consider the boundary ∂K of the convex compact set K (which is different from K if X is finite dimensional and the interior of K is non-empty).

Theorem 9.10. *Consider a convex compact set K with a non-empty interior, together with an upper hemi-continuous set-valued map C from $K \times [0, 1]$ to X, with non-empty, convex, closed values.*

Suppose that

(i)   *the set-valued map $x \to C(x, 0)$ satisfies the tangential condition;*
(ii) $\forall \lambda \in [0, 1[, \ \forall x \in \partial K, \ 0 \notin C(x, \lambda).$ (46)

Then

$$\exists \bar{x} \in K \text{ such that } 0 \in C(\bar{x}, 1). \qquad (47)$$

Proof. We shall suppose that the conclusion (47) is false and derive a contradiction.

We set $A := \partial K$, which is a closed subset of K and introduce the subset

$$B := \{x \in K | \exists \lambda \in [0, 1] \text{ satisfying } 0 \in C(x, \lambda)\}. \qquad (48)$$

The set B is non-empty, since it contains the equilibria of $x \to C(x, 0)$. It is closed (since C is upper hemi-continuous) and disjoint from A (if $x \in A$ and $t \in [0, 1[$, assumption (46)(ii) implies that $x \notin B$; if $x \in A$ and $t = 1$, then $x \notin B$, since $C(\cdot, 1)$ has no zeros).

Next we introduce a continuous function ϕ from X to $[0, 1]$, which is equal to 0 on A and 1 on B

$$\phi(x) := \frac{d(x, A)}{d(x, A) + d(x, B)},$$

together with the set-valued map D defined by

$$D(x) := C(x, \phi(x)). \qquad (49)$$

D is clearly upper hemi-continuous with non-empty, convex, closed values. It coincides with $C(x, 0)$ on A and consequently satisfies the assumptions of Theorem 9.4. Thus, the set-valued map D has a critical point $\bar{x} \in K$ such that $0 \in D(\bar{x}) = C(\bar{x}, \phi(\bar{x}))$. This now implies that $\bar{x} \in B$, whence that $\phi(\bar{x}) = 1$ and so $0 \in C(\bar{x}, 1)$. It follows that $C(\cdot, 1)$ has a critical point, which is the desired contradiction. $\qquad \square$

In particular, we obtain the following result:

Theorem 9.11. *Suppose that K is a convex compact subset with a non-empty interior and that C and D are two upper hemi-continuous set-valued maps from K to X with non-empty, convex, closed values.*

Suppose that

$$C \text{ satisfies the tangential condition} \qquad (50)$$

and that

$$\forall \mu \geq 0, \ \forall x \in \partial K, \ 0 \notin C(x) + \mu D(x). \qquad (51)$$

Then the set-valued map D has a zero $\bar{x} \in K$.

Proof. We apply the previous theorem with $C(x, t) = (1 - t)C(x) + tD(x)$. $\qquad \square$

Let us take a finite-dimensional space X and $x_0 \in \text{Int } K$. Then the mapping $C(x) = x - x_0$ satisfies the tangential condition. Thus, we have the following theorem:

Theorem 9.12. *Suppose that x_0 is a point in the interior of a convex compact subset K of X and that D is an upper hemi-continuous set-valued map from K to X with non-empty closed values. Suppose further that*

$$\forall \mu \geq 0, \ \forall x \in \partial K, \ x_0 \notin x + \mu D(x). \qquad (52)$$

Then D has a zero $\bar{x} \in K$.

9.11 Quasi-variational Inequalities

We shall now prove a theorem which reconciles Ky Fan's Inequality and Kakutani's Fixed-point Theorem. This result will be useful in the theory of non-cooperative games.

Theorem 9.13. *We suppose that*

$$K \text{ is a convex compact subset of a Hilbert space } X \qquad (53)$$

and that

$$C : K \to X \text{ is an upper hemi-continuous set-valued map}$$
$$\text{with non-empty, convex, closed values.} \qquad (54)$$

We consider a function $\phi : K \times K \to \mathbb{R}$ *satisfying*

(i) $\forall y \in K, \quad x \to \phi(x, y)$ *is lower semi-continuous*
(ii) $\forall x \in K, \quad y \to \phi(x, y)$ *is concave*
(iii) $\sup_{y \in K} \phi(y, y) \leq 0.$ \qquad (55)

We suppose further that the set-valued map C *and the function* ϕ *are related by the property*

$$\{x \in K | \alpha(x) := \sup_{y \in C(x)} \phi(x, y) \leq 0\} \text{ is closed.} \qquad (56)$$

Then there exists a solution $\bar{x} \in K$ *of the* **quasi-variational inequality**:

(i) $\bar{x} \in C(\bar{x})$
(ii) $\sup_{y \in C(\bar{x})} \phi(\bar{x}, y) \leq 0.$ \qquad (57)

Remark. Assumptions (53) and (54) are those of Kakutani's Theorem and assumptions (53) and (55) are those of Theorem 8.6 (Ky Fan's Inequality). Assumption (56) is an assumption of consistency between C and ϕ.

Proof. We shall argue by reduction to the absurd. If the conclusion is false, for all $x \in K$ we would have either $\alpha(x) > 0$ or $x \notin C(x)$. To say that $x \notin C(x)$ implies that there exists $p \in X^*$ such that $\langle p, x \rangle - \sigma(C(x), p) > 0$. We set

(i) $V_0 := \{x \in K | \alpha(x) > 0\}$
(ii) $V(p) := \{x \in K | \langle p, x \rangle - \sigma(C(x), p) > 0\}.$ \qquad (58)

The negation of the conclusion may be expressed in the form

$$K \subset V_0 \cup \bigcup_{p \in X^*} V(p). \qquad (59)$$

Assumptions (54) and (55)(i) imply that the sets V_0 and $V(p)$ are open. Since K is compact, it follows that there exist $p_1, \ldots, p_n$ such that

$$K \subset V_0 \cup \bigcup_{i=1}^{n} V(p_i) \qquad (60)$$

and that there exists a *continuous partition of unity* $\{g_0, g_1, \ldots, g_n\}$ subordinate to this covering.

Next we introduce the function $\psi : K \times K \to \mathbb{R}$ defined by

$$\psi(x, y) = g_0(x)\phi(x, y) + \sum_{i=1}^{n} g_i(x)\langle p_i, x - y \rangle. \tag{61}$$

This function ψ is lower semi-continuous in x (by virtue of (55)(i)) and concave in y (by virtue of (55)(ii)). Since K is convex and compact (by virtue of (53)) and since $\sup_y \psi(y, y) \leq 0$ (by virtue of (55)(ii)), Theorem 8.6 (Ky Fan's Inequality), implies that there exists $\bar{x} \in X$ satisfying

$$\sup_{y \in K} \psi(\bar{x}, y) \leq 0. \tag{62}$$

We shall contradict this inequality by proving that there exists $\bar{y} \in K$ such that

$$\psi(\bar{x}, \bar{y}) > 0. \tag{63}$$

We take:

(i)   any $\bar{y} \in C(\bar{x})$ if $\alpha(\bar{x}) \leq 0$;

(ii) $\bar{y} \in C(\bar{x})$ satisfying $\phi(\bar{x}, \bar{y}) \geq \alpha(\bar{x})/2$ if $\alpha(\bar{x}) > 0$ (64)

(the choice of $\bar{y}$ is free).

Since $g_0, g_1, \ldots, g_n$ is a partition of unity, $g_i(\bar{x}) > 0$ for at least one index $i = 0, 1, \ldots, n$. The inequality (63) then follows from the following assertions:

(i) $g_0(\bar{x}) > 0$ implies that $\phi(\bar{x}, \bar{y}) > 0$

(ii) $g_i(\bar{x}) > 0$ implies that $\langle p_i, \bar{x} - y \rangle > 0$.   (65)

Let us now prove these assertions. If $g_0(\bar{x}) > 0$, then $\bar{x} \in V_0$ and consequently, $\alpha(\bar{x}) > 0$. Thus, $\phi(\bar{x}, \bar{y}) \geq \alpha(\bar{x})/2$. If $g_i(\bar{x}) > 0$, then $\bar{x} \in V(p_i)$ and consequently, $\langle p_i, \bar{x} \rangle > \sigma(C(\bar{x}), p_i) \geq \langle p_i, \bar{y} \rangle$, since $\bar{y} \in C(\bar{x})$. Thus, $\langle p_i, \bar{x} - \bar{y} \rangle > 0$.   □

Remark. It is useful to give sufficient conditions implying assumption (56). One such is that the function $\alpha : x \to \alpha(x) = \sup_{y \in C(x)} \phi(x, y)$ be lower semi-continuous. For, Proposition 9.3 implies that if the set-valued map C is lower semi-continuous then so too is α.

Theorem 9.14. *Suppose that C is a **continuous** set-valued map from a convex compact subset K into itself, with non-empty, convex, closed values. Suppose that ϕ is a function satisfying assumptions (55) which is lower semi-continuous in both variables. Then there exists a solution $\bar{x} \in K$ of the quasi-variational inequality (57).*

9.12 Shapley's Generalisation of the Three-Poles Lemma

We know that the story began in 1910 with the Brouwer Fixed Point Theorem. It was proved later in 1926 via the *Three Polish Lemma*, the three Poles being

Knaster, Kuratowski and Mazurkiewicz, which allowed them to derive the Fixed Point Theorem in a simpler way. Knaster saw the connection between Sperner's Lemma and the fixed point theorem, Mazurkiewicz provided a proof corrected by Kuratowski. The extension to Banach spaces was proved in 1930 by their colleague Schauder.

Von Neumann did need the set-valued version of this Fixed Point Theorem in game theory, which was proved by Kakutani in 1941.

Lemma 9.1 (Three-Poles or K–K–M lemma). *Consider n closed subsets F_i of the simplex $M^n := \{x \in \mathbb{R}^n_+ \mid \sum_{i=1}^n x_i = 1\}$ satisfying the condition*

$$\forall x \in M^n, \quad x \in \bigcup_{\{i \mid x_i > 0\}} F_i. \tag{66}$$

Then

$$\bigcap_{i=1}^n F_i \neq \emptyset. \tag{67}$$

At the time, this lemma was proved from Sperner's lemma on the simplicial decomposition and thus stemmed from the area of combinatorics. Shapley generalised it in 1973 and we shall deduce the Three-Poles Lemma from his theorem.

However, before we do so, some indications as to how to prove Brouwer's Theorem from this lemma would not go amiss.

Proof of Brouwer's Theorem. Let D be a continuous mapping of the simplex M^n into itself. We associate this with the sets F_i defined by

$$F_i := \{x \in M^n \mid x_i \geq D_i(x)\} \tag{68}$$

which are closed since D is continuous. Condition (66) is satisfied, otherwise there would exist $x \in M^n$ such that for all indices i, $x_i < D_i(x)$. Since x and $D(x)$ belong to M^n, we obtain the contradiction $1 < 1$ by summing these inequalities.

The Three-Poles Lemma then implies that there exists a point $\bar{x} \in M^n$ belonging to the intersection of the F_i, in other words satisfying

$$\forall i = 1, \dots, n, \quad \bar{x}_i \geq D_i(\bar{x}). \tag{69}$$

The inequality cannot be nontrivial since, otherwise, taking the sum, we would again obtain the contradiction $1 < 1$. Thus, $\bar{x}_i = D_i(\bar{x})$ for all i, and consequently, $\bar{x}$ is a fixed point of the continuous mapping D. □

Let us now denote the set of n elements by $N := \{1, \dots, n\}$. With any subset T of N, we associate the sub-simplex M^T defined by

$$M^T := \{x \in M^n \mid \forall i \in T, \ x_i \neq 0\}. \tag{70}$$

The characteristic functions $c_T \in \{0, 1\}^n$ of the subsets $T \subset N$ are defined by

$$c_T(i) = 1 \text{ if } i \in T, \qquad c_T(i) = 0 \text{ if } i \notin T. \tag{71}$$

Theorem 9.15 (K–K–M–S Theorem). *Every non-empty subset $T \subset N$ is associated with a closed subset (possibly empty) $F_T \subset M^n$ in such a way that the condition*

$$\forall T \neq \emptyset, \quad M^T \subset \bigcup_{S \subset T} F_S \tag{72}$$

is satisfied. Then there exist non-negative scalars $m(T)$ such that

(i)
$$c_N = \sum_{T \neq \emptyset} m(T) c_T$$

(ii)
$$\bigcap_{\{T \mid m(T) > 0\}} F_T \neq \emptyset. \tag{73}$$

Proof of Lemma 9.1. We apply Theorem 9.15 with $F_S := F_i$ when $S = \{i\}$ and $F_S = \emptyset$ if $|S| := \text{card}(S) \geq 2$. Condition (66) implies assumption (72), whilst the conclusion (73)(i) implies that $m(i) = 1$ for all $i = 1, \dots, n$ and the conclusion (73)(ii) implies that the intersection of the F_i is non-empty.   □

Proof of Theorem 9.15. This is a consequence of Theorem 9.4 applied to the set-valued map $G : M^n \to \mathbb{R}^n$ defined by

$$G(x) := \frac{1}{|N|} c_N - \overline{\text{co}} \left\{ \frac{1}{|S|} c_S \right\}_{F_S \ni x}. \tag{74}$$

It is easy to prove that G is upper semi-continuous with convex compact values. It remains to show that it satisfies the tangential condition:

$$\forall x \in M^n, \quad G(x) \cap T_{M^n}(x) \neq \emptyset. \tag{75}$$

We denote the set of indices i such that $x_i > 0$ by T. Assumption (72) implies that there exists a subset $R \subset T$ such that x belongs to F_R. Thus,

$$y := \frac{1}{|N|} c_N - \frac{1}{|R|} c_R \in G(x). \tag{76}$$

But y also belongs to the tangent cone to M^n at x, which is described by formula (44) of Chapter 4. In fact, $\sum_{i=1}^n y_i$ is equal to zero, and for all i such that $x_i = 0$, we have $y_i = \frac{1}{|N|} c_N(i) - \frac{1}{|R|} c_R(i) = \frac{1}{|N|} \geq 0$, since i does not belong to T. Thus, the tangential condition (75) is satisfied. Whence, Theorem 9.4 implies that there exists $\bar{x} \in M^n$ such that $0 \in G(\bar{x})$, in other words, such that

$$c_N = \sum_{\{T \mid F_T \ni \bar{x}\}} \frac{\lambda(T) |N|}{|T|} c_T \quad \text{and} \quad \bar{x} \in \bigcap_{F_T \ni \bar{x}} F_T. \tag{77}$$

□

Remark. We note that property (73)(i) may be written in the form

$$\forall i, 1, \dots, n, \quad \sum_{T \ni i} m(T) = 1. \tag{78}$$

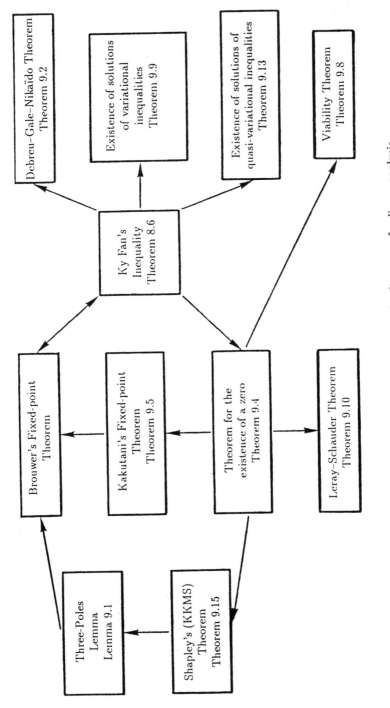

Fig. 9.3. Relationships between the main theorems of nolinear analysis.

Definition 9.9. *A family $\mathcal{B}$ of non-empty subsets $T \subset N$ such that*

$$c_N = \sum_{T \in \mathcal{B}} m(T) c_T \quad where \quad m(T) > 0 \;\; \forall T \in \mathcal{B} \tag{79}$$

is called a **balanced family** *and the vector $m = (m(T))_{T \in \mathcal{B}}$ satisfying (78) or (79) is called a* **balancing**.

Theorem 9.15 may be reformulated by saying that assumption (72) implies that there exists a balanced family $\mathcal{B}$ such that $\bigcap_{T \in \mathcal{B}} F_T \neq \emptyset$.

10. Introduction to the Theory of Economic Equilibrium

10.1 Introduction

We shall describe two ways of explaining the role of prices in the problem of *decentralisation* of consumer choice (in the static framework only). This is taken to mean that knowledge of prices enables each consumer to make his own choice, in accordance with his own objectives, without knowing the global state of the economy and in particular, without knowing the choice of other consumers, all the while respecting the scarcity constraints.

It was Adam Smith who, more than two centuries ago, originated this concept of decentralisation. He introduced this paradoxical and mysterious property in a poetic way. Here is the famous quotation from his book, *The Wealth of Nations*, published in 1778.

'Every individual endeavours to employ his capital so that its produce may be of greatest value. He generally neither intends to promote the public interest, nor knows how much he is promoting it. He intends only his own security, only his own gain. And he is in this led by an invisible hand to promote an end which was no part of his intention. By pursuing his own interest, he frequently thus promotes that of society more effectually than when he really intends to promote it.'

But Adam Smith did not state what this famous hand manipulated and *a fortiori* put forward a rigorous argument to justify its existence.

It was a century later that Léon Walras suggested that this invisible hand acted on prices via the demand functions, using them to provide economic agents with sufficient information to guarantee the consistency of their actions whilst respecting the scarcity constraints.

This concept of *economic equilibrium* which we owe to Walras is not the only thing we owe to him. For, it was Léon Walras who, from his first publication in 1859, which refuted the ideas of Proudhon, suggested that mathematical methods could be useful in economic theory. Originality often consists of a new way of viewing the world rather than of discoveries and inventions which arouse the interest of contemporaries. Walras introduced mathematical rigour into an area which at that time had not benefitted from detailed work for a number of centuries. He did this outside of (and against) all customs, despite major difficulties, alone and unaided, without the encouragement and moral support of his colleagues, whether mathematicians or economists. He did it, because in his heart, he was able to recognise the perspectives involved before he even started. It should also be noted that the Parisian scientific community at that time (like that of today), guided by snobbism, custom and prestige, did not allow Léon Walras to take root. It was the University of Lausanne which was recompensed by having offered him the chair of economics in which he was succeeded by Vilfredo Pareto who shared with Walras the conviction of the applicability of mathematics to the social sciences.

It was in 1874 that Léon Walras introduced his concept of economic equilibrium in *Élements d'économie politique pure*, as the solution of a system of a nonlinear equations. The fact that there were the same number of unknowns as equations gave him sufficient optimism about the final outcome to affirm the existence of a solution.

But this required the tools of nonlinear analysis, which were developed following the proof of Brouwer's theorem in 1910. It took another century of maturation before the works of Wald and von Neumann (in the 1930s), Arrow and Debreu (1954), Gale, Nikaïdo and many others, produced the rigorous results which we shall describe in the greatly-simplified framework of exchange economies

10.2 Exchange Economies

We begin by describing an economy by introducing l types of elementary commodities, each with a unit of measurement, so that it is possible to talk about x units of an elementary commodity. An elementary commodity is described not only by its physical properties, but also by other characteristics such as its location and/or the date when it will be available and, in case of uncertainty, the event which will take place, etc.

Services may also be viewed as elementary commodities as long as they can be quantified by units of measurement.

A commodity (or a 'complex' or 'basket' of commodities) consists of a vector $x \in \mathbb{R}^l$ which describes the quantity x_h of each elementary commodity $h = 1, \ldots, l$.

The description of an exchange economy involves

$$\text{a subset } M \subset \mathbb{R}^l \text{ of available commodities} \tag{1}$$

together with n consumers. We shall describe two consumer models, the first of which could be called the classical Walrasian model. In both cases, the description of the ith consumer involves

$$\text{the consumption set } L_i \subset R^l \tag{2}$$

This is interpreted as the set of commodities which the ith consumer needs. If x belongs to L_i, then the hth component x_h represents the consumer's demand for the elementary commodity h if $x_h > 0$ and $|x_h|$ represents the supply of this elementary commodity if x_h is negative.

The fundamental question then arises: can consumers share an available commodity?

This leads us to introduce the following concept of allocation.

Definition 10.1. *An* **allocation** $x \in (\mathbb{R}^l)^n$ *is a sequence of n commodities $x_i \in L_i$ such that their sum $\sum_{i=1}^n x_i$ is available.*

We denote the set of allocations by:

$$K := \left\{ x \in \prod_{i=1}^n L_i \mid \sum_{i=1}^n x_i \in M \right\} \tag{3}$$

Assuming that the set of allocations K is non-empty, we must describe reasonable mechanisms which enable consumers to choose allocations.

The two mechanisms which we shall describe are *decentralised mechanisms*.

By this, we mean mechanisms which do not require each consumer to know the set M of available commodities and the behaviour and the choices of the other consumers, but which only require each consumer to know his own particular environment and to have access to common information about the state of the economy.

In the two models to be described, *this common information will take the form of a price* (or price system) which is perhaps best viewed as an adaptive control.

As far as we are concerned, a price is a linear form $p \in \mathbb{R}^{l*}$ which associates a commodity $x \in \mathbb{R}^l$ with its value $\langle p, x \rangle \in \mathbb{R}$, expressed in monetary units.

Since an elementary commodity h is represented by the hth vector $e^h := (0, \dots, 0, 1, 0, \dots, 0)$ of the canonical basis of $\mathbb{R}^l$, the components $p^h := \langle p, e^h \rangle$ of the price p represent what is usually called the price of the commodity h.

We denote the *price simplex* by

$$M^l := \left\{ p \in \mathbb{R}_+^{l\,*} \mid \sum_{h=1}^l p_h = 1 \right\} \tag{4}$$

We could have taken a different normalisation rule, for example, by taking a reference commodity $\omega \in \overset{\circ}{\mathbb{R}}_+^l$, called the *currency* whose value is always 1; this amounts to only considering prices $p \in \mathbb{R}^{l*}$ such that $\langle p, \omega \rangle = 1$. For simplicity, we shall take $\omega := \frac{1}{l}\mathbb{1}$.

10.3 The Walrasian Mechanism

In the case of the Walrasian mechanism, we view the consumer i as *an automaton which associates a subset of consumptions $D_i(p, r) \subset L_i$ with each price* *system $p \in \mathbb{R}^{l*}$ and each income r*. In other words, a consumer is described by

$$\text{a correspondence } D_i : M^l \times \mathbb{R} \to L_i \tag{5}$$

We interpret the support function

$$\sigma_M(p) := \sup_{y \in M} \langle p, y \rangle \tag{6}$$

of the set M of available commodities as the *collective-income function* which is the maximum value of the commodities available for each price p.

The essential assumption of this mechanism is that the collective income is shared between the n consumers:

$$\text{there exist } n \text{ functions } r_i : M^l \to \mathbb{R} \text{ such that } \textstyle\sum_{i=1}^n r_i(p) = \sigma_M(p). \tag{7}$$

With this in place, for each price $p \in \mathbb{R}^l$, the income of each consumer is $r_i(p)$ and he is thus led to choose a consumption x_i in the set $D_i(p, r_i(p))$. This choice is *decentralised*; it depends only on the price p (via r_i) and is independent of the choice of other consumers.

There is clearly a consistency problem.

Is there a price $\bar{p}$ such that the sum of the consumptions $\sum_{i=1}^n \bar{x}_i \in \sum_{i=1}^n D_i(\bar{p}, r_i(\bar{p}))$ is available (in other words belongs to M) or such that the consumptions $\bar{x}_i \in D_i(\bar{p}, r_i(\bar{p}))$ form an allocation?

Definition 10.2. *We shall say that a price $\bar{p} \in M^l$ is a* **Walrasian equilibrium price** *if it is a solution of the inclusion*

$$0 \in \sum_{i=1}^n D_i(\bar{p}, r_i(\bar{p})) - M. \tag{8}$$

We shall call the correspondence E from M^l to $\mathbb{R}^l$ defined by

$$E(p) := \sum_{i=1}^n D_i(p, r_i(p)) - M \tag{9}$$

the **excess-demand** *correspondence.*

Consequently, the Walrasian equilibrium prices are the zeros of the excess-demand correspondence. These are the prices which Adam Smith's invisible hand should be able to propose to the market – by solving the inclusion (8).

Remark. To avoid misunderstandings, it is useful to stress that the partition $\sum_{i=1}^n r_i(p) = \sigma_M(p)$ of the collective income is *given* in the model and is not a solution. In other words, there are as many Walrasian equilibrium prices as partitions $r(p) = \sum_{i=1}^n r_i(p)$. This model is *neutral* as far as any question of the justice of the partition of the collective income between the players is concerned.

We can solve the existence problem for Walrasian equilibrium prices using one of the many theorems for the existence of zeros of correspondences. In addition, we have to find such a theorem with assumptions which are susceptible to a reasonable economic interpretation. This is possible. We shall show that simple (decentralised) budgetary rules guarantee the existence of an equilibrium.

Collective Walras Law

The demand correspondences D_i should satisfy the condition

$$\forall p \in M^l, \ \ \forall x_i \in D_i(p, r_i), \ \ \left\langle p, \sum_{i=1}^{n} x_i \right\rangle \leq \sum_{i=1}^{n} r_i. \tag{10}$$

In other words, this law *forbids the set of consumers from spending more (in terms of monetary units) then their total income.*

The collective Walras law provides for transfer of income between consumers. A stronger, decentralised law is given below.

Walras Law

Every correspondence D_i satisfies the condition

$$\forall p \in M^l, \ \ \forall x \in D_i(p, r), \ \ \langle p, x \rangle \leq r. \tag{11}$$

We note that the Walras law is independent of the set M of available commodities.

Theorem 10.1. *We make the following assumptions:*

The consumption set M is convex and may be written as $M = M_0 - \mathbb{R}_+^l$, where M_0 is compact. (12)

The demand correspondences $D_i : M^l \times \mathbb{R} \to L_i$ are upper hemi-continuous with convex, compact values and satisfy the collective Walras law. (13)

The income functions r_i are continuous. (14)

Then there exists a Walrasian equilibrium.

Proof. We apply Theorem 9.2 (Debreu–Gale–Nikaïdo) to the correspondence $C : M^l \to \mathbb{R}^l$ defined by

$$C(p) := M_0 - \sum_{i=1}^{n} D_i(p, r_i(p)) \tag{15}$$

which is clearly upper hemi-continuous.

It follows from (12) and (13) that $C(p) - \mathbb{R}_+^l$ is convex and closed. Since

$$\forall p \in M^l, \ \ \sum_{i=1}^{n} r_i(p) = \sigma(M_0 - \mathbb{R}_+^l, p) = \sigma(M_0, p), \tag{16}$$

it follows from the collective Walras law (10) that

$$\sigma(C(p), p) = \sigma(M_0, p) - \sup_{x_i \in D_i(p, r_i(p))} \sum_{i=1}^{m} \langle -p, x_i \rangle$$

$$\geq \sigma(M_0, p) - \sum_{i=1}^{n} r_i(p)$$

$$= \sigma(M,p) - \sum_{i=1}^{n} r_i(p)$$

$$= 0.$$

Thus, there exists $\bar{p} \in M^l$ such that $0 \in C(\bar{p}) - \mathbb{R}_+^l = -E(\bar{p})$, in other words, a Walrasian equilibrium price. □

One important particular case is that in which

$$M := w - \mathbb{R}_+^l \tag{17}$$

is the set of commodities less than the available commodity $w \in \mathbb{R}^l$.

Corollary 10.1. *We suppose that assumptions (13) and (14) are in force and that*

$$w = \sum_{i=1}^{n} w_i \text{ is allocated to the } n \text{ consumers} \tag{18}$$

Then there exists a Walrasian equilibrium price $\bar{p}$ and consumptions $\bar{x}_i \in D_i(\bar{p}, \langle \bar{p}, w_i \rangle)$ such that $\sum_{i=1}^{n} \bar{x}_i \le \sum_{i=1}^{n} w_i$.

In fact, Léon Walras, and the neoclassical economists following him assumed that the demand functions and correspondence arose from the maximisation of a utility function under the budgetary constraints:

$$D_i(p,r) = \left\{ x \in L_i, \langle p, x \rangle \le r | u_i(x) = \sup_{\langle p,y \rangle \le r} u_i(y) \right\} \tag{19}$$

Of course, these demand correspondences satisfy the Walras law. Assumptions needed to return to the case of Theorem 10.1 are imposed on the utility functions. So as not to overload the description with technical complications and above all because it is not clear that the maximisation of utility functions according to *Homo economicus* is compatible with the teachings of cognitive psychology, we shall not develop this point of view any further.

We have seen that the Walrasian equilibrium prices are the zeros of the excess-demand correspondence E.

It is tempting (as in physics) to consider these zeros as the stationary solutions

$$0 \in E(\bar{p}) \tag{20}$$

of the dynamical system (multi-valued)

$$p'(t) \in E(p(t)) \tag{21}$$

(again called a differential inclusion).

The algorithm thus defined (called **Walras tâtonnement**) cannot be implemented outside the stationary state, since in this case, 0 does not belong to $E(p(t))$ and the sum of the corresponding demands $x_i(t) \in D_i(p(t), r_i(p(t))$ does not necessarily belong to the set M of available commodities.

We shall introduce another mathematical description of the consumers and another concept of equilibrium which may be viewed as a stationary state of a dynamical system. Although dynamic considerations are beyond the limited scope of this book, the mechanism which we shall describe avoids the criticisms made of the Walrasian equilibrium. These criticisms (which were made too quickly) do not relate to the concept of price decentralisation, but to an excessively-specific mathematical translation which has become a dogma. The fact that this attempt to respond to criticisms has been called the theory of disequilibrium has added a great deal of confusion to an already complicated situation. The notion of equilibrium is very specific, its converse could be almost anything!

10.4 Another Mechanism for Price Decentralisation

We consider an exchange economy described by a subset M of available commodities together with n consumers whose consumption sets $L_i \subset \mathbb{R}$ are given.

Here, we represent each consumer i, not by a demand function or correspondence, but by a continuous function

$$c_i : L_i \times M^l \rightarrow \mathbb{R}^l \qquad (22)$$

called a *change function*

This function associates the commodity $x \in L_i$ and the price $p \in M^l$ with the change $c_i(x, p)$ that the automaton consumer wishes to make to the composition of the commodity x. If the hth component $c_i(x, p)_h$ is positive, he will increase his consumption of the elementary commodity h and if $c_i(x, p)_h = 0$ he will conserve his elementary commodity h.

Definition 10.3. *In this context, an equilibrium is defined by an* **allocation** $\overline{x} \in K$ *and a price* $\overline{p} \in M^l$ *such that*

$$\forall i = 1, \ldots, n, \quad c_i(\overline{x}_i, \overline{p}) = 0. \qquad (23)$$

In other words, the equilibrium price $\overline{p}$ is such that it stimulates each consumer i to conserve his consumption $\overline{x}_i$.

The choice of such an equilibrium allocation is again *decentralised*; it depends only on the price p and the personal consumption $x_i \in L_i$ of each consumer, and does not depend either on the choice of the other consumers or on the set M of available commodities.

To solve the equations (23), we must chose an existence theorem with assumptions which are susceptible to a reasonable economic interpretation. As in the case of the Walrasian mechanism, we shall show that simple (and decentralised) budgetary rules guarantee the existence of an equilibrium.

10.5 Collective Budgetary Rule

The change functions $c_i : L_i \times M^l \to \mathbb{R}^l$ satisfy

$$\forall x \in \prod_{i=1}^{n} L_i, \quad \forall p \in M^l, \quad \left\langle p, \sum_{i=1}^{n} c_i(x_i, p) \right\rangle \le 0. \tag{24}$$

In other words, this rule states that for each price p, the *total value* of the deficits $\langle p, c_i(x_i, p) \rangle$ incurred by each agent should not be *negative or zero*.

The collective budgetary rule allows for transfers of deficits between the consumers. It is possible to forbid such transfers and to require consumers to obey the following more restrictive (but decentralised) rule:

$$\forall x \in L_i, \quad \forall p \in M^l, \quad \langle p, c_i(x_i, p) \rangle \le 0. \tag{25}$$

We obtain the following existence theorem.

Theorem 10.2. *We make the following assumptions:*

The consumption set M is convex and may be written as $M = M_0 - \mathbb{R}_+^l$, where M_0 is compact. $\tag{26}$

$\forall i = 1, \dots, n$ *the consumption sets are convex, closed, bounded below and satisfy* $0 \in \text{Int} \left(\sum_{i=1}^{n} L_i - M \right).$ $\tag{27}$

For each $i = 1, \dots, n$, the change functions satisfy

(i) $\forall x \in L_i, \quad p \to c_i(x, p)$ *is affine*

(ii) $\forall x \in L_i, \quad \forall p \in M^l, \quad c_i(x, p) \in T_{L_i}(x)$ $\tag{28}$

together with the collective budgetary rule

$$\forall x \in \prod_{i=1}^{n} L_i, \quad \forall p \in M^l, \quad \left\langle p, \sum_{i=1}^{m} c_i(x_i, p) \right\rangle \le 0. \tag{29}$$

Then there exists an equilibrium allocation $\overline{x} = (\overline{x}_1, \dots, \overline{x}_n) \in K$ and price $\overline{p} \in M^l$.

Proof. We apply Theorem 9.8 with

$$X := (\mathbb{R}^l)^n, \quad Y = \mathbb{R}^l, \quad Ax := \sum_{i=1}^{n} x_i, \quad P := M^l,$$

$$L := \prod_{i=1}^{n} L_i \text{ and } c(x, p) = (c_i(x_i, p))_{i=1,\dots,n}.$$

Assumptions (34) of Chapter 9 follow from assumptions (28).

Since $M = M_0 - \mathbb{R}_+^l$, where M_0 is compact, and since the sets L_i are contained in cones $\xi_i + \mathbb{R}_+^l$, it follows that the set of allocations

$$K := \left\{ x \in \prod_{i=1}^{n} L_i \mid \sum_{i=1}^{n} x_i \in M \right\} \tag{30}$$

is compact. In fact, since M is contained in a cone $\omega - \mathbb{R}_+^l$, the set K is contained in $\prod_{i=1}^n [\xi_i, \omega - \sum_{j \neq i} \xi_j]$ where $[y, z]$ denotes the set of $x \in \mathbb{R}^l$ such that $y \leq x \leq z$.

Clearly 0 belongs to $\text{Int}(A(L) - M)$. Since $M = M_0 - \mathbb{R}_+^l$, it follows that for all $y \in M$, the normal cone $N_M(y)$ is contained in $\mathbb{R}_+^l$, which is generated by M^l. Thus, assumptions (35) of Chapter 9 are satisfied. The collective budgetary rule clearly implies assumption (36) of Chapter 9. It remains to check that the solutions $\bar{x} \in K := L \cap A^{-1}(M)$ and $\bar{p} \in M^l$ of the equation $c(\bar{x}, \bar{p}) = 0$ are the desired equilibria. $\square$

Example. The functions of the form

$$c(x, p) := \theta(x)(\langle p, f(x) \rangle g(x) - \|f(x)\| \|g(x)\| p - h(x)) \tag{31}$$

where

(i) $\theta : L \to \mathbb{R}_+$ and $h : L \to R_+^l$ are positive

(ii) f and g are defined from L to $\mathbb{R}_+^l$ (32)

are affine with respect to p and satisfy the budgetary rule, since the Cauchy–Schwarz inequality implies that

$$\langle p, c(x, p) \rangle \leq -\theta(x) \langle p, h(x) \rangle \leq 0 \tag{33}$$

when p runs through M^l.

If we also suppose that

$$\forall x \in L, \quad f(x) \in -\mathbb{R}_+^l \tag{34}$$

we obtain the inequalities

$$\forall x \in L, \quad \forall p \in M^l, \quad \langle f(x), c(x, p) \rangle \geq 0 \tag{35}$$

again by virtue of the Cauchy–Schwarz inequality.

If we take

$$f = g \quad \text{and} \quad \langle f, h \rangle = 0 \tag{36}$$

we obtain

$$\forall x \in L, \quad \forall p \in M^l, \quad \langle f(x), c(x, p) \rangle = 0. \tag{37}$$

If for the mapping f we take the gradient ∇u of a utility function, the conditions (36) (or (37)) express the fact the changes $c(x, p)$ are directions which change (or leave invariant) the level of utility.

Example. Another example of a change function may be constructed from:

a twice-continuously-differentiable function w_i defined on a neighbourhood of L_i (38)

by setting

$$c_i(x, p)_h = \sum_{k=1}^l \frac{\partial^2 w_i(x)}{\partial x_h \partial x_k} p_k. \tag{39}$$

If we assume that

$$\sum_{i=1}^{n} w_i \text{ is concave (or } \forall i, w_i \text{ is concave)} \tag{40}$$

the collective budgetary rule (or the budgetary rule, respectively) is satisfied. An equilibrium is defined by the conditions

(i) $\qquad\qquad\qquad \overline{x} \in K$

(ii) $\qquad \forall i = 1, \ldots, n, \ \ \forall h = 1, \ldots, \ \sum_{k=1}^{l} \dfrac{\partial^2 w_i(\overline{x})}{\partial x_h \partial x_k} \overline{p}_k = 0.$ $\tag{41}$

Remark. Since the equilibrium allocations are the zeros of the correspondence C defined by

$$\forall x \in L, \ \ C(x) := \{c(x, p)\}_{p \in M^l} \tag{42}$$

it is tempting to view them as the stationary points of the differential inclusion

$$\frac{dx}{dt} \in C(x(t)) \tag{43}$$

which may be written explicitly as the system of differential equations

$$x_i'(t) = c_i(x_i(t), p(t)), \qquad i = 1, \ldots, n \tag{44}$$

in which the price p appears as a control.

The problem then is to know if there exists a function $t \to p(t)$ with values in the set of prices M^l, such that the trajectories of the differential system (44) at any given time form allocations of the set M of available commodities, in other words they satisfy

(i) $\qquad\qquad \forall t \geq 0, \ \forall i = 1, \ldots, n, \ x_i(t) \in L_i$

(ii) $\qquad\qquad \forall t \geq 0, \ \sum_{i=1}^{n} x_i(t) \in M$ $\tag{45}$

It can be shown that, under the assumptions of Theorem 10.2 there exists such a function $p(\cdot)$. If we set

$$\forall x \in K, \ \ \Pi(x_1, \ldots, x_n) := \left\{ p \in M^l \mid \sum_{i=1}^{n} c_i(x_i, p) \in T_M\left(\sum_{i=1}^{n} x_i\right) \right\} \tag{46}$$

the price $p(t)$ is linked to the allocations $X(t)$ by the feedback relation

$$\forall t \geq 0, \ p(t) \in \Pi(x_1(t), \ldots, x_n(t)) \tag{47}$$

It is at this level that one can put one's finger on the difference between the two concepts of equilibrium. The Walrasian equilibrium price is the stationary state of a dynamical system $p'(t) \in E(p(t))$ *involving prices, which cannot be implemented*; whilst the second concept for equilibrium allocation is the stationary state of a dynamical system $x'(t) \in C(x(t))$ involving the *commodities consumed*, which is *viable*, in the sense that at any given time, the $x(t)$ are allocations. In this case, the price evolves according to the feedback law (46). This being the case, the two models translate the same idea of price decentralisation allowing each consumer to find an allocation.

Remark. In this context, one could equally well propose a decentralised planning model which would allow one to move around in the set of allocations.

Suppose there exists a continuous mapping $x \in K \to p(x_1, \dots, x_n) \in M^l$ such that

$$\forall x \in K, \ p(x_1, \dots, x_n) \in \Pi(x_1, \dots, x_n). \tag{48}$$

Such a mapping is called a *continuous selection* of the correspondence Π.

In the language of planning, this is interpreted by surmising that the planning office knowing the allocation $x \in K$ is able to associate it with a price $p(x_1, \dots, x_n)$ which is an element of $\Pi(x_1, \dots, x_n)$.

Knowing this price system, the consumers modify their consumption by solving the system of differential equations

$$x_i'(t) = c_i(x_i(t), p(x_1(t), \dots x_i(t), \dots, x_n(t))) \ (i = 1, \dots, n) \tag{49}$$

the solutions of which at any given time satisfy the viability conditions

(i) $$\forall i = 1, \dots, n, \ \forall t \geq 0, \ x_i(t) \in L_i$$

(ii) $$\forall t \geq 0, \ \sum_{i=1}^{n} x_i(t) \in M. \tag{50}$$

Table 10.1. Comparison between Walrasian and viable equilibrium

Process:	Walrasian	Viable
Description of the behavior of consumers	demand functions $d_i(p,r)$ $x_i = d_i(p, r_i(p))$	change functions $c_i(x_i, p)$ $x_i'(t) = c_i(x_i(t), p(t))$
Demand map derived from change function		$x \in D_i(p) :=$ if and only if $c_i(x, p) = 0$
Derivation from utility function	$d_i(p, r)$ maximizes U_i under $\langle p, x \rangle \leq r$	$c_i(x, p)$ $= U_i'(x) - p$ $= \partial \left(U_i|_{\langle p,x \rangle \leq r} \right)(x)$
Equilibrium: stationarity and (static) viability	$\forall\, i,\ \bar{x}_i = d_i(\bar{p}, r_i(\bar{p}))$ such that $\displaystyle\sum_{i=1}^{n} \bar{x}_i \in M$	$\forall\, i,\ c_i(\bar{x}_i, \bar{p}) = 0$ such that $\displaystyle\sum_{i=1}^{n} \bar{x}_i \in M$
Budget rule	$\langle p, d_i(p, r) \rangle \leq r$	$\langle p, c_i(x, p) \rangle \leq 0$
(dynamic) Viability		$\exists\, p(t)$ such that $\forall\, t \geq 0,\ \displaystyle\sum_{i=1}^{n} x_i(t) \in M$
Characterization of the viability		$\forall\, (x_1, \ldots, x_n),$ $\Pi(x_1, \ldots, x_n) \neq \emptyset$
Regulation law		$p(t) \in \Pi(x_1(t), \ldots, x_n(t))$

11. The Von Neumann Growth Model

11.1 Introduction

In 1945, J.von Neumann proposed a general economic-equilibrium model. This
model is of historical interest, because at that time it was the only economic
model which could be used to prove existence theorems for economic equilibrium. Another remarkable aspect of this model is that it was aimed at growth
models. At any rate, a whole area of the economic literature has developed the
points of view discovered by von Neumann. This will also provide us with the
opportunity to prove the Perron–Frobenius theorem on the existence of positive
eigenvectors of positive matrices and to study the surjectivity properties of M
matrices.

11.2 The Von Neumann Model

We shall begin by studying von Neumann's model, which is largely concerned
with the production sector. We suppose that there are m commodities to produce and consume and that for this there are n production processes which
consume these commodities as inputs and produce them as outputs.

Each production process is implemented with a certain level of activity. The
state of the economy is then described by a vector $x \in \mathbb{R}^n$ the component x_i of
which denotes the level of activity at which the ith production process operates.
These levels of activity are positive or zero and are normalised, for example, by
imposing that x belongs to

$$M^n := \left\{ x \in \mathbb{R}^n_+ \mid \sum_{i=1}^n x_i = 1 \right\}$$

We assume that we are dealing with an economy with constant yields, in
which inputs and outputs depend linearly on the levels of activity. In other
words, the economy is described by a pair of matrices F and G from $\mathbb{R}^n$ to $\mathbb{R}^m$.
The coefficient f_{ij} of the matrix F represents the quantity of the commodity
i consumed by the producer j operating at the unit level of activity, whilst
the coefficient g_{ij} of the matrix G represents the quantity of the commodity i
produced.

Let us consider the commodity i. Its total consumption is

$$(Fx)_i = \sum_{j=1}^{n} f_{ij} x_j \tag{1}$$

and its total production is

$$(Gx)_i = \sum_{j=1}^{n} g_{ij} x_j \tag{2}$$

Suppose that the production process is implemented over a period of time. Thus, we assume that the consumption of the commodity i at the end of the period of production is lower than the production of this commodity at the beginning of the period:

$$Fx^1 \leq Gx^0 \tag{3}$$

We shall say that there is *balanced growth* if the levels of activity increase at the same rate, in other words, if there exists α such that

$$x^1 = (1 + \alpha)x^0. \tag{4}$$

If such an α exists it is called the *balanced-growth rate*. Conditions (3) and (4) imply that x^0 is a solution of the inequalities

$$(1 + \alpha)fx^0 \leq Gx^0. \tag{5}$$

We now consider the transposes of the matrices F and G. We interpret the dual $\mathbb{R}^{m*}$ (identified with $\mathbb{R}^m$) as the space of prices $p = (p^1, \dots, p^m)$, where the component p^i represents the unit price of the commodity i. These unit prices are assumed to be non-negative and are normalised with the assumption that

$$p \in M^m := \left\{ p \in \mathbb{R}_+^m \mid \sum_{i=1}^{m} p^i = 1 \right\}$$

Thus, the images F^*p and G^*p denote the value of the consumptions (inputs) and the productions (outputs). The problem is now to find prices such that the value of the outputs at the beginning of the period does not exceed that of the inputs at the end of the period:

$$F^*p^1 \geq G^*p^0. \tag{6}$$

We assume that the price p^1 is related to the price p^0 by

$$p^0 = \frac{1}{1+\rho} p^1 \quad \text{where } \rho = \frac{p^1 - p^0}{p^0} \tag{7}$$

where ρ is interpreted as the interest rate. Conditions (5) and (6) imply that p^0 is the solution of the inequalities

$$(1 + \rho)F^*p^0 \geq G^*p^0. \tag{8}$$

We shall solve problems (7) and (8) and show that the interest rate ρ and the balanced-growth rate coincide.

Theorem 11.1 (von Neumann). *Consider two matrices F and G from $\mathbb{R}^n$ to $\mathbb{R}^m$ satisfying*

(i) the coefficients g_{ij} of G are non-negative, but not all zero;
(ii) $\forall i = 1, \ldots, m,\ \sum_{j=1}^n g_{ij} > 0$;
(iii) $\forall j = 1, \ldots, n,\ \sum_{i=1}^m f_{ij} > 0$. $\hspace{2cm}$ (9)
Then there exist $\overline{x} \in M^n$, $\overline{p} \in M^m$ and $\delta > 0$ such that

$$(i) \hspace{4cm} \delta F\overline{x} \le G\overline{x}$$
$$(ii) \hspace{4cm} \delta F^* \overline{p} \ge G^* \overline{p}$$
$$(iii) \hspace{3.5cm} \delta \langle \overline{p}, F\overline{x} \rangle = \langle \overline{p}, G\overline{x}. \rangle \hspace{1.5cm} (10)$$

Moreover, δ is maximal in the sense that if $\tilde{x} \in M^n$ is a solution of $\lambda F\tilde{x} \le G\tilde{x}$ then $\lambda \le \delta$. Also, for all $\mu > \delta$ and all $y \in \operatorname{Int} \mathbb{R}_+^m$, there exists $\hat{x} \in \mathbb{R}_+^n$ such that

$$\mu F\hat{x} - G\hat{x} \le y. \hspace{2cm} (11)$$

We shall deduce this theorem from a more general result due to Ky Fan, in which the linearity assumptions are replaced by convexity assumptions.

We recall that we have previously described the $n-1$ simplex of $\mathbb{R}^n$ by

$$M^n := \left\{ x \in \mathbb{R}_+^n \mid \sum_{i=1}^n x_i = 1 \right\}$$

Theorem 11.2. *Consider two mappings F and G from M^n to $\mathbb{R}^m$ satisfying*

(i) the components f_i of f are convex and lower semi-continuous;
(ii) the components g_i of g are concave, positive and lower semi-continuous;
(iii) $\exists \tilde{p} \in M^m$ such that $\forall x \in M^n,\ \langle \tilde{p}, F(x) \rangle > 0$;
(iv) $\exists \tilde{x} \in M^n$ such that $\forall i = 1, \ldots, n,\ g_i(\tilde{x}) > 0$. $\hspace{1.5cm}$ (12)
(a) Then there exist $\delta > 0$, $\overline{x} \in M^n$ and $\overline{p} \in M^m$ such that

$$(i) \hspace{3cm} \forall i = 1, \ldots, n,\ \delta f_i(\overline{x}) \le g_i(\overline{x})$$
$$(ii) \hspace{3cm} \forall x \in M^n,\ \langle G(x) - \delta F(x), \overline{p} \rangle \le 0$$
$$(iii) \hspace{2.5cm} \forall i = 1, \ldots, n,\ \overline{p}_i(\delta f_i(\overline{x}) - g_i(\overline{x})) = 0 \hspace{1cm} (13)$$

(b) The number $\delta > 0$ is defined by

$$\frac{1}{\delta} = \sup_{p \in M^m} \inf_{x \in M^n} \frac{\langle p, F(x) \rangle}{\langle p, G(x) \rangle}$$
$$= \inf_{x \in M^n} \sup_{p \in M^m} \frac{\langle p, F(x) \rangle}{\langle p, G(x) \rangle}. \hspace{1.5cm} (14)$$

If $\lambda > 0$ and $x \in M^n$ satisfy the inequalities $\lambda f_i(x) \le g_i(x),\ \forall i = 1, \ldots, n$, then $\lambda \le \delta$.

(c) For all $\mu > \delta$ and for all $y \in \text{Int}(\mathbb{R}^m_+)$, there exist $\beta > 0$ and $\hat{x} \in M^n$ such that

$$\forall i = 1, \ldots, n, \quad \mu f_i(\hat{x}) - g_i(\hat{x}) \le \beta y_i \tag{15}$$

Proof. We begin by defining the number δ by

$$\frac{1}{\delta} := \sup_{p \in M^m} \inf_{x \in M^n} \frac{\langle p, F(x) \rangle}{\langle p, G(x) \rangle} \tag{16}$$

which is positive and finite by virtue of assumptions (12)(ii)–(iv).

Next we consider the mapping $\delta F - G$ from M^n to $\mathbb{R}^m$ the components $\delta f_i - g_i$ of which are convex and lower semi-continuous. Propositions 1.8 and 2.6 imply that

$$\text{the subset } (\delta F - G)(M^n) + \mathbb{R}^m_+ \text{ is closed and convex.} \tag{17}$$

(a) We note that

$$0 \in (\delta F - G)(M^n) + \mathbb{R}^m_+ \tag{18}$$

Otherwise, following the separation theorem (Theorem 2.4), there would exist $p_0 \in \mathbb{R}^m$ and $\varepsilon > 0$ such that, $\forall x \in M^n$,

$$\delta \langle p_0, F(x) \rangle - \langle p_0, G(x) \rangle + \inf_{v \in \mathbb{R}^m_+} \langle p_0, v \rangle \ge \varepsilon.$$

This implies that $p_0 \in \mathbb{R}^m_+$; whence, after dividing by $\sum_{i=1}^m p_{0i}$ and setting $\bar{p}_0 = p_0 / \sum_{i=1}^m p_{0i}$, we obtain the inequality

$$\forall x \in M^n, \quad \delta \langle p_0, F(x) \rangle - \langle p_0, G(x) \rangle \ge \varepsilon \Big/ \sum_{i=1}^n p_{0i}$$

But the definition of δ in (16) implies that

$$\forall x \in M^n, \quad \delta \langle p_0, F(x) \rangle - \langle p_0, G(x) \rangle \le 0.$$

Thus we have obtained a contradiction.

(b) Consequently, the inclusion (18) implies that there exists $\bar{x} \in M^n$ such that

$$\delta F(\bar{x}) \le G(\bar{x}). \tag{19}$$

Taking the scalar product of this inequality in $\mathbb{R}^m$ with $p \in M^m$, we obtain

$$\sup_{p \in M^m} \frac{\langle p, F(\bar{x}) \rangle}{\langle p, G(\bar{x}) \rangle} \le \frac{1}{\delta}$$

whence we deduce the minimax equation (14).

Since M^m is compact and $p \to \dfrac{\langle p, F(\bar{x}) \rangle}{\langle p, G(\bar{x}) \rangle}$ is continuous, there exists $\bar{p} \in M^m$ such that

$$\frac{1}{\delta} = \frac{\langle \overline{p}, F(\overline{x}) \rangle}{\langle \overline{p}, G(\overline{x}) \rangle} = \inf_{x \in M^n} \frac{\langle \overline{p}, F(x) \rangle}{\langle \overline{p}, G(x) \rangle}.$$

Next we consider $\tilde{x} \in M^n$ and $\lambda > 0$ such that $\lambda F(\tilde{x}) \leq G(\tilde{x})$. Since, for all $p \in M^m$, $\lambda \langle p, F(\tilde{x}) \langle \leq \rangle p, G(\tilde{x}) \rangle$, we obtain

$$\frac{1}{\delta} \leq \sup_{p \in M^m} \frac{\langle p, F(\tilde{x}) \rangle}{\langle p, G(\tilde{x}) \rangle} \leq \frac{1}{\lambda}$$

(c) Suppose now that $\mu > \delta$ and that $y \in \text{Int}(\mathbb{R}^m_+)$. We define the number β by

$$\beta := \sup_{p \in M^m} \inf_{x \in M^n} \frac{\langle p, \mu F(x) - G(x) \rangle}{\langle p, y \rangle}$$

which is strictly positive since the inequality

$$\frac{1}{\mu} < \frac{1}{\delta} = \inf_{x \in M^n} \frac{\langle \overline{p}, F(x) \rangle}{\langle \overline{p}, G(x) \rangle}$$

implies that

$$\beta \geq \inf_{x \in M^n} \frac{\langle \overline{p}, \mu F(x) - G(x) \rangle}{\langle \overline{p}, y \rangle} > 0.$$

We shall prove that

$$\beta y \in (\mu F - G)(M^n) + \mathbb{R}^m_+ \tag{20}$$

If this is not the case, then, since this set is convex and closed, the separation theorem (Theorem 2.4) implies that there exist $p_1 \in \mathbb{R}^m$ and $\varepsilon > 0$ such that

$$\beta \langle p_1, y \rangle + \varepsilon \leq \langle p_1, \mu F(x) - G(x) \rangle + \inf_{v \in \mathbb{R}^m_+} \langle p_1, v \rangle.$$

This implies that $p_1 \in \mathbb{R}^m_+$ and that $\overline{p}_1 := p_1 / \sum_{i=1}^m p_{1i}$ satisfies the inequality:

$$\beta + \varepsilon \leq \inf_{x \in m^n} \frac{\langle \overline{p}_1, \mu F(x) - G(x) \rangle}{\langle \overline{p}_1, y \rangle} \leq \beta$$

which is a contradiction.

The inclusion (20) implies that there exists $\hat{x} \in M^n$ such that

$$\mu F(\hat{x}) - G(\hat{f}) \leq y \tag{21}$$

from which we deduce the minimax equation

$$\beta = \inf_{x \in M^n} \sup_{p \in M^m} \frac{\langle p, \mu F(x) - G(x) \rangle}{\langle p, y \rangle} \tag{22}$$

and the existence of $\hat{q} \in M^m$ such that

$$\beta = \frac{\langle \hat{q}, \mu F(\hat{x}) - G(\hat{x}) \rangle}{\langle \hat{q}, y \rangle} \tag{23}$$

This completes the proof of Theorem 11.2 $\square$

Proof of Theorem 11.1. We note that the assumptions (9)(ii) and (iii) imply assumptions (12)(iii) and (iv) with $\tilde{p} = \frac{1}{m}(1, 1, \ldots, 1)$ and $\tilde{x} = \frac{1}{n}(1, 1, \ldots, 1)$. $\square$

11.3 The Perron–Frobenius Theorem

When the dimensions n and m are equal, boundary conditions on F imply that the solutions $\bar{x}$ and $\hat{x}$ of the inequalities

$$\delta F(\bar{x}) \leq G(\bar{x})$$

and

$$\mu F(\hat{x}) - G(\hat{x}) \leq \beta y$$

are in fact equalities

$$\delta F(\bar{x}) = G(\bar{x})$$

and

$$\mu F(\hat{x}) - G(\hat{x}) = \beta y.$$

Theorem 11.3. *Suppose that F is a mapping from M^n to $\mathbb{R}^n$ satisfying*

(i) the components f_i of F are convex and lower semi-continuous;
(ii) $\exists \tilde{p} \in M^n \cap \text{Int}(\mathbb{R}^n_+)$ such that $\forall x \in M^n$, $\langle \tilde{p}, F(x) \rangle > 0$;
(iii) if $x_i = 0$ then $f_i(x) \leq 0$.   (24)
Suppose also that G is another mapping from M^n to $\mathbb{R}^n$ satisfying

(i) the components g_i of G are concave and upper semi-continuous;
(ii) $\forall x \in M^n$, $\forall i = 1, \ldots, n$, $g_i(x) > 0$.   (25)

Consider the number δ defined by (14). Then there exist $\bar{x} \in M^n \cap \text{Int}(\mathbb{R}^n_+)$ and $\bar{p} \in M^n \cap \text{Int}(\mathbb{R}^n_+)$ such that

(i) $$\delta F(\bar{x}) = G(\bar{x})$$
(ii) $$\forall x \in M^n, \quad \langle \bar{p}, G(x) - \delta F(x) \rangle \leq 0.$$ (26)

If $\mu > \delta$ and $y \in \text{Int}(\mathbb{R}^n_+)$ are given, there exist $\beta > 0$ and $\hat{x} \in M^n \cap \text{Int}(\mathbb{R}^n_+)$ such that

$$\mu F(\hat{x}) - G(\hat{x}) = \beta y$$ (27)

Proof. We let e^j denote the jth element of the canonical basis of $\mathbb{R}^n$. The boundary condition (24)(iii) implies that

$$\forall k \neq i, \quad f_k(e^i) \leq 0$$ (28)

which, together with the positivity condition (24)(ii), implies that

$$\forall i = 1, \ldots, n, \quad f_i(e^i) \geq 0$$ (29)

since, because $\tilde{p} \in \text{Int}(\mathbb{R}^n_+)$,

$$\tilde{p}_i f_i(e^i) + \sum_{k \neq i} \tilde{p}_k f_k(e^i) = \langle \tilde{p}, F(e^i) \rangle \geq 0.$$

(a) Let us now consider the solutions $\overline{x} \in M^n$ and $\overline{p} \in M^n$ of the system

(i) $\qquad\qquad\qquad \delta F(\overline{x}) \leq G(\overline{x})$

(ii) $\qquad\qquad\qquad \forall x \in M^n, \ \langle \overline{p}, G(x) - \delta F(x) \rangle \leq 0$

(iii) $\qquad\qquad\qquad \langle \overline{p}, G(\overline{x}) - \delta F(\overline{x}) \rangle = 0$ $\qquad\qquad\qquad$ (30)

which exist by virtue of Theorem 11.2.

We note firstly that $\overline{p}$ belongs to the interior of $\mathbb{R}^n_+$. Taking $x := e^i$ in (30)(ii) and using the fact that $g_k(e^i) > 0$ for all k, it follows that

$$\delta \overline{p}_i f_i(e^i) + \delta \sum_{k \neq i} \overline{p}_k f_k(e^i) \geq \langle \overline{p}, G(e^i) \rangle > 0.$$

Inequalities (28) and (29) show that $\overline{p}_i > 0$. We set $\overline{z} := G(\overline{x}) - \delta F(\overline{x})$ which belongs to $\mathbb{R}^n_+$ by virtue of (30)(i). Property (30)(iii) may now be written as $\langle \overline{p}, \overline{z} \rangle = 0$.

Since the components $\overline{p}_i$ are strictly positive, it follows that $\overline{z} = 0$ and consequently that $\delta F(\overline{x}) = G(\overline{x})$. Finally, since the components $g_i(\overline{x})$ are strictly positive, the same is true of the components $f_i(\overline{x})$ of $F(\overline{x})$ (since $\delta > 0$). The boundary condition (24)(iii) then implies that $\overline{x}_i > 0$ for all $i = 1, \ldots, n$.

The proof of the second part of the theorem is completely analogous. $\qquad \square$

The mapping $F := 1$ clearly satisfies the conditions (24). Thus, we obtain the following corollary on the eigenvalues of concave, positive operators.

Corollary 11.1. *Suppose G is a mapping from M^n to $\text{Int}(\mathbb{R}^n_+)$ the components of which are concave and upper semi-continuous. The number δ defined by*

$$\frac{1}{\delta} := \sup_{p \in M^n} \inf_{x \in M^n} \frac{\langle p, x \rangle}{\langle p, G(x) \rangle} \qquad\qquad (31)$$

is strictly positive and if there exist $\tilde{x} \in M^n$ and $\lambda > 0$ such that $\lambda x \leq G(x)$ then $\lambda \leq \delta$.

There exist $\overline{x} \in M^n \cap \text{Int}(\mathbb{R}^n_+)$ and $\overline{p} \in M^n \cap \text{Int}(\mathbb{R}^n_+)$ such that

(i) $\qquad\qquad\qquad \delta \overline{x} = G(\overline{x})$

(ii) $\qquad\qquad\qquad \forall x \in M^n, \ \langle \overline{p}, Gx - \delta x \rangle \leq 0.$ $\qquad\qquad$ (32)

If $\mu > \delta$ and $y \in \text{Int}(\mathbb{R}^n_+)$ are given, there exist $\beta > 0$ and $\hat{x} \in M^n \cap \text{Int}(\mathbb{R}^n_+)$ which are solutions of the equation

$$\mu \hat{x} - G\hat{x} = \beta y. \qquad\qquad (33)$$

When G is a positive matrix we obtain the Perron–Frobenius theorem.

Theorem 11.4 (Perron–Frobenius). *Let G be a matrix with strictly positive coefficients.*

(a) G has a strictly positive eigenvalue δ and an associated eigenvector $\bar{x}$ the components of which are strictly positive.

(b) δ is the only eigenvalue associated with an eigenvector of M^n.

(c) δ is greater than or equal to the absolute value of any other eigenvalue of G.

(d) The matrix $\mu - G$ is invertible and $(\mu - G)^{-1}$ is positive if and only if $\mu > \delta$.

Proof. (a) The existence of $\delta > 0$, $\bar{x} \in M^n \cap \mathrm{Int}(\mathbb{R}^n_+)$ and $\bar{p} \in M^n \cap \mathrm{Int}(\mathbb{R}^n_+)$ such that $\delta\bar{x} = G\bar{x}$ and $G^*\bar{p} - \delta\bar{p} \leq 0$ follows from Corollary 11.1. In fact, we have the equality $G^*\bar{p} - \delta\bar{p} = 0$, since

$$\langle G^*\bar{p} - \delta\bar{p}, \bar{x} \rangle = \langle \bar{p}, G\bar{x} - \delta\bar{x} \rangle = 0$$

and since the components $\bar{x}_i$ of $\bar{x}$ are strictly positive.

(b) Suppose $x \in M^n$ and μ are such that $Gx = \mu x$, It follows that

$$\mu \langle \bar{p}, x \rangle = \langle \bar{p}, Gx \rangle = \langle G^*\bar{p}, x \rangle = \delta \langle \bar{p}, x \rangle.$$

Since $\langle \bar{p}, x \rangle$ is strictly positive (because x belongs to M^n and $\bar{p} \in \mathrm{Int}(\mathbb{R}^n_+)$), the previous equality implies that $\mu = \delta$.

(c) Suppose that λ is an eigenvalue of G and that $z \in \mathbb{R}^n$ is an associated eigenvector. The equalities

$$\lambda z_i = \sum_{j=1}^n g_{ij} z_j \qquad (i = 1, \ldots, n)$$

imply the inequalities

$$|\lambda||z_i| \leq \sum_{j=1}^n g_{ij}|z_j|.$$

If $|z|$ denotes the vector with components $|z_i|$, it follows that $|\lambda||z| \leq G|z|$, which implies that $|\lambda| \leq \delta$.

(d) We know that when $\mu > \delta$ the matrix $\mu - G$ is invertible, since δ is the largest eigenvalue. We also know that for all $y \in \mathrm{Int}(\mathbb{R}^n_+)$, the solution $(\mu - G)^{-1}y$ belongs to $\mathrm{Int}(\mathbb{R}^n_+)$, by virtue of the second part of Corollary 11.1. This implies that $(\mu - G)^{-1}$ is positive.

Conversely, suppose that $(\mu - G)$ is invertible and that $(\mu - G)^{-1}$ is positive. The inequality $\mu \leq \delta$ cannot hold, since this would imply the inequalities

$$\mu\bar{x} \leq \delta\bar{x} = G\bar{x} \quad \text{where } \bar{x} \in M^n$$

and thus also

$$-\bar{x} = (\mu - G)^{-1}(G\bar{x} - \mu\bar{x}) \in \mathbb{R}^n_+$$

since $G\bar{x} - \mu\bar{x}$ is a positive vector. Thus μ is strictly larger than δ.   □

11.4 Surjectivity of the M matrices

We shall use the following more general result to show that the M matrices which we define below are surjective.

Theorem 11.5. *We consider a mapping H from $\mathbb{R}_+^n$ to $\mathbb{R}^n$ satisfying*

(i) the components h_i of H are convex, positively homogeneous and lower semi-continuous;
(ii) $\exists b \in \mathbb{R}$ such that $\forall x \in \mathbb{R}_+^n$, $bx_i > h_i(x)$;
(iii) $\forall x \in M^n$, $\exists q \in M^n$ such that $\langle q, H(x) \rangle > 0$. $\qquad(34)$

   Then

$$\forall y \in \mathrm{Int}(\mathbb{R}_=+^n), \quad \exists x \in \mathrm{Int}\mathbb{R}_+^n \ \text{ such that } \ H(x) = y. \qquad(35)$$

Proof. We choose a number μ strictly larger than the number b used in assumption (34)(ii). We associate H with the mapping $G := \mu - H$. Corollary 11.1 implies that there exist $\delta > 0$ and $\overline{x} \in M^n$ satisfying

$$\delta\overline{x} = G\overline{x} = \mu\overline{x} - H(\overline{x}), \quad \overline{x} \in M^n \cap \mathrm{Int}(\mathbb{R}_+^n).$$

Since $(\mu - \delta)\overline{x} = H(\overline{x})$, assumption (34)(iii) implies that

$$(\mu - \delta)\langle \overline{q}, \overline{x} \rangle = \langle \overline{q}, H(\overline{x}) \rangle > 0.$$

Since $\langle \overline{q}, \overline{x} \rangle$ is strictly positive, it follows that $\mu > \delta$. Again by virtue of Corollary 11.1, we can associate any $y \in \mathrm{Int}(\mathbb{R}_+^n)$ with a strictly positive number β and $\hat{x} \in M^n \cap \mathrm{Int}(\mathbb{R}_+^n)$ such that

$$H\hat{x} = (\mu - G)\hat{x} = \beta y$$

Then $\overline{x} := \hat{x}/\beta$ is the desired solution. $\qquad\square$

We now deduce the surjectivity theorem for the M matrices.

Definition 11.1. *A matrix $H := (h_{ij})$ from $\mathbb{R}^n$ to itself is called an **M matrix** if the following two conditions are satisfied*

$$\forall i \neq j, \ h_{ij} \leq 0 \qquad(36)$$

$$\forall x \in M^n, \ \exists q \in M^n \ \text{ such that } \ \langle q, Mx \rangle > 0. \qquad(37)$$

Theorem 11.6. *Suppose H is a matrix from $\mathbb{R}^n$ to $\mathbb{R}^n$ satisfying (36). The following conditions are equivalent:*

(a) H is an M matrix;
(b) H is invertible and H^{-1} is positive;
(c) H^* is invertible and H^{*-1} is positive.

Proof. The implication (a)⇒(b) follows from the previous theorem and the implication (b)⇒(c) is clear. It remains to show that (c) implies (a).

Suppose $p \in \mathrm{Int}(\mathbb{R}^n_+)$ is the solution of the equation $H^*p = \mathbf{1}$ where $\mathbf{1}$ is the vector with components all equal to 1. Then, for all $x \in M^n$,

$$\langle p, Hx \rangle = \langle H^*p, x \rangle = \sum_{i=1}^{n} x_i = 1.$$

Property (37) is satisfied and this implies that H is an M matrix.   □

Remark. There is another criterion for the surjectivity of a matrix: any positive-definite matrix satisfying

$$\forall x \neq 0, \quad \langle Hx, x \rangle > 0$$

is invertible.

12. n-person Games

12.1 Introduction

The fundamental concepts of two-person games extend to n-person games.

The ith player is denoted by $i = 1, \ldots, n$. Each player i may play a strategy x^i in a strategy set E^i.

We denote the set of multistrategies $x := (x^1, \ldots, x^n)$ by

$$E := \prod_{i=1}^{n} E^i \tag{1}$$

12.2 Non-cooperative Behaviour

Let us put ourselves in the place of the ith player. From his point of view, the set of multistrategies is considered to be the product of the set E^i of strategies which he may choose and the set

$$E^{\hat{i}} := \prod_{j \neq i} E^j \tag{2}$$

of strategies $x^{\hat{i}} = (x^1, \ldots, x^n)$ of the other players, over which he has no control in the absence of cooperation. Thus, from the ith player's point of view, the set of multistrategies $x := (x^i, x^{\hat{i}})$ may be written as the set

$$E := E^i \times E^{\hat{i}} \tag{3}$$

The choice of the players' strategies may be determined using decision rules.

Definition 12.1. *A decision rule of the ith player is a correspondence C^i from $E^{\hat{i}}$ to E^i which associates the multistrategies $x^{\hat{i}} \in E^{\hat{i}}$ determined by the other players with a strategy set $C^i(x^{\hat{i}})$.*

Once each of the n players i has been described in terms of the decision rules C^i, as in the case of two-person games, we single out the *consistent multistrategies* .

Definition 12.2. *Consider an n-person game described by n decision rules C^i from $E^{\hat{i}}$ to E^i. We shall say that a multistrategy $x \in E$ is **consistent** if*

$$\forall i = 1, \ldots, n, \quad x^i \in C^i(x^{\hat{i}}) \tag{4}$$

In other words, the set of consistent multistrategies is the set of fixed points of the correspondence $\mathbf{C}$ from E to E defined by

$$\mathbf{C}(x) := \prod_{i=1}^{n} C^i(x^{\hat{i}}) \tag{5}$$

Thus, Kakutani's fixed-point theorem immediately provides an existence theorem for consistent multistrategies.

Theorem 12.1. *Suppose that the n strategy sets are convex, compact subsets and that the n decision rules C^i are upper semi-continuous. with non-empty, convex, closed values. Then there exists a consistent multistrategy.*

Proof. We apply Kakutani's theorem (Theorem 9.5) to the correspondence $\mathbf{C}$ defined by (5) from the convex, compact set E into itself, which is clearly upper semi-continuous with non-empty, convex, closed values.

12.3 n-person Games in Normal (Strategic) Form

We shall suppose now that the decision rules of the n players are determined by loss functions.

Definition 12.3. *A game in normal (strategic) form is a game in which the behaviour of the ith player is defined by a **loss function** $f^i : E \to \mathbb{R}$ that evaluates the loss $f^i(x)$ inflicted on the ith player by each multistrategy x.*

A game described in strategic form may be summarised by the multiloss mapping $\mathbf{f} : E \to \mathbb{R}^n$ defined by

$$\forall x \in E, \quad \mathbf{f}(x) := (f^1(x), \ldots, f^n(x)) \in \mathbb{R}^n. \tag{6}$$

The associated decision rules are defined by

$$\overline{C}^i(x^{\hat{i}}) := \{x^i \in E^i | f^i(x^i, x^{\hat{i}}) = \inf_{y^i \in E^i} f^i(y^i, x^{\hat{i}})\}. \tag{7}$$

Definition 12.4. *The decision rules $\overline{C}^i$ associated with the loss functions f_i by (7) are called the **canonical decision rules**, A multistrategy $\overline{x} \in E$ which is consistent for the canonical decision rules is called a **non-cooperative equilibrium** (or Nash equilibrium).*

This definition leads to the following characterisation. We introduce the function $\phi : E \times E$ to $\mathbb{R}$ defined by

$$\phi(x, y) := \sum_{i=1}^{n} (f^i(x^i, x^{\hat{i}}) - f^i(y^i, x^{\hat{i}})). \tag{8}$$

Proposition 12.1. *The following assertions are equivalent:*

(a) $\overline{x} \in E$ is a non-cooperative equilibrium;
(b) $\forall i = 1, \ldots, n, \ \forall y^i \in E^i, \ f^i(\overline{x}^i, \overline{x}^{\hat{i}}) \leq f^i(y^i, \overline{x}^{\hat{i}})$;
(c) $\forall y \in E, \ \phi(\overline{x}, y) \leq 0.$ $\qquad\qquad$ (9)

The equivalence of (9)(a) and (9)(b) follows immediately from the definitions. The implication (9)(b)$\Rightarrow$(9)(c) is obtained by adding the inequalities

$$f^i(\overline{x}^i, \overline{x}^{\hat{i}}) - f^i(y^i, \overline{x}^{\hat{i}}) < leq0. \qquad (10)$$

To prove that (9)(c) implies (9)(b), we fix i and take $y := (y^i, \overline{x}^{\hat{i}})$. The inequality $\phi(\overline{x}, y) \leq 0$ may be written as

$$f_i(\overline{x}^i, \overline{x}^{\hat{i}}) - f_i(y^i, \overline{x}^{\hat{i}}) + \sum_{j \neq i} f_j(\overline{x}^j, \overline{x}^{\hat{j}}) - f_j(y^j, \overline{x}^{\hat{j}}) \leq 0. \qquad (11)$$

Now, $x = \{x^j, x^{\hat{j}}\} = \{y^j, x^{\hat{j}}\}$ whenever $j \neq i$. Thus, (11) implies that

$$\forall y^i \in E^i, \ f_i(\overline{x}^i, \overline{x}^{\hat{i}}) \leq f_i(y^i, \overline{x}^{\hat{i}}).$$

Theorem 12.2 (Nash). *We suppose that*

$$\forall i \in N, \ \text{the sets } E^i \text{ are convex and compact} \qquad (12)$$

and that

$$\forall i \in N, \text{ the functions } f_i \text{ are continuous}$$
$$\text{and the functions } y^i \to f_i(y^i, x^{\hat{i}}) \text{ are convex} \qquad (13)$$

Then there exists a non-cooperative equilibrium.

Proof. The theorem follows from Ky Fan's theorem (Theorem 8.6).
    We have introduced the set E and the function ϕ defined by

(i) $$E = \prod_{i=1}^{n} E^i$$

(ii) $$\phi(x, y) = \prod_{i=1}^{n} (f_i(x^i, x^{\hat{i}}) - f_i(y^i, x^{\hat{i}})). \qquad (14)$$

The set E is convex and compact, since it is the product of convex, compact sets E^i (by assumption (12)). Moreover, assumptions (13) clearly imply that the functions $x \to \phi(x, y)$ are continuous and that the functions $y \to \phi(x, y)$ are concave. Thus, Ky Fan's theorem implies that there exists $\overline{x} = \{\overline{x}^i, \overline{x}^{\hat{i}}\} \in E$ such that

$$\sup_{y \in E} \phi(\overline{x}, y) \leq \sup_{y \in E} \phi(y, y) = 0 \qquad (15)$$

since $\phi(y, y) = 0$ for all y.
    Proposition 12.1 may then be applied. $\qquad\qquad\square$

12.4 Non-cooperative Games with Constraints (Metagames)

Here, we consider a game defined by both decision rules $C^i : E^{\hat{i}} \to E^i$ and loss functions $f^i : E \to \mathbb{R}$.

We associate these with the *canonical decision rules* defined by

$$\overline{C}^i(x^{\hat{i}}) := \left\{ x^i \in C^i(x^{\hat{i}}) | f^i(x^i, x^{\hat{i}}) = \inf_{y^i \in C^i(x^{\hat{i}})} f^i(y^i, x^{\hat{i}}) \right\} \tag{16}$$

We shall say that the consistent multistrategies are social equilibria of the metagame. It is easy to adapt the proof of Theorem 12.1 to this new game.

For this, we set

(i)
$$E := \prod_{i=1}^{n} E^i$$

(ii)
$$\phi(x, y) := \sum_{i=1}^{n} (f^i(x^i, x^{\hat{i}}) - f^i(y^i, x^{\hat{i}}))$$

(iii)
$$\mathbf{C}(x) := \prod_{i=1}^{n} C^i(x^{\hat{i}}). \tag{17}$$

Proposition 12.2. *The following assertions are equivalent*

(a) $\overline{x} \in E$ *is a social equilibrium;*
(b) $\forall i = 1, \ldots, n$, $\overline{x}^i \in C^i(\overline{x}^{\hat{i}})$ *and* $\forall y^i \in C^i(\overline{x}^{\hat{i}})$, $f^i(\overline{x}^i, \overline{x}^{\hat{i}}) \le f^i(y^i, \overline{x}^{\hat{i}})$;
(c) $\overline{x} \in \mathbf{C}(\overline{x})$ *and* $\forall y \in \mathbf{C}(x)$, $\phi(\overline{x}, y) \le 0$. $\tag{18}$

Proof. The proof is left as an exercise. The existence of a social equilibrium then follows from Theorem 9.14.

Theorem 12.3 (Arrow–Debreu–Nash). *We suppose that*

$$\forall i \in N, \text{ the sets } E^i \text{ are convex and compact} \tag{19}$$

and that

$$\forall i \in N, \text{ the correspondences } C^i \text{ from } \prod_{j \ne i} E^j \text{ to } E^i$$
$$\text{are continuous with non-empty, convex, closed values.} \tag{20}$$

Lastly, we assume that

$$\forall i \in N, \quad \text{the functions } f^i \text{ are continuous}$$
$$\text{and the functions } y^i \to f^i(y^i, x^{\hat{i}}) \text{ are convex.} \tag{21}$$

Then there exists a social equilibrium.

Proof. The set E is convex and compact. The correspondence $\mathbf{C}$ is clearly continuous with non-empty, convex, closed values.

The function ϕ is continuous and for all $x \in E$, the function $y \to \phi(x, y)$ is concave. Moreover, $\phi(y, y) = 0$ for all $y \in E$.

Theorem 9.14 implies that there exists $\bar{x} \in X$ such that

(i) $$\bar{x} = \{\bar{x}^1, \ldots, \bar{x}^n\} \in \mathbf{C}(\bar{x}) = \prod_{i=1}^{n} C^i(\bar{x}^{\hat{i}})$$

(ii) $$\phi(\bar{x}, y) \leq 0, \quad \forall y \in C(\bar{x}) \tag{22}$$

Proposition 12.2 may then be applied. $\qquad\square$

12.5 Pareto Optima

As in the case of two-person games, we single out Pareto optima when the players are permitted to exchange information and to collaborate.

Definition 12.5. *A multistrategy $\bar{x} \in E$ is said to be Pareto optimal if there are no other multistrategies $x \in E$ such that*

$$\forall i = 1, \ldots, n, \quad f^i(x) < f^i(\bar{x}). \tag{23}$$

We also saw in the case of two-person games that there may be a number of Pareto optima. There thus arises the problem of choosing these optima. For example, one might attribute a weight $\lambda^i \geq 0$ to each player.

If the players accept this weighting, they may agree to collaborate and to minimise the weighted function

$$f_\lambda(x) := \sum_{i=1}^{n} \lambda^i f^i(x) \tag{24}$$

over E. If the vector λ with components λ^i is not zero, we note that *any multistrategy $\bar{x} \in E$ which minimises $f_\lambda(x)$ is a Pareto minimum.* For, if this were not the case, there would exist x satisfying the inequalities (23). Multiplying these by $\lambda_i \geq 0$ and summing them, we obtain the contradiction $f_\lambda(x) < f_\lambda(\bar{x})$.

If the n players could be made to agree on a weighting λ, we would no longer have a game problem proper but a simple optimisation problem. However, it is interesting to know the conditions under which any Pareto optimum may be obtained by minimising the function f_λ associated with a weighting λ which is borne in some way by this Pareto optimum. This question has a positive answer if we apply convexity assumptions.

Proposition 12.3. *Suppose that the strategy sets E^i are convex and that the loss functions $f^i : E \to \mathbb{R}$ are convex. Any Pareto optimum $\bar{x}$ may be associated with a non-zero weighting $\lambda \in \mathbb{R}^n$ such that $\bar{x}$ minimises the function f_λ over E.*

Proof. Proposition 2.6 implies that

$$\mathbf{f}(E) + \mathring{\mathbb{R}}^n_+ \text{ is convex} \tag{25}$$

We then note that an element $\overline{x} \in E$ is a Pareto minimum if and only if

$$\mathbf{f}(\overline{x}) \notin \mathbf{f}(E) + \mathring{\mathbb{R}}^n_+ \tag{26}$$

Thus, we may use the large separation theorem (Theorem 2.5) to see that there exists $\lambda \in \mathbb{R}^n, \lambda \neq 0$, such that

$$\langle \lambda, \mathbf{f}(\overline{x}) \rangle = \inf_{\substack{x \in E \\ u \in \mathring{\mathbb{R}}^n_+}} (\langle \lambda, \mathbf{f}(x) \rangle + \langle \lambda, u \rangle).$$

It follows that λ is positive and that $\overline{x}$ minimises $x \to f_\lambda(x) = \langle \lambda, \mathbf{f}(x) \rangle$ on E.
$\square$

Remark. A Pareto minimum also minimises other functions.

For example, we introduce the *virtual minimum* $\boldsymbol{\alpha}$ defined by its components

$$\alpha^i := \inf_{x \in E} f^i(x). \tag{27}$$

We shall say that the game is *bounded below* if $\forall i = 1, \dots, n, \ \alpha^i > -\infty$.

In this case, we take $\beta^i < \alpha^i$ for all i and set $\beta := (\beta^1, \dots, \beta^n) \in \mathbb{R}^n$.

Proposition 12.4. *Suppose that the game is bounded below.*

An element $\overline{x} \in E$ is a Pareto minimum if and only if there exists $\lambda \in \mathring{\mathbb{R}}^n_+$ such that $\overline{x}$ minimises the function g_λ defined by

$$g_\lambda(x) := \max_{i=1,\dots,n} \frac{1}{\lambda^i}(f^i(x) - \beta^i) \tag{28}$$

over E.

Proof. (a) If $\overline{x} \in E$ minimises g_λ on E and is not a Pareto minimum, we could find $x \in E$ satisfying the inequalities (23). Subtracting β_i, and multiplying by $\frac{1}{\lambda^i}$ and taking the maximum of the two terms, we obtain the contradiction $g_\lambda(x) < g_\lambda(\overline{x})$.

(b) If $\overline{x}$ is a Pareto minimum, we take $\lambda^i = f^i(\overline{x}) - \beta^i > 0$ such that $g_\lambda(\overline{x}) = 1$. If there were an $x \in E$ such that $g_\lambda(x) < g_\lambda(\overline{x})$, then we would have $\max_{i=1,\dots,n} \left(\frac{f^i(x) - \beta^i}{f^i(\overline{x}) - \beta^i} \right) < 1$ which would imply the inequalities (23). $\square$

We can also define conservative strategies for the players. We set

$$f^{i\sharp}(x^i) := \sup_{x^{\hat{i}} \in E^{\hat{i}}} f^i(x^i, x^{\hat{i}}) \tag{29}$$

We shall say that $x^{i\sharp} \in E^i$ is a conservative strategy for the ith player if

$$f^{i\sharp}(x^{i\sharp}) = \inf_{x^i \in E^i} \sup_{x^{\hat{i}} \in E^{\hat{i}}} f^i(x^i, x^{\hat{i}}). \tag{30}$$

and we shall say the number $v_i^{\sharp}$ defined by

$$v_i^{\sharp} := \inf_{x^i \in E^i} \sup_{x^{\hat{\imath}} \in E^{\hat{\imath}}} f^i(x^i, x^{\hat{\imath}}). \tag{31}$$

is the conservative value of the game. As in the case of two-person games, this conservative value may be used as a threat, by refusing to accept any multistrategy x such that

$$f^i(x) > v_i^{\sharp} \tag{32}$$

since by playing a conservative strategy $x^{i\sharp}$ the loss $f^i(x^{i\sharp}, x^i)$ is strictly less than $f^i(x)$.

Suppose that

$$\forall i = 1, \ldots, n, \quad v_i^{\sharp} > \alpha^i. \tag{33}$$

(the conservative value is strictly greater than the virtual minimum).

Consider the function $g_0 : E \to \mathbb{R}$ defined by

$$g_0(x) = \max_{i=1,\ldots,n} \frac{f^i(x) - \alpha^i}{v_i^{\sharp} - \alpha^i} \tag{34}$$

Proposition 12.4 (with $\beta^i = \alpha^i$ and $\lambda^i = v_i^{\sharp} - \alpha^i$) implies that

$$\text{if } x_0 \in E \text{ minimises } g_0 \text{ on } E, \text{ then } x_0 \text{ is a Pareto minimum.} \tag{35}$$

If $d := \min_{x \in E} g_0(x)$, it follows that x_0 minimises g_0 on E if and only if

$$\forall i, \ldots, n, \quad f^i(x_0^i) \leq (1 - d)v_i^{\sharp} + d\alpha_i. \tag{36}$$

This property suggests that such choices of Pareto optima should be viewed as *best compromise* solutions.

Other methods of selection by optimisation involve minimising functions

$$x \to s\left(\frac{f^1(x) - \alpha^1}{v_1^{\sharp} - \alpha^1}, \ldots, \frac{f^n(x) - \alpha^n}{v_n^{\sharp} - \alpha^n} \right) \tag{37}$$

on E, where the function s satisfies the following increasing property

$$\text{if } a^i > b^i \text{ for all } i, \text{ then } s(a) > s(b). \tag{38}$$

It is easy to show that any $\overline{x} \in E$ which minimises (37) is a Pareto minimum. We also note that the function (37) remains invariant whenever the loss functions f^i are replaced by functions $a_i f^i + b_i$, where $a_i > 0$.

We also say that by replacing the functions f^i by the functions g^i

$$g^i(x) = \frac{f^i(x) - \alpha^i}{v_i^{\sharp} - \alpha^i} \tag{39}$$

we have 'normalised' the same game. For the normalised game the virtual minimum is zero and the conservative value is 1.

12.6 Behaviour of Players in Coalitions

We denote the set of n players by N and write $S \subset N$ to denote a coalition of players. As a member of a coalition S, a player $i \in S$ *will modify his behaviour.* For example, we suppose that he cooperates with the players of the coalition S and that he does not cooperate with the players of the adverse coalition $\hat{S} := N\backslash S$. In other words, the player $i \in S$, as a member of the coalition S, assumes that the players $j \in \hat{S}$ of the adverse coalition will maximise his loss.

For this, we set

$$E^S := \prod_{i \in S} E^i, \quad E^{\hat{S}} := \prod_{j \notin S} E^j \tag{40}$$

and it is convenient to make the following identifications:

(i) $E = E^S \times E^{\hat{S}}, \quad x = (x^S, x^{\hat{S}})$ where $(ii) x^S \in E^s, \quad x^{\hat{S}} \in E^{\hat{S}}$

(ii) $f^i(x) = f^i(x^S, x^{\hat{S}}).$ \hfill (41)

Thus, the behaviour of the player i as a member of the coalition S is described by the loss function $f_S^{i\sharp} : E^S \to \mathbb{R}$ by

$$f_S^{i\sharp}(x^S) := \sup_{y^{\hat{S}} \in E^{\hat{S}}} f^i(x^S, y^{\hat{S}}). \tag{42}$$

When S reduces to a single player i, this definition is compatible with that of $f^{i\sharp}$ given in (29).

We let c_S denote the operator from $\mathbb{R}^n$ to itself defined by

$$(c_S \cdot r)_i := \begin{cases} r_i & \text{if } i \in S \\ 0 & \text{if } i \notin S \end{cases} \tag{43}$$

and set

$$\mathbb{R}^S := c_S \cdot \mathbb{R}^n, \quad \mathbb{R}_+^S := c_S \cdot \mathbb{R}_+^n, \quad \mathring{\mathbb{R}}_+^S := c_S \cdot \mathring{\mathbb{R}}_+^n. \tag{44}$$

Thus, the behaviour of the players i of the coalition S is described by the multiloss mapping $\mathbf{f}_S^\sharp$ from E^S to $\mathbb{R}^S$ defined by

$$\mathbf{f}_S^\sharp(x^S)^i := \begin{cases} f_S^{i\sharp}(x^S) & \text{if } i \in S \\ 0 & \text{if } i \notin S \end{cases} \tag{45}$$

Consider now a multistrategy $x \in E$ and a player i. As a member of the whole coalition, he incurs a loss $f^i(x)$. As a member of the coalition S, his loss is $f_S^{i\sharp}(x^S)$ in the worst case.

If all the players i of the coalition S can find a strategy $y^S \in E^S$ such that

$$\forall i \in S, \quad f_S^{i\sharp}(y^S) < f^i(x) \tag{46}$$

they will reject the multistrategy x, form a coalition S and choose the multistrategy y^S. Consequently, for the multistrategy x to be *accepted* by all the players, it must be the case that *whatever the coalition S, there is no multi-strategy $y^S \in E^S$ such that the inequalities (46) are satisfied.*

This leads us to make the following definition.

Definition 12.6. *We shall say that the* **core** *of the game is the set of multi-strategies $x \in E$ which are* **accepted** *by all non-empty coalitions $S \subset N$, in the sense that*

$$\forall S, \text{ it is not the case that } (\exists y^S \in E^S | \forall i \in S, f_S^{i\sharp}(y^S) < f^i(x)). \tag{47}$$

This concept is not void, since we shall prove the following theorem due to Scarf (1971).

Theorem 12.4. *We suppose that the strategy sets E^i are convex and compact and that the loss functions $f^i : E \to \mathbb{R}$ are convex (with respect to all the variables) and lower semi-continuous. Then the core of the game is non-empty.*

In fact, for clarity, we shall get rid of the strategy sets and loss functions and retain only their images in the space of multilosses.

For this, we set

$$\forall S \subset N, \ W(S) := \mathbf{f}_S^\sharp(E^S) \subset \mathbb{R}^S. \tag{48}$$

It then follows that x belongs to the core of the game if and only if its multiloss $r := \mathbf{f}(x)$ satisfies

(i) $\qquad\qquad\qquad r \in W(N)$

(ii) $\qquad\qquad\qquad \forall S \subset N, \ c_S \cdot r \notin W(S) + \mathring{\mathbb{R}}_+^S \tag{49}$

We then note that the problem (49) is equivalent to the problem

(i) $\qquad\qquad\qquad r \in W(N) + \mathbb{R}_+^N$

(ii) $\qquad\qquad\qquad \forall S \subset N, \ c_S \cdot r \notin W(S) + \mathring{\mathbb{R}}_+^S \tag{50}$

Any solution of (49) is clearly a solution of (50), let us consider a solution r of this latter problem. Then r may be written as $r_0 + r_1$ where $r_0 \in W(N)$ and $r_1 \in \mathbb{R}_+^n$ and we note that r_0 satisfies the conditions (49)(ii); for otherwise, there would exist S and $r_2 \in W(S) + \mathbb{R}_+^S$ such that $\forall i \in s, \ r_i = r_{0i} + r_{1i} > r_{2i} + r_{1i} > r_{2i}$, in other words $r \in W(S) + \mathring{\mathbb{R}}_+^S$. $\qquad\square$

By setting $V(S) := W(S) + \mathbb{R}_+^S$, we may thus assume that the sets $V(S)$ satisfy the property $V(S) = V(S) + \mathbb{R}_+^S$.

12.7 Cooperative Games Without Side Payments

In 1971, Scarf deduced the above theorem from a famous theorem which he proved in 1963 on the non-emptiness of a game defined solely by the multiloss sets $V(S)$ of each coalition S. With Nash's theorem (1951), this is one of the two major theorems of game theory. For a long time it remained a very difficult theorem to prove, until, in 1973 Shapley gave a simple proof based on Theorem 9.15, which he conceived and proved just for that purpose.

Thus, you will appreciate that the discovery of these theorems and their original proofs required considerable effort on the part of their authors (Nash, Scarf, Shapley) not to mention originality and intelligence. If they now feature in this master's level book it is because, little by little, over two decades, work to understand these results and developments in nonlinear analysis have enabled us to unravel Ariadne's threads and to find direct approaches. Much exploration of unknown territory was needed to find better signposted paths leading without excessive difficulty to an understanding of these theorems.

Definition 12.7. *A cooperative n-person game without side payments is described by the introduction for every non-empty subset S of N of a non-empty subset $V(S)$ of $\mathbb{R}^S$ satisfying*

$$V(S) = V(S) + \mathbb{R}_+^S \qquad (51)$$

A multiloss $r \in \mathbb{R}^N$ is said to be accepted if, for any coalition S, there is no $r^S \in V(S)$ such that $r_i^S < r^i$ for all players i in S. The core of the game is the set of multilosses of $V(N)$ accepted (by all the coalitions).

We denote the set of multilosses r accepted by all coalitions by $\mathcal{A}$. Since $V(S) + \mathring{\mathbb{R}}_+^S$ is the set of multilosses which are not accepted (thus rejected) by the coalition S, we note that the set $\mathcal{A}$ of accepted multilosses may be written in the form

$$\mathcal{A} := \bigcap_{S \subset N} \mathrm{comp}(c_S \cdot)^{-1}(V(S) + \mathring{\mathbb{R}}_+^S) \qquad (52)$$

This is a closed set. Thus, the core of the game is equal to

$$C(N) := V(N) \cap \mathcal{A} \qquad (53)$$

We must show that this intersection is non-empty. The idea is to consider the subcores

$$C(T) := (c_T \cdot)^{-1} V(T) \cap \mathcal{A} \qquad (54)$$

of the multilosses of the coalition T accepted by all the others. We shall prove the following theorem:

Theorem 12.5. *Suppose that the subsets $V(T)$ are closed and bounded below. Then there exists a balanced family $\mathcal{B}$ of coalitions such that*

$$\cap_{T \in \mathcal{B}} C(T) \neq \emptyset \qquad (55)$$

We shall prove a theorem to show that the core is non-empty under the following assumptions.

Definition 12.8. *We shall say that a game is balanced , if for any balanced family $\mathcal{B}$, we have*

$$\bigcap_{T \in \mathcal{B}} (c_T \cdot)^{-1} V(T) \subset V(N). \tag{56}$$

Theorem 12.6 (Scarf). *Suppose that the game is balanced and that the subsets $V(T)$ are closed and bounded below, Then the core of the game is non-empty.*

Proof of Theorem 12.5. The idea is simple, namely to apply Theorem 9.15. We cannot do this directly since the subsets $C(T)$ are not contained in the simplex. But $C(T)$ is contained in $V(T) \cap \text{comp}(V(T) + \mathring{\mathbb{R}}_+^T)$ which is the Pareto surface of $V(T)$ and which is intuitively isomorphic to the simplex M^T. We shall explain all this.

First we note that

$$\mathcal{A} \subset \prod_{i=1}^{n}] - \infty, v_i] \quad \text{where} \quad v_i := \inf V(\{i\}) \tag{57}$$

since, for coalitions with one player

$$\text{comp}(c_{\{i\}} \cdot)^{-1}(V(\{i\}) + \mathring{\mathbb{R}}_+) = \{r \in \mathbb{R}^r | r_i \leq v_i\}.$$

Normalising by the condition

$$\forall i \in N, \quad v_i := \inf V(\{i\}) = 0 \tag{58}$$

so that $\mathcal{A} \subset -\mathbb{R}_+^n$ changes nothing in the game.

Then, since the sets $V(T)$ are bounded below, there exists a finite number $\alpha_i \geq 0$ such that

$$-\alpha_i := \inf_{S \ni i} \inf_{\substack{r \in V(S) \\ r \leq 0}} r_i \tag{59}$$

Consequently, the set $\mathcal{A}$ of the accepted multilosses is bounded:

$$\mathcal{A} \subset \prod_{i=1}^{n}] - \infty, -\alpha_i] \tag{60}$$

We set

$$\rho := n \sup_{i=1,\dots,n} \alpha_i > 0. \tag{61}$$

We change nothing in the game by taking $\rho = 1$.

Let us now consider an element x of the simplex and the straight line $x + \mathbb{R}\mathbf{1}$. Since the sets $V(S)$ satisfy the condition

$$V(S) = V(S) + \mathbb{R}_+^S$$

it follows that

$$\text{if } \sigma \leq \tau \text{ then } \{x - \sigma \mathbf{1} \in \mathcal{A} \Rightarrow x - \tau \mathbf{1} \in \mathcal{A}\}. \tag{62}$$

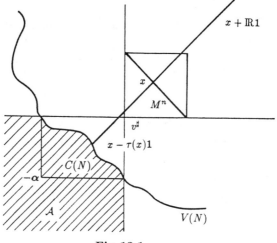

Fig. 12.1.

Moreover, we note that if $x - \tau\mathbf{1}$ belongs to $\mathcal{A}$, then τ is positive.

Thus, we may associate every $x \in M^n$ with the number $\tau(x) \geq 0$ defined by

$$\tau(x) := \inf\{\tau | x - \tau\mathbf{1} \in \mathcal{A}\}. \tag{63}$$

Let us for the moment accept the following lemma,

Lemma 12.1 *The mapping τ from M^n to $\mathbb{R}_+$ is continuous.*

We then define the subsets

$$F_T := \{x \in M^n | c_T \cdot (x - \tau(x)\mathbf{1}) \in V(T)\}. \tag{64}$$

Since $V(T)$ is closed and τ is continuous, the subsets F_T are closed. Suppose for a moment that there exist a balanced family $\mathcal{B}$ and $\bar{x} \in M^n$ such that

$$\bar{x} \in \bigcap_{T \in \mathcal{B}} F_T \tag{65}$$

Then, by the construction of τ and the F_T we have proved Theorem 12.5, since

$$\bar{x} - \tau(\bar{x})\mathbf{1} \in \bigcap_{T \in \mathcal{B}} C(T) \tag{66}$$

Thus, we must prove (65), for which we must apply Theorem 9.15, that is to say verify the assumption

$$\forall T \subset N, \ M^T \subset \bigcup_{S \subset T} F_S \tag{67}$$

Consider firstly the case where $T = N$. We have

$$M^n \subset \bigcup_{S \subset N} F_S \tag{68}$$

In fact, we may write

$$\tau(x) = \max_{S \subset N} \tau_S(x) \tag{69}$$

where

$$\tau_S(x) := \inf\{\tau | (x - \tau \mathbf{1}) \in \text{comp}\, c_S{}^{-1} \cdot (V(S) + \mathring{\mathbb{R}}_+^S)\} \tag{70}$$

Thus, we may associate any element $x \in M^n$ with a coalition S such that $\tau(x) = \tau_S(x)$, whence such that $x - \tau(x)\mathbf{1} = x - \tau_S(x)\mathbf{1} \in (c_S \cdot)^{-1}(V(S))$.

Consider now the case when $T \neq N$. Since $M^T \subset M^n \subset \bigcup_{S \subset N} F_S$, it is sufficient to prove that

$$\text{if } M^T \cap F_S \neq \emptyset, \text{ then } S \subset T \tag{71}$$

to verify that the assumption (67) is satisfied.

Thus, we take x in $M^T \cap F_S$. We shall show that $\forall i \in S$, $x_i > 0$, which will imply that $S \subset T$. Since x belongs to M^T and since $T \neq N$, there exists a player $i_0 \in T$ for which

$$x_{i_0} > \frac{1}{n} \tag{72}$$

Moreover, since $x - \tau(x)\mathbf{1}$ belongs to $\mathcal{A}$, then

$$\forall i = 1, \ldots, n, \quad x_i - \tau(x) \leq 0. \tag{73}$$

These two inequalities imply that

$$\frac{1}{n} < \tau(x) \tag{74}$$

Since x also belongs to F_S, then $c_s \cdot (x - \tau(x)\mathbf{1})$ belongs to $V(S)$, whence, following the definition of α_i in (59), we obtain the inequalities

$$\forall i \in S, \quad -\alpha_i \leq x_i - \tau(x) < x_i - \frac{1}{n}. \tag{75}$$

Since $n \sup \alpha_i = 1$, it follows that

$$\forall i \in S, \quad x_i > 0 \tag{76}$$

and thus that $S \cap \text{comp}\, T = \emptyset$, in other words, $S \subset T$. Assumption (67) is satisfied, there exist x and $\mathcal{B}$ such that (65) is satisfied and Theorem 12.5 holds and has been proved. □

Proof of Lemma 12.1. We let $\mathcal{R}$ denote the complement of $\mathcal{A}$ which satisfies the property $\mathcal{R} + \mathring{\mathbb{R}}_+^n = \mathcal{R}$. We consider the cone $x - \tau(x)\mathbf{1} + \mathring{\mathbb{R}}_+^n$, which is contained in $\mathcal{R}$ and the cone $x - \tau(x)\mathbf{1} - \mathbb{R}_+^n$ which is contained in $\mathcal{A}$.

Consider a sequence x_η converging to x.

We set

(i) $\qquad\qquad t_n := \sup\{t | x_n - t\mathbf{1} \in x - \tau(x)\mathbf{1} + \mathring{\mathbb{R}}_+^n\}$

(ii) $\qquad\qquad s_n := \inf\{s | x_n - s\mathbf{1} \in x - \tau(x)\mathbf{1} - \mathbb{R}_+^n\}$ $\qquad$ (77)

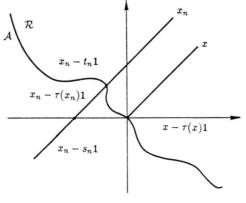

Fig. 12.2.

It then follows that

$$t_n \leq \tau(x_n) \leq s_n \tag{78}$$

It suffices now to note that s_n and t_n tend to $\tau(x)$ as x_n tends to x. □

Proof of Theorem 12.6 (Scarf). Following Propositions 1.8 and 2.6, we know that the subsets $V(S) := \mathbf{f}_S^{\sharp}(E^S) + \mathbb{R}_+^S$ satisfy

$$V(S) \text{ is closed, convex and bounded below.} \tag{79}$$

Thus, Theorem 12.5 applies to the $V(S)$. □

To prove Theorem 12.4 from Theorem 12.6, we must show that this game is balanced.

This will follow from Proposition 12.5, below. We recall that we defined *balancings* (Definition 9.9) as vectors $m = (m(S))_{S \subset N}$ satisfying

$$m(S) \geq 0, \quad \forall i = 1, \dots, n, \quad \sum_{S \ni i} m(S) = 1 \tag{80}$$

or alternatively

$$m(S) \geq 0, \quad c_N := \sum_{S \subset N} m(S) c_S \tag{81}$$

where c_S denotes the characteristic function of the coalition S.

Proposition 12.5. *Suppose that the strategy sets E^i and the loss functions f^i are convex. Then ,for any balancing $m = (m(S))_{S \subset N}$ and any family of multistrategies $x^S \in E^S$, we have the inequalities*

$$\forall i \in N, \quad f^i \left(\sum_{S \subset N} m(S) x^S \right) \leq \sum_{S \ni i} m(S) f_S^{i\sharp}(x^S). \tag{82}$$

Proof of Theorem 12.4. It is sufficient to show that the game is balanced. For this, we take $r \in \cap_{S \in \mathcal{B}} (c_S \cdot)^{-1} V(S)$ where $\mathcal{B}$ is a balanced family associated with the balancings $m(S) > 0$. Thus, we may associate every $S \in \mathcal{B}$ with a multistrategy $x^S \in E^S$ such that

$$\forall S \in \mathcal{B}, \quad c_S \cdot r \geq \mathbf{f}_S^\sharp(x^S). \tag{83}$$

We take $x^N := \sum_{S \subset N} m(s)x^S$, which is a multistrategy in E. Since $\sum_{S \ni i} m(S) = 1$, it follows from Proposition 12.5 that, for all $i \in N$,

$$r_i = \sum_{S \ni i} m(S)r_i \geq \sum_{S \ni i} m(S)f_S^{i\sharp}(x^S) \geq f^i(x^N) \tag{84}$$

whence that $r \in \mathbf{f}(x^N) + \mathbb{R}_+^n \subset V(N)$. $\qquad\qquad\square$

Proof of Proposition 12.5. Consider a balancing $m = (m(S))_{S \subset N}$ and two players i and j. We observe firstly that

$$\sum_{\substack{T \ni j \\ T \not\ni i}} m(T) = \sum_{\substack{S \ni i \\ S \not\ni j}} m(S) \tag{85}$$

since

$$\begin{aligned}
1 &= \sum_{S \ni i} m(S) = \sum_{\substack{S \ni i \\ S \not\ni j}} m(S) + \sum_{S \supset \{i,j\}} m(S) \\
&= \sum_{T \ni j} m(T) = \sum_{\substack{T \ni j \\ T \not\ni i}} m(T) + \sum_{T \supset \{i,j\}} m(T)
\end{aligned}$$

Now we consider multistrategies $x^S \in E^S$. We shall associate these with multistrategies $y_i \in E^i$ defined by

$$\forall j \neq i, \quad y_i^j := \left(\sum_{T \ni j, T \not\ni i} m(T)x^{T,j} \right) \bigg/ \left(\sum_{S \ni i, S \not\ni j} m(S) \right). \tag{86}$$

From equation (48) y_i^j belongs to E^j since the latter is convex. If $S \ni i$ and $j \neq i$ we set

$$y_i^{\hat{S}} = (y_i^j)_{j \in \hat{S}} \in E^{\hat{S}}. \tag{87}$$

We note that for $j \neq i$,

$$\begin{aligned}
\sum_{S \ni i} m(S)(x^S, y_i^{\hat{S}})^j &= \sum_{S \supset \{i,j\}} m(S)x^{S,j} + \sum_{\substack{S \ni i \\ S \not\ni j}} m(S)y_i^j \\
&= \sum_{S \supset \{i,j\}} m(S)x^{S,j} + \sum_{\substack{S \ni i \\ S \not\ni j}} \left(\sum_{\substack{T \ni j \\ T \not\ni i}} m(T)x^{T,j} \bigg/ \sum_{\substack{S \ni i \\ S \not\ni j}} m(S) \right) \\
&= \sum_{S \supset \{i,j\}} m(S)x^{S,j} + \sum_{\substack{T \ni j \\ T \not\ni i}} m(T)x^{T,j} \\
&= \sum_{T \ni j} m(T)x^{T,j}
\end{aligned}$$

if $j = i$, then, clearly,

$$\sum_{S \ni i} m(S)(x^S, y_i^{\hat{S}})^i = \sum_{S \ni i} m(S)x^{S,i}.$$

Thus, we have established the equality

$$\forall i \in N, \quad \sum_{S \ni i} m(S)(x^S, y_i^{\hat{S}}) = \sum_{S \subset N} m(S)x^S. \tag{88}$$

Since $\sum_{S \ni i} m(s) = 1$, the convexity of the loss functions f^i implies that

$$\begin{aligned}
f^i(\sum_{S \subset N} m(S)x^S) &= f^i(\sum_{S \ni i} m(S)(x^S, y_i^{\hat{S}})) \\
&\leq \sum_{S \ni i} m(S)f^i(x^S, y_i^{\hat{S}}) \\
&\leq \sum_{S \ni i} m(S)f_S^{i\sharp}(x^S)
\end{aligned}$$

$\square$

Suppose that the players in the coalition S decide to combine their losses. The best worst-loss of the coalition S is then defined by

$$w(S) := \inf_{x^S \in E^S} \sum_{i \in S} f_S^{i\sharp}(x^S). \tag{89}$$

Corollary 12.1 *We suppose that the assumptions of Proposition 12.5 are in force. Then*

$$w(N) \leq \sum_{S \subset N} m(S)w(S). \tag{90}$$

Proof. Following Proposition 12.5

$$\begin{aligned}
\sum_{i \in N} f^i(\sum_{S \subset N} m(S)x^S) &\leq \sum_{i \in N}\sum_{S \ni i} m(S)f_S^{i\sharp}(x^S) \\
&= \sum_{S \subset N} m(S)\sum_{i \in S} f_S^{i\sharp}(x^S).
\end{aligned}$$

Next we choose $x^S \in E^S$ such that

$$\sum_{i \in S} f_S^{i\sharp}(x^S) \leq w(S) + \varepsilon \Big/ \sum_{S \subset N} m(S).$$

Then $x^N := \sum_{S \subset N} m(S)x^S \in E$ and we have

$$\begin{aligned}
w(N) &\leq \sum_{i \in N} f^i(\sum_{S \subset N} m(S)x^S)) \\
&\leq \sum_{S \subset N} m(S)\sum_{S \subset N} f_S^{i\sharp}(x^S) \\
&\leq \sum_{S \subset N} m(S)w(S) + \varepsilon
\end{aligned}$$

$\square$

12.8 Evolutionary Games

We regard the probability simplex

$$M^n := \left\{ x \in \mathbb{R}^n_+ \mid \sum_{i=1}^{n} x_i = 1 \right\}$$

as the set of mixed strategies of the evolutionary game. Such games provide equilibria of dynamical systems which we shall built. We begin with systems for which we know the growth rates $g_i(\cdot)$ of the evolution without constraints (also called "specific growth rates"):

$$\forall \; i = 1, \ldots, n, \quad x_i'(t) = x_i(t) g_i(x(t))$$

We set $g(x) := (g_1(x_1), \ldots, g_n(x_n))$ and $x \circ g(x) := (x_1 g_1(x_1), \ldots, x_n g_n(x_n))$. If the map $x \in M^n \to x \circ g(x) \in \mathbb{R}^n$ does not satisfy the tangential condition

$$\forall \, x \in M^n, \quad x \circ g(x) \in T_{M^n}(x)$$

which boils down to

$$\forall \, x \in M^n, \quad \sum_{i=1}^{n} x_i g_i(x) = 0$$

thanks to formula (44) of chapter 4, we cannot us theorem 9.4 for obtaining the exitence of an equilibrium. But we can correct this situation by subtracting to each initial growth rate the common "feedback control $\tilde{u}(\cdot)$" (also called "global flux" in many applications) defined as the weighted mean of the specific growth rates

$$\forall \; x \in M^n, \quad \tilde{u}(x) := \sum_{j=1}^{n} x_j g_j(x)$$

because

$$\forall \; x \in M^n, \quad \sum_{i=1}^{n} x_i(g_i(x) - \tilde{u}(x)) := 0$$

Hence, we replace the initial dynamical system by

$$\begin{cases} \forall \; i = 1, \ldots, n, \; x_i'(t) = x_i(t)(g_i(x(t)) - \tilde{u}(x(t))) \\[2mm] = x_i(t)(g_i(x(t)) - \sum_{j=1}^{n} x_j(t) g_j(x(t))) \end{cases}$$

called *replicator system* (or system *under constant organization*) by the biologists who introduced these games.

Remark. There are other methods for correcting a dynamical system to make a given closed subset a viability domain. A general method consists in projecting the dynamics onto the tangent cone (see variational inequalities of chapter 9.) Here, we have taken advantage of the particular nature of the simplex.

An equilibrium α of the replicator system is thus a solution to the system

$$\forall\, i = 1, \ldots, n, \quad \alpha_i(g_i(\alpha) - \tilde{u}(\alpha)) = 0$$

Such an equilibrium does exist, thanks to Theorem 9.3. These equations imply that either $\alpha_i = 0$ or $g_i(\alpha) = \tilde{u}(\alpha)$ or both, and that $g_{i_0}(\alpha) = \tilde{u}(\alpha)$ holds true for at least one i_0. We shall say that an *equilibrium α is nondegenerate* if

$$\forall\, i = 1, \ldots, n, \quad g_i(\alpha) = \tilde{u}(\alpha)$$

Equilibria α which are strongly positive (this means that $\alpha_i > 0$ for all $i = 1, \ldots, n$) are naturally non degenerate. We shall say that an equilibrium α is *evolutionary stable* if and only if the property

$$\exists\, \eta > 0 \quad \text{such that } x \neq \alpha, \ \sum_{i=1}^{n} g_i(x)(\alpha_i - x_i) > 0$$

holds true in a neighborhood of α. Let us consider the function V_α defined on the simplex M^n by

$$V_\alpha(x) := \prod_{i=1}^{n} x_i^{\alpha_i} := \prod_{i \in I_\alpha} x_i^{\alpha_i}$$

where we set $0^0 := 1$ and $I_\alpha := \{i = 1, \ldots, n \mid \alpha_i > 0\}$. Such an equilibrium is called evolutionary stable because

1. α is the unique maximizer of V_α on the simplex M^{n1},

2. starting from an equilibrium $\alpha \in M^n$, the solution $x(\cdot)$ to the replicator system satisfies
$$t \to V_\alpha(x(t)) \text{ is increasing}$$

   since

$$\frac{d}{dt} V_\alpha(x(t)) = V_\alpha(x(t)) \sum_{i=1}^{n} (\alpha_i(t) - x_i) g_i(x(t)) > 0$$

in a neighborhood of α.

Therefore, whenever the equilibrium is evolutionary stable, $x(t)$ converges to the equilibrium α. Indeed, the Cauchy-Schwarz inequality implies that

$$\left(\sum_{i=1}^{n} x_i g_i(x) \right)^2 \leq \left(\sum_{i=1}^{n} x_i \right) \left(\sum_{j=1}^{n} x_i g_i(x)^2 \right) = \sum_{i=1}^{n} x_i g_i(x)^2$$

---

[1]This follows from the concavity of the function $\varphi := \log$: Setting $0 \log 0 = 0 \log \infty = 0$, we get

$$\sum_{i=1}^{n} \alpha_i \log \frac{x_i}{\alpha_i} = \sum_{\alpha_i > 0} \alpha_i \log \frac{x_i}{\alpha_i} \leq \log\left(\sum_{\alpha_i > 0} x_i \right) \leq \log 1 = 0$$

so that

$$\sum_{i=1}^{n} \alpha_i \log x_i \leq \sum_{i=1}^{n} \alpha_i \log \alpha_i$$

and thus, $V_\alpha(x) \leq V_\alpha(\alpha)$ with equality if and only if $x = \alpha$.

and thus,

$$\forall\, t \geq 0, \ \sum_{i=1}^{n} g_i(x(t))x_i'(t) \,\geq\, 0$$

Therefore, whenever $\alpha \in M^n$ is a nondegenerate equilibrium,

$$\begin{cases} \dfrac{d}{dt}V_\alpha(x(t)) \,=\, \sum_{i \in I_\alpha} \dfrac{\partial}{\partial x_i}V_\alpha(x(t))x_i'(t) \\[2mm] \qquad =\, V_\alpha(x(t))\sum_{i \in I_\alpha} \alpha_i \dfrac{x_i'(t)}{x_i(t)} \end{cases}$$

and

$$\sum_{i=1}^{n} \alpha_i \frac{x_i'(t)}{x_i(t)} \,=\, \sum_{i=1}^{n}(\alpha_i - x_i(t))g_i(x(t))$$

Example: Replicator systems for linear growth rates. The main class of examples is provided by linear growth rates

$$\forall\, i = 1, \ldots, n, \ \ g_i(x) \,:=\, \sum_{j=1}^{n} a_{ij}x_j$$

Let A denote the matrix the entries of which are the above a_{ij}'s. Hence the global flux can be written

$$\forall\, x \in M^n, \ \ \tilde{u}(x) \,=\, \sum_{k,l=1}^{n} a_{kl}x_k x_l \,=\, < Ax, x >$$

Therefore, first order replicator systems can be written

$$\forall\, i = 1, \ldots, n, \ \ x_i'(t) \,=\, x_i(t)\Big(\sum_{j=1}^{n} a_{ij}x_j(t) - \sum_{k,l=1}^{n} a_{kl}x_k(t)x_l(t)\Big)$$

Such systems have been investigated independently in   — *population genetics* (allele frequencies in a gene pool)   — *theory of prebiotic evolution* of self replicating polymers (concentrations of polynucleotides in a dialysis reactor) — *sociobiological studies* of evolutionary stable traits of animal behavior (distributions of behavioral phenotypes in a given species)   — *population ecology* (densities of interacting species). In population genetics, *Fisher-Wright-Haldane's model* regards the state $x \in M^n$ as the frequencies of alleles in a gene pool and the matrix $A := (a_{ij})_{i,j=1,\ldots,n}$ as the *fitness matrix*, where a_{ij} represents the fitness of the genotype (i, j). In this case, the matrix A is obviously symmetric and we denote by

$$\tilde{u}(x) :=< Ax, x > \quad \textit{the average fitness}$$

In the theory of *prebiotic evolution*, the state represents the concentrations of polynucleotides. It is assumed in *Eigen-Schuster's "hypercycle"* that the growth rate of the ith-polynucleotide is proportional to the concentration of the preceding one:

$$\forall\, i = 1, \ldots, n, \quad g_i(x) = c_i x_{i-1} \text{ where } x_{-1} := x_n$$

In other words, the growth of polynucleotide i is catalyzed by its predecessor by Michaelis-Menten type chemical reactions. The feedback $\tilde{u}(x) = \sum_{i=1}^{n} c_i x_i x_{i-1}$ can be regarded as a selective pressure to maintain the concentration. The equilibrium α of such a system is equal to

$$\forall\, i = 1, \ldots, n, \quad \alpha_i = \frac{1}{c_{i+1}} \left(\sum_{j=1}^{n} \frac{1}{c_j} \right)^{-1} \text{ where } c_{n+1} := c_1$$

First order replicator systems also offer a quite interesting model of *dynamic game* theory proposed in 1974 by J. Maynard-Smith to explain the evolution of genetically programmed behaviors of individuals of an animal species. We denote by $i = 1, \ldots, n$ the n possible "strategies" used in interindividual competition in the species and denote by a_{ij} the "gain" when strategy i is played against strategy j. The state of the system is described by the "mixed strategies" $x \in M^n$, which are the probabilities with which the strategies are implemented. Hence the growth rate $g_i(x) := \sum_{j=1}^{n} a_{ij} x_j$ is the gain obtained by playing strategy i against the mixed strategy x and $\tilde{u}(x) := \sum_{i,j=1}^{n} a_{ij} x_i x_j$ can be interpreted as the *average gain*. So the growth rate of the strategy i in the replicator system is equal to the difference between the gain of i and the average gain (a behavior which had been proposed in 1978 by Taylor and Jonker.) In ecology, the main models are elaborations of the *Lotka-Volterra equations*

$$\forall\, i = 1, \ldots, n, \quad x_i'(t) = x_i(t) \left(a_{i0} + \sum_{j=1}^{n} a_{ij} x_j(t) \right)$$

where the growth rate of each species depend in an affine way upon the number of organisms of the other species. A very simple transformation replaces this system by a first order replicator system. We compactify $\mathbb{R}_+^n$ by introducing homogeneous coordinates. We set $x_0 := 1$ and we introduce the map

$$\forall\, i = 0, \ldots, n, \quad y_i := \frac{x_i}{\sum_{j=1}^{n} x_j}$$

from $\mathbb{R}_+^n$ onto S^{n+1}, the inverse of which is defined by $x_i := y_i/y_0$. We set $a_{0j} = 0$ for all j, so that Lotka-Volterra's equation becomes

$$\forall\, i = 1, \ldots, n, \quad y_i' = \frac{y_i}{y_0} \left(\sum_{j=0}^{n} a_{ij} y_j - \sum_{k,l=1}^{n} a_{kl} y_l y_k \right)$$

because

$$y_i' = \frac{x_i'}{\sum x_j} - \frac{x_i \sum x_j'}{(\sum x_j)^2} = x_i \left(\sum_{j=0}^{n} a_{ij} x_j \right) y_0 - x_i \left(\sum_{k,l=0}^{n} a_{kl} x_l x_k \right) y_0^2$$

This is, up to the multiplication by $\frac{1}{y_0}$, i.e., up to a modification of the time scale, a $(n+1)$-dimensional first order replicator system. So, first-order replicator systems appear as a common denominator underlying these four biological processes.

13. Cooperative Games and Fuzzy Games

13.1 Introduction

Let us consider a set N of n players and the set $\mathcal{D}(N)$ of subsets of players S. Cooperative games are those which take into account not only the behaviour of the players but also that of *coalitions of players*. Thus, we require a completely different formalism from that used for non-cooperative games. From the beginning, theorists of cooperative games have wrestled with difficulties resulting from the finite nature of $\mathcal{D}(N)$. The structure of this set is too weak and the results relating to it are either trivial or very difficult. Several attempted approaches have involved increasing the number of players by various means. For example, one such approach involved taking the interval $[0, 1]$ as the set of players (the interval is called the continuum of players). This technique, which was first used by R. Aumann, is one which physicists have used since the invention of differential calculus.

13.2 Coalitions, Fuzzy Coalitions and Generalised Coalitions of n Players

We denote the set of n players by N.

The first definition of a coalition which comes to mind is that of a subset of players $S \subset N$. Thus there are 2^n coalitions. However, although the number of coalitions rapidly becomes important, it remains finite which prevents us from using analytical techniques.

In defining mixed strategies, we saw a first example of the 'convexification' of a finite set.

We shall study a natural way of 'convexifying' the set $\mathcal{D}(N)$ of coalitions of n players. For this, we identify the set of coalitions $\mathcal{D}(N)$ with the set $\{0, 1\}^n$ with the aid of the set characteristic functions

$$S \in \mathcal{D}(N) \rightarrow c_S \in \{0, 1\}^n \tag{1}$$

where c_S, the characteristic function of S, is defined by

$$c_S(i) := \begin{cases} 1 & \text{if } i \in S \\ 0 & \text{if } i \notin S \end{cases} \tag{2}$$

Since $\{0,1\}^n$ is a subset of $\mathbb{R}^n$, we can take its *convex hull*, which is the cube $[0,1]^n$. We shall call any element c of $[0,1]^n$ which is defined by

$$c : i \in N \to c(i) \in [0,1] \tag{3}$$

a *fuzzy subset* of N. The number $c(i) \in [0,1]$ is called the *level of membership* (of i) of the fuzzy subset c.

This concept of the fuzzy set was introduced in 1965 by L.A. Zadeh. Since then, it has been wildly successful, above all in many areas outside the mathematical community. In this age of anti-scientific reaction, the adjective 'fuzzy' must have raised some people's hopes of being able to escape from the constraints of *rigour* to which mathematicians are subjected. Whilst the latter are wary of fuzziness could this not offer a way of avoiding the punishing logical consistency of scientific reasoning, without a bad conscience? Did not the author of *Caroline chérie* and *Les corps tranquilles* (J. Laurent) recently entitle his last novel *Les sous-ensembles flous* (Fuzzy Subsets)?

Beyond such anecdotes – and the rather unkind reflections on snobbery which they may evoke – it is useful to reflect on the power of words and the harm which may result from word play. The success of catastrophe theory outside mathematics must be associated with this phenomenon. At this time of collective pessimism and end-of-the-world atmosphere, was not scientific support for this made even more legitimate when the originator of catastrophe theory won the Fields medal, the mathematician's equivalent of the Nobel prize? However, all this is as nothing compared with the alarming word play around the concept of entropy, which takes a cowardly advantage of the difficulty of this notion. Must pervading ideologies be no longer viable for the second law of thermodynamics to be raised to a quasi-religious statute?

That this is no exaggeration is proved by J. Rivkin's book, the title of which, *Entropy*, is repeated three times in flamboyant colours. To see this, we need only read the titles of the last paragraphs:

Entropy: a new world in view, Toward a new economic theory, Third world development, Domestic redistribution of wealth, A new infrastructure for the solar age, Values and institutions in an entropic society, Reformulating science, Reformulating education, A second Christian education, Facing the entropy crisis, From despair to hope.

This is so scandalous that a famous professor of mathematics at the École Polytechnique has formulated the third law of thermodynamics: "Any sophistical explanation of the second law of thermodynamics is a foolish affirmation".

Since we have interpreted any subset of N as a coalition of players, we shall interpret any fuzzy subset c of $[0,1]^n$ as a *fuzzy coalition* of players and each number $c(i)$ as the *level of participation* of player i in the fuzzy coalition c. Player i participates fully in c if $c(i) = 1$, does not participate at all if $c(i) = 0$ and participates in a fuzzy way if $c(i) \in]0,1[$.

The interest of the concept of fuzzy coalitions in political games is clear for all to see!

Since the set of fuzzy coalitions is the convex hull of the set of coalitions any fuzzy coalition may be written in the form

$$c = \sum_{S \in \mathcal{D}(N)} m(S)c_S \text{ where } m(S) \geq 0, \quad \sum_{S \in \mathcal{D}(N)} m(S) = 1. \tag{4}$$

The levels of participation of the players are then defined by

$$c(i) = \sum_{S \ni i} m(S) \qquad (i = 1, \dots, n). \tag{5}$$

Thus, if $m(S)$ is interpreted as the probability of the formation of the coalition S, the level of participation of player i is the sum of the probabilities of the formation of the coalitions S to which i belongs.

But why stop there? Why not model non-cooperative behaviour of the ith player by a negative level of participation?

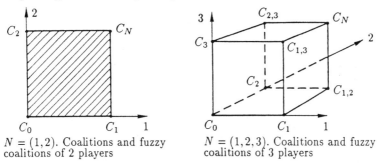

$N = (1,2)$. Coalitions and fuzzy coalitions of 2 players

$N = (1,2,3)$. Coalitions and fuzzy coalitions of 3 players

Fig. 13.1. Coalitions and fuzzy coalitions of players. (a) $N = (1,2)$, 2 players, (b) $N = (1,2,3)$, 3 players.

Definition 13.1 *A* **generalised coalition** *of n players is defined to be any element of the cube $[-1, +1]^n$ of functions $c : N \to [-1, +1]$ which associate each player i with his* **level of participation** *$c(i) \in [-1, +1]$.*

A positive level of participation is interpreted as cooperative participation of the player i in the coalition, whilst a negative level of participation is interpreted as non-cooperative participation of the ith player in the generalised coalition.

We can also enrich the description of the players by representing each player i by what psychologists call his 'behaviour profile '. Let us explain this.

We consider q 'behavioural qualities' $k = 1, \dots, q$, each with a unit of measurement. For example, $k = 1$: intelligence, $k = 2$: patience, $k = 3$: creativity, etc. We also suppose that a behavioural quantity can be measured (evaluated) in terms of a real number (positive or negative) of units. A behaviour profile is a vector $a = (a_1, \dots, a_q) \in \mathbb{R}^q$ which specifies the quantities a_k of the q qualities k attributed to the player. Thus, instead of representing each player by a letter of the alphabet, he is described as an element of the vector space $\mathbb{R}^q$.

We then suppose that each player may implement all, none, or only some of his behavioural qualities when he participates in a social coalition. For example, suppose that we have retained the two qualities $k = 1$: intelligence and $k = 2$: patience. A player may implement these two qualities in different ways, according as to whether he is participating in the Fraternal Society for Social Psychology, the Anglers' Association or the Association of Belote Players. In the first case, we assume that he implements the two qualities in full, while in the second case he does not use his intelligence at all but proves his patience to the full and in third case he uses half of the potential of each of these two qualities.

This translates to the statement that the level of participation (or, in the terminology of social science 'the degree of actualisation') of his behaviour profile is (1,1) in the first case, (0,1) in the second case and (1/2,1/2) in the third case.

We note (and this is important) that the level of participation is independent of the behaviour profile.

More precisely, we introduce the following concept:

Definition 13.2 *Consider n players represented by their behaviour profiles in $\mathbb{R}^q$. Any matrix $C = (C_i^k)$ describing the levels of participation $C_i^k \in [-1, +1]$ of the behavioural qualities k for the n players i is called a* **social coalition**.

qualities players	quality 1	...	quality k	...	quality q
player 1	C_1^1		C_1^k		C_1^q
$\vdots$					
player i	C_i^1		C_i^k		C_i^q
$\vdots$					
player n	C_n^1		C_n^k		C_n^q

The q rows represent the levels of behavioural participation of the q players.

The set of all social coalitions which it is possible to construct is the qn-dimensional hypercube $[-1, +1]^{qn}$.

If the players are described by their behaviour profiles $a^i = (a_1^i, \ldots, a_k^i, \ldots a_q^i) \in \mathbb{R}^q$, the qualities brought into play by a social coalition $C \in [-1, +1]^{qn}$ are equal to

$$\forall i = 1, \ldots, n, \quad (C_i^k a_k^i)_{k=1,\ldots,q} \in \mathbb{R}^q. \tag{6}$$

For example, let us consider three players and two qualities (intelligence and patience) and the social coalition:

	1	2
Xavier	1/6	1
Yvette	3/4	1/4
Zoe	1	0

Xavier brings 1/6 of his intelligence and all his patience to bear. Yvette uses 3/4 of her intelligence and 1/4 of her patience. Zoe shows all her intelligence but is not patient (nor is she impatient).

Suppose that the behaviour profiles of our three players are:

$$
\begin{aligned}
a^X &= (6, -3) &&\text{(Xavier is intelligent and quick tempered)} \\
a^Y &= (4, 4) &&\text{(Yvette is intelligent and patient)} \\
a^Z &= (-3, 11) &&\text{(Zoe is not very smart but has the patience of an angel)}
\end{aligned}
$$

The behaviour profiles *effectively* implemented by this social coalition will be:

$$
\begin{aligned}
b^X &= (1, -3) = \left(\frac{1}{6} \cdot 6, 1 \cdot (-3)\right) \\
b^Y &= (3, 1) = \left(\frac{3}{4} \cdot 4, \frac{1}{4} \cdot 4\right) \\
b^Z &= (-3, 0) = (1 \cdot (-3), 0 \cdot 11).
\end{aligned}
$$

We note that a social coalition of n players is simply a generalised coalition of nq subplayers (i, k) formed by the kth quality of player i.

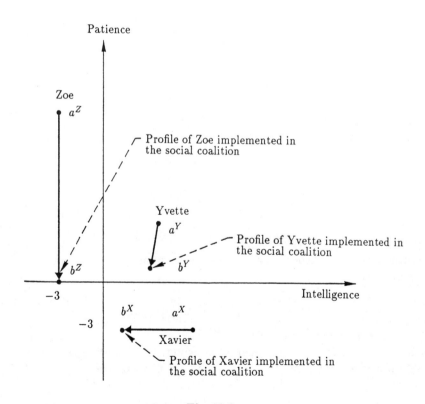

Fig. 13.2.

13.3 Action Games and Equilibrium Coalitions

Consider n players $i = 1, \ldots, n$. We suppose that the behaviour of the ith player entails *acting on the environment* to transform it.

The environment is described by:

$$\text{a convex closed subset } L \text{ of a finite-dimensional vector space } X \qquad (7)$$

and the action of the ith player is described by

$$\text{a } continuous \text{ mapping } f_i \text{ from } L \text{ to } X. \qquad (8)$$

We also suppose that

the action of a generalised coalition $c \in [-1, +1]^n$ on the environment

is described by the continuous mapping $\sum_{i=1}^{n} c_i f_i$ from L to X. $\qquad (9)$

Definition 13.3. *We shall say that a state of the environment $\bar{x} \in L$ and a generalised coalition $\bar{c} \in [-1, +1]^n$ form an* **equilibrium** *if*

$$\sum_{i=1}^{n} \bar{c}_i f_i(\bar{x}) = 0. \qquad (10)$$

Such a state of the environment $\bar{x}$ is not modified by the action of the generalised coalition $\bar{c} \in [-1, +1]^n$.

Theorem 13.1. *Suppose that L is compact and that*

$$\forall x \in L, \ \exists c \in [-1, +1]^n \ such \ that \ \sum_{i=1}^{n} c_i f_i(x) \in T_L(x). \qquad (11)$$

Then there exists an equilibrium state $\bar{x}$ and coalition $\bar{c} \in [-1, +1]^n$.

Proof. We apply Theorem 9.4 to the set-valued map C defined on the convex compact set L by

$$C(x) := \left\{ \sum_{i=1}^{n} c_i f_i(x) \right\}_{c \in [-1, +1]^n} \qquad (12)$$

which is clearly upper semi-continuous with convex compact values. The assumption (11) says that the tangential condition is satisfied. There exists a state of the environment $\bar{x} \in L$ such that $0 \in C(\bar{x})$ which, by virtue of (12), implies the existence of a generalised coalition $\bar{c}$ satisfying (10). $\qquad \square$

We shall complicate this model slightly by assuming that any state of the environment $x \in L$ inflicts a loss $\langle p_i, x \rangle$ on each player $i = 1, \ldots, n$, where $p_i \in X^*$ is a linear form on X.

By interpreting $\mathbb{R}^n$ as the space of multilosses of the n players, we suppose that the survival of the n players requires that the multilosses be confined to a set M.

Thus, we require that a state of the environment should obey the additional constraints:

$$((\langle p_i, x \rangle))_{i=1,\ldots,n} \in M. \tag{13}$$

Theorem 13.2. *We suppose that assumptions (7) and (8) are satisfied and that*

$$K := \{x \in L | (13) \text{ holds}\} \quad \text{is compact.} \tag{14}$$

We also suppose that

$$\forall x \in L, \text{ the matrix of coefficients } \langle p_i, f_k(x) \rangle \text{ is negative semi-definite.} \tag{15}$$

Then there exists an equilibrium state of the environment $\bar{x} \in L$ satisfying the constraints (13) and a generalised (equilibrium) coalition $\bar{c}$:

(i) $\qquad\qquad \bar{x} \in L \text{ and } ((\langle p_i, \bar{x} \rangle))_{i=1,\ldots,n} \in M$

(ii) $\qquad\qquad \sum_{i=1}^{n} \bar{c}_i f_i(\bar{x}) = 0. \tag{16}$

Proof. We use Theorem 9.8, where $Y = \mathbb{R}^n$ and the operator A is defined by:

$$Ax := ((\langle p_i, x \rangle))_{i=1,\ldots,n} \tag{17}$$

and where the role of the parameters p is played by the generalised coalitions $c \in [-1, +1]^n =: P$. Assumption (36) of Chapter 9 in Theorem 9.8 follows from assumption (15) since

$$\left\langle c, A\left(\sum_{k=1}^{n} c_k f_k(x)\right) \right\rangle = \sum_{i,k=1}^{n} \langle p_i, f_k(x) \rangle c_i c_k \leq 0.$$

The other assumptions of Theorem 9.8 are clearly satisfied. Thus, we deduce the existence of $\bar{x}$ and $\bar{c}$ satisfying the conditions (16). $\qquad\qquad \square$

Remark. $\langle p_i, f_k(x) \rangle$ may be interpreted as the marginal loss inflicted on the player i by the action of the player k on the state of the environment x. Assumption (15) implies that for each player i, the marginal loss $\langle p_i, f_i(x) \rangle$ which follows from his own action is not positive.

Remark. These models may be given a dynamic interpretation by considering the equilibrium states $\bar{x} \in L$ as the stationary points of the differential equation

$$x'(t) = \sum_{i=1}^{n} c_i(t) f_i(x(t)) \tag{18}$$

where the generalised coalitions $c(t)$ play the role of control parameters.

13.4 Games with Side Payments

A *game with side payments* is described by a *loss function* v (called a characteristic function) defined on the set of coalitions:

$$v : \mathcal{D}(N) \rightarrow \mathbb{R}. \tag{19}$$

The problem here is *how to partition the loss $v(N)$ of the set of players amongst all the players,*

$$\text{find } s = (s_1, \ldots, s_n) \in \mathbb{R}^n \text{ such that } \sum_{i=1}^{n} s_i = v(N). \tag{20}$$

We interpret the elements $s \in \mathbb{R}^n$ as multilosses.

The goal which game-theory specialists gave themselves was to find a *fair distribution* of the loss $v(N)$ taking into account the results of the cooperation of the players as described (a priori) by the characteristic function of the game. We shall of course give examples of cooperative games with side payments.

But, before that, we shall define the notions of *fuzzy games* and *generalised games with side payments* which are described by loss functions defined on the sets $[0, 1]^n$ and $[-1, +1]^n$ of fuzzy and generalised coalitions, respectively.

Since the number $+1$ which we have chosen to define set characteristic functions and fuzzy coalitions is arbitrary, it is clear that this function should depend only on the *relative* levels of participation; in other words, it should be positively homogeneous.

Definition 13.4. *A generalised sharing game with side payments is defined to be a function v from $\mathbb{R}^n$ to $\mathbb{R} \cup \{+\infty\}$ satisfying*

(i)   *v is positively homogeneous;*
(ii)   *v is Lipschitz in the neighbourhood of $c_N = (1, \ldots, 1)$;*
(iii)   *$v(c_S) < +\infty$ for any coalition $S \subset N$.* $\tag{21}$

We shall say that the subset

$$M := \{s \in \mathbb{R}^n | \forall c \in \mathbb{R}^n, \ \langle c, s \rangle \leq v(c)\} \tag{22}$$

is the set of multilosses *accepted* by all the coalitions of players since, for any generalised coalition c, the loss $\sum_{i=1}^{n} s_i c_i = \langle c, s \rangle$ imputed to the coalition c in a pro rata fashion, based on the levels of participation of the players does not exceed the loss $v(c)$ attributed to this coalition a priori.

The conjugate function v^* defined by

$$v^*(s) = \sup_{c \in \mathbb{R}^n} (\langle c, s \rangle - v(c)) \tag{23}$$

is the *indicator function of the set M of accepted multilosses.*

We shall impose a number of axioms, which must be respected by any rule for sharing out $v(c_N)$ with a priori knowledge of the losses inflicted on any generalised coalition c.

A *share-out rule* is, by definition, a set-valued map which associates with any game v a subset $S(v)$ of multilosses in $\mathbb{R}^n$.

Efficiency (Pareto) Axiom. This simply says that the multilosses s of $S(v)$ form a partition of $v(c_N)$:

$$\forall s \in S(v), \ \sum_{i=1}^{n} s_i = v(c_N). \tag{24}$$

Symmetry Axiom (or Axiom of A Priori Justice). Consider a permutation $\theta : N \to N$ of the set of n players, which defines the order in which these players play. The action of θ on the function v is defined by:

$$(\theta * v)(c) := v(c_{\theta^{-1}(1)}, \dots, c_{\theta^{-1}(n)}) \tag{25}$$

and the action of θ on the multiloss $s \in \mathbb{R}^n$ is defined by

$$(\theta * s)_i = s_{\theta(i)} \ \forall i = 1, \dots, n. \tag{26}$$

The symmetry axiom states that the share-out rule does not depend on the order in which the players are called to play, in the sense that

$$\text{for any permutation } \theta, \ S(\theta * v) = \theta * S(v). \tag{27}$$

Atomicity Axiom. If $P := (S_1, \dots, S_m)$ is a *partition* of the set N of n players into m non-empty subsets of players S_j, any game v with n players may be associated with the game $P \Box v$ of m players defined by

$$(P \Box v)(d_1, \dots, d_m) := v(c_1, \dots, c_n) \text{ where } c_k = d_j \text{ when } k \in S_j. \tag{28}$$

Any n-loss $s \in \mathbb{R}^n$ is associated with the m-loss $P \Box s \in \mathbb{R}^m$ by the formula

$$(P \Box s)_j := \sum_{k \in S_j} s_k \ j = 1, \dots, m. \tag{29}$$

The atomicity axiom states that

$$S(P \Box v) = P \Box S(v). \tag{30}$$

Dummy Player Axiom. Consider a superset $M \supset N$ of m players and an n-person game v. This is associated with an m-player game $\pi_M \Delta v$ by the following formula. If C denotes the projection of $\mathbb{R}^m$ onto $\mathbb{R}^n$, we set

$$(\pi_M \Delta v)(d) := v(C \cdot d), \ \forall d \in \mathbb{R}^m \tag{31}$$

and

$$(\pi_M \Delta s)_j := \begin{cases} s_j & \text{if } j \in N \\ 0 & \text{if } j \notin N \end{cases} \quad j = 1, \ldots, m. \tag{32}$$

In other words, the players in M who do not belong to N are dummy.

The dummy player axiom ensures that the redundant players receive nothing:

$$S(\pi_M \Delta v) = \pi_M \Delta S(v). \tag{33}$$

Before we introduce a share-out rule in the general case (which demands a knowledge of the generalised gradients described in Chapter 6), we shall begin with *two special cases* which extend the concepts of the *Shapley value* and the *core* to the case of generalised games.

Example 1. Shapley Value of Regular Games. If v is continuously differentiable at $c_N = (1, \ldots, 1)$, we shall say that the game described by v is *regular* and we define the *Shapley value* of the regular game v to be the gradient of v at c_N:

$$S(v) := \nabla v(c_N) \in \mathbb{R}^n. \tag{34}$$

The loss $s_i := S(v)_i$ attributed to the ith player is the *marginal loss which he incurs by belonging to the coalition of all the players*. It in some way measures the role of the player i as a *pivot*.

The Shapley value defines a share-out rule. In fact, since v is positively homogeneous, we know that, setting $s := S(v)$,

$$\sum_{i=1}^{n} s_i = \langle s, c \rangle = \langle \nabla v(c_N), c_N \rangle = v(c_N).$$

It is easy to check that the symmetry, atomicity and dummy player axioms hold for the Shapley value. In fact, we associate a permutation θ with the matrix $A = (\alpha_i^j)$ defined by

$$\alpha_i^j = 1 \text{ if } j = \theta^{-1}(i) \text{ and } \alpha_i^j = 0 \text{ if } j \neq \theta^{-1}(i). \tag{35}$$

Since

$$\theta * v = v \circ A, \quad \theta * s = A^* s \text{ and } Ac_N = c_N, \tag{36}$$

it follows that

$$S(\theta * v) = \nabla(v \circ A)(c_N) = A^* \nabla v(Ac_N) = \theta * S(v). \tag{37}$$

Similarly, we associate any partition $P = (S_1, \ldots, S_m)$ with the linear operator B from $\mathbb{R}^m$ to $\mathbb{R}^n$ defined by

$$(Bd)_i = d_j \text{ whenever } i \in A_j. \tag{38}$$

Since

$$P \Box v = v \circ B, \quad P \Box s = B^* s \text{ and } Bc_M = c_N \tag{39}$$

it follows that

$$S(P \Box v) = \nabla(v \circ B)(c_M) = B^* \nabla v(Bc_M) = B^* S(v). \tag{40}$$

Finally, we associate any superset $M \supset N$ with the matrix C from $\mathbb{R}^m$ to $\mathbb{R}^n$ which is the projection of $\mathbb{R}^m$ onto $\mathbb{R}^n$. Since

$$\pi_M \Delta v = v \circ C, \quad \pi_M \Delta s = C^* s \quad \text{and} \quad Cc_M = c_N, \tag{41}$$

it follows that

$$S(\pi_M \Delta v) = \nabla(v \circ C)(c_M) = C^* \nabla v(Cc_M) = \pi_M \Delta S(v). \tag{42}$$

Example 2. Weighting Game. We associate any weighting $k = (k_1, \ldots, k_n) \in \mathbb{R}^n$ with the function γ_k defined by

$$\gamma_k(c) := \left(\prod_{i=1}^{n} c_i^{k_i} \right)^{1/|k|} \quad \text{where } |k| := k_1 + \ldots + k_n \tag{43}$$

(where, by convention, $0^0 = 1$). We note that

$$S(\gamma_k) = \left(\frac{k_i}{|k|} \right)_{i=1,\ldots,n} \tag{44}$$

since $\frac{\partial}{\partial c_i} \gamma_k(c) = \frac{k_i}{|k|} \gamma_k(c) c_i$.

In this weighting game, the Shapley value leads us to share $v(c_N) = 1$ proportionally according to the weight of each player.

We are now in a position to characterise the Shapley value by a system of axioms.

Proposition 13.1. *Let $\mathcal{V}$ be the set of games with side payments generated by the weighting games γ_k as k runs through the set N^n of integer vectors.*

*Then the mapping $S : v \in \mathcal{V} \to S(v) \in \mathbb{R}^n$ is the unique **linear** operator which satisfies the efficiency, symmetry and atomicity axioms.*

Proof. Suppose ϕ is a *linear* mapping from $\mathcal{V}$ to $\mathbb{R}^n$ which satisfies the efficiency, symmetry and atomicity axioms.

Consider an integer vector $k = (k_1, \ldots, k_n) \in N^n$ and the games γ_k and $\gamma_1^{|k|}$ with n and $|k|$ players, defined respectively by

$$\gamma_k(c) = \left(\prod_{i=1}^{n} c_i^{k_i} \right)^{1/|k|} \quad \text{and} \quad \gamma_1^{|k|}(d) = \left(\prod_{j=1}^{|k|} d_j \right)^{1/|k|}. \tag{45}$$

The efficiency and symmetry axioms imply that

$$\forall j = 1, \ldots, k, \quad \phi(\gamma_1^{|k|})_j = \frac{1}{|k|}.$$

If P is the partition of the set of k players into n subsets S_1 of k_1 players, $\ldots$, S_n of k_n players, we note that $\gamma_k = P \square \gamma_1^{|k|}$. The atomicity axiom then implies that

$$\phi(\gamma_k)_i = \sum_{j \in S_i} \phi(\gamma_1^{|k|})_j = \sum_{j \in S_i} \frac{1}{|k|} = \frac{k_i}{|k|} = S(\gamma_k)_i.$$

Since S and ϕ are linear and coincide on the basis of M formed by the games γ_k, they are equal. $\square$

Example 3. Core of Subadditive Games. We shall say that the generalised game described by a function v is *subadditive* if

$$\forall c_1, c_2, \quad v(c_1 + c_2) \leq v(c_1) + v(c_2). \tag{46}$$

Since v is positively homogeneous, this is equivalent to saying that

$$\text{the loss function is convex.} \tag{47}$$

We note that such games are a translation of the idea that *unity makes for strength*. In fact, if S and T are two disjoint coalitions, $c_S + c_T$ is the characteristic function of $S \cup T$ and inequality (46) leads to the inequality

$$v(c_{S \cup T}) \leq v(c_S) + v(c_T) \tag{48}$$

which states that the loss incurred by the union of two disjoint coalitions is less than or equal to the sum of the losses incurred by each coalition separately.

Proposition 13.2. *Suppose that the function v which describes the game is subadditive. Then the subdifferential $\partial v(c_N)$ is the set (non-empty) of accepted multilosses $s \in M$ into which $v(C_N)$ may be partitioned.*

$$\partial v(c_N) = \left\{ s \in M \mid \sum_{i=1}^{n} s_i = v(C_N) \right\}. \tag{49}$$

Proof. To say that Δ belongs to $\partial v(c_N)$ is equivalent to saying that for all $c \in \mathbb{R}^n$,

$$v(c_N) - v(c) \leq \langle s, c_N - c \rangle. \tag{50}$$

Taking $c = \lambda c_N$, this inequality implies that

$$(1 - \lambda)(v(c_N) - \langle s, c_N \rangle) \leq 0.$$

Choosing $\lambda = +1/2$ and $\lambda = -1/2$, it follows that, on the one hand,

$$v(c_N) = \langle s, c_N \rangle = \sum_{i=1}^{n} s_i \tag{51}$$

and, on the other hand, taking into account (50),

$$\forall c \in \mathbb{R}^n, \quad \langle s, c \rangle \leq v(c) \qquad (\text{i.e. } s \in M). \tag{52}$$

Conversely, (51) and (52) clearly imply (50). $\qquad\qquad\qquad\qquad\square$

We define the *core* of the subadditive game v to be the subdifferential of v at c_N

$$S(v) := \partial v(c_N) \subset \mathbb{R}^n. \tag{53}$$

This defines a natural share-out rule. If the game is both regular and subadditive, then the core consists of the single Shapley value, since in this case

$$S(v) = \partial v(c_N) = \{\nabla v(c_N)\}. \tag{54}$$

We note that the Shapley value and the core of a generalised game are two special cases of a single concept, which we shall now define.

Definition 13.5 *Suppose we have a generalised game with side payments defined by a function v. We define the* **solution** *of the game to be the* **generalised gradient** *of v at the characteristic function c_N of the set of all the players*

$$S(v) := \partial v(c_N) \subset \mathbb{R}^n. \tag{55}$$

Thus, this coincides with the Shapley value when the game is regular and with the core when the game is subadditive.

The solution $S(v)$ defines a share-out rule.

Theorem 13.3. *The solution $S(v)$ satisfies the efficiency, symmetry and dummy player axioms, together with the properties*

(i) $\qquad\qquad\qquad\qquad \forall \lambda \geq 0, \quad S(\lambda v) = \lambda S(v)$

(ii) $\qquad\qquad\qquad\qquad S(v_1 + v_2) \subset S(v_1) + S(v_2). \tag{56}$

Proof. This follows from the properties of generalised gradients. The efficiency axiom follows from the fact that v is positively homogeneous (see Proposition 6.11, formula (45)). Since the matrices A associated with the permutations θ by (35) and the projections C from $\mathbb{R}^m$ onto $\mathbb{R}^n$ associated with the extensions $M \supset N$ are surjective, using properties (36) and (41) we obtain

$$S(\theta * v) = \partial(v \circ A)(c_N) = A^* \partial v(A c_N) = \theta * S(v) \tag{57}$$

and

$$S(\pi_N \Delta v) = \partial(v \circ c)(c_N) = A^* \partial v(C c_N) = \pi_M \Delta S(v). \tag{58}$$

Formulae (57) and (58) now imply the symmetry axiom and the dummy player axiom. $\qquad\qquad\qquad\qquad\square$

Remark. The properties (39) and Proposition 6.11 imply that

$$S(P \square v) \subset P \square S(v). \tag{59}$$

Equality (the atomicity axiom) only occurs under additional regularity assumptions. For example, when v is subadditive, Corollary 4.3 implies that the core of a game satisfies the atomicity axiom

$$S(P \Box v) = P \Box S(v) \quad \text{for all partitions } P \tag{60}$$

together with the additivity property

$$S(v_1 + v_2) = S(v_1) + S(v_2). \tag{61}$$

The solution scheme also satisfies all the properties of the generalised gradient described in Chapter 6.

The notion of the solution $S(v) := \partial v(c_N)$ makes the coalition of all players c_N play a privileged role. We note that for any coalition $c \in \text{Int Dom } v$, the generalised gradient $\partial v(c)$ provides a subset of multilosses $s \in \mathbb{R}^n$ which partition the losses of the coalition c, since

$$\forall s \in \partial v(c), \quad \langle c, s \rangle = v(c). \tag{62}$$

The converse question then arises. Does a given multiloss s belong to the generalised gradient $\partial v(c)$ of a coalition c?

Theorem 13.4. *Suppose P is a convex compact subset contained in the interior of the domain of the function v. Any **accepted** multiloss $s \in M$ may be associated with a generalised coalition $\bar{c} \in P$ such that*

$$s \in \partial v(\bar{c}) + P^-. \tag{63}$$

Proof. We apply Theorem 8.6 (Ky Fan) to the function ϕ defined by

$$\phi(c, d) := \langle d, s \rangle - D_c v(c)(d) \tag{64}$$

which is concave in d and lower semi-continuous in c (see Theorem 6.1) and which satisfies $\phi(c, c) = \langle c, s \rangle - D_c v(c)(c) = \langle c, s \rangle - v(c) \leq 0$ when s belongs to the set M of accepted multilosses. Since the set P is compact, Ky Fan's Inequality implies that there exists $\bar{c} \in P$ such that

$$\forall d \in P, \quad \langle d, s \rangle \leq D_c v(\bar{c})(d) = \sigma(\partial v(\bar{c}), d).$$

This implies that $s \in \partial v(\bar{c}) + P^-$.   □

Example 4. Core of Market Games. Consider a (Hilbert) strategy space X and

$$\begin{array}{l} n \text{ nontrivial, convex, lower semi-continuous} \\ \text{loss functions from } X \text{ to } \mathbb{R} \cup \{+\infty\}. \end{array} \tag{65}$$

Suppose we have

$$n \text{ vectors } y_i \in X. \tag{66}$$

Suppose that $c \in \mathbb{R}^n_+$ is a fuzzy coalition. We define the set $K(c)$ of *allocations of the fuzzy coalition* c by

$$K(c) := \left\{ x \in \prod_{i=1}^n \mid \sum_{i=1}^n x_i = \sum_{i=1}^n c_i y_i \right\}. \tag{67}$$

This means that in participating in the fuzzy coalition with a level of participation c_i, the ith player (consumer) offers $c_i y_i$. The loss function of the coalition c will be defined by

$$f(c, x) := \sum_{c_i > 0} c_i f_i \left(\frac{x_i}{c_i} \right). \tag{68}$$

Thus, the *minimum loss function* of the fuzzy coalition c is defined by

$$v(c) := \inf_{x \in K(c)} \sum_{c_i > 0} c_i f_i \left(\frac{x_i}{c_i} \right). \tag{69}$$

We shall say that v describes a fuzzy market game.

Lemma 13.1. *Suppose that*

$$\forall i = 1, \dots, n \; y_i \in \text{Int} \, (\text{Dom} \, f_i). \tag{70}$$

Then we may write

$$v(c) := \sup_{p \in X^*} \left(\sum_{c_i > 0} c_i (\langle p, y_i \rangle - f_i^*(p)) \right) \tag{71}$$

and the supremum is attained at a point $\bar{p}_c \in X^*$.

Proof. We apply Corollary 5.2 with $X = X^n, Y := X$,

$$A_c x = \sum_{i=1}^n c_i x_i, \quad y_c = - \sum_{i=1}^n c_i y_i, \quad f_c(x) := \sum_{i=1}^n c_i f_i(x_i) \text{ and } M = \{0\}.$$

Assumptions (70) imply that

$$-y_c \in \text{Int} \, (A_c \, \text{Dom} \, f_c)$$

for any fuzzy coalition c.

Thus, we have

$$
\begin{aligned}
v(c) &= - \inf_{p \in X^*} (f_c^*(-A_c^* p) - \langle p, y_c \rangle) \\
&= - \inf_{p \in X^*} \left(\sum_{c_i > 0} c_i f_i^*(-p) + \langle p, \sum_{c_i > 0} c_i y_i \rangle \right) \\
&= \sup_{p \in X^*} \sum_{c_i > 0} c_i (\langle p, y_i \rangle - f_i^*(p))
\end{aligned}
$$

and there exists $\bar{q}$ at which the maximum of this problem is attained.   □

Whence, it follows that the function v is *convex*. Proposition 4.4 enables to calculate $\partial v(c_N)$; in other words, the core of the fuzzy market game.

Proposition 13.3. *Suppose that the assumptions (70) are in force, whence*

$$\forall i = 1, \ldots, n, \quad y_i \in \text{Int}\,(\text{Dom}\, f_i).$$

We set

(i) $$P(N) := \left\{ p \in P | v(C_N) = \sum_{i=1}^{n} (\langle p, y_i \rangle - f_i^*(p)) \right\}$$

(ii) $$\sigma_i(p) := \langle p, y_i \rangle - f_i^*(p) \ \text{and}\ \sigma(p) = (\sigma_1(p), \ldots, \sigma_n(p)). \qquad (72)$$

Then the core of the fuzzy market game is equal to

$$S(v) = \overline{\text{co}}\{\sigma(p)\}_{p \in P(N)}. \qquad (73)$$

13.5 Core and Shapley Value of Standard Games

Consider a standard game with side payments defined by a function w from $\mathcal{D}(N)$ to $\mathbb{R}$.

We may associate this with a solution scheme whenever we have a means associating w with a fuzzy game or with a generalised game πw by taking $S(\pi w)$.

Thus, there are *as many solution schemes as methods of extending a game to a fuzzy game*. We shall introduce two methods which will lead to the concepts of the core and the Shapley value.

Core of a Standard Game

Consider a game defined by a function w from $\{0, 1\}^n$ to $\mathbb{R}$. We associate this with the set M of accepted multilosses defined by

$$M := \left\{ s \in \mathbb{R}^n | \forall S \subset N, \ \sum_{i \in S} s_i \le w(S) \right\}. \qquad (74)$$

Definition 13.6. *We define the* **core** *$C(w)$ of the game w to be the set of accepted multilosses $s \in M$ such that $\sum_{i=1}^{n} s_i = w(N)$.*

This notion is compatible with that of Definition 12.2. The game with side payments w described by the function w is associated with the game without side payments defined by the sets

$$V(S) := \left\{ r \in \mathbb{R}^S | \sum_{i \in S} r_i \ge w(S) \right\}.$$

This suggests associating w with a fuzzy subadditive game which has the same set of accepted multilosses.

We shall say that the generalised game πw defined by

$$\pi w(c) := \sup_{s \in M} \langle c, s \rangle \qquad (75)$$

is the 'subadditive covering' of the game w, which is the support function of the set of accepted multilosses of the game w.

We always have the inequalities $\pi w(c_S) \leq w(S)$ for any coalition S.

We shall say that a game is *balanced* if

$$w(N) = \pi w(c_N). \qquad (76)$$

Proposition 13.4. *The core $C(w)$ of a game is non-empty if and only if it is balanced. In this case,*

$$C(w) = S(\pi w).$$

Proof. (a) If $s \in M$ belongs to $C(w)$, then

$$\pi w(c_N) \geq \langle c_N, s \rangle = w(N).$$

Thus, the game is balanced since we know that $\pi w(C_N) \leq w(N)$.

(b) If the game is balanced, the set $S(\pi w)$ is non-empty and clearly coincides with $C(w)$. □

We note that the set M is of the form $A^{-1}(w - P)$, where $A : \mathbb{R}^n \to \mathbb{R}^{2n-1}$ is defined by $A_S = (\langle c^T, s \rangle)_{T \subset N}$, $w := (w(T))_{T \subset N}$ and $P := \mathbb{R}_+^{2n-1}$. Then formula (70) of Chapter 4 allows us to write the support function πw of the set M in the form

$$\pi w(c) = \inf_{\substack{A^* m = c \\ u \in P}} (\langle m, w \rangle - \langle m, u \rangle).$$

Since $A^* m = \sum_{T \subset N} m(T) c^T$, we obtain

$$\pi w(c) = \inf_{m(T) \geq 0} \sum_{T \subset N} m(T) w(T)$$

$$c = \sum_{T \subset N} m(T) c_T.$$

In particular, taking $c = c_N$, the formula becomes

$$\pi w(c_N) = \inf \{ \sum_{T \subset N} m(T) w(T), \; m \text{ is a balancing} \}$$

(see Definition 9.9).

To say that the game is balanced is then to say that

$$w(N) \leq \sigma_{T \subset N} m(T) w(T) \qquad \text{for all balancings } m.$$

Example. The following corollary may be deduced from Theorem 12.1

Corollary 13.1. *Consider a game defined in strategic form by n convex strategy sets and n convex loss functions and associate it with the cooperative game defined by*

$$w(T) := \inf_{x^T \in E^T} \sum_{i \in T} f_T^{i\sharp}(x^T).$$

Then its core is non-empty.

Shapley Value of a Standard Game

We shall use an extension χ which associates every standard game w with a regular game χw. We associate every coalition $S \subset N$ with its characteristic function $c_S \in \{0,1\}^n$ and the *weighting* function

$$\gamma_S(c) := \gamma_{c_S}(c) := \left(\prod_{i \in S} c_i \right)^{1/|S|} \quad \text{where } |S| = \text{card}(S). \tag{77}$$

We associate every coalition S with the functionals

$$\alpha_s(w) := \sum_{T \subset S} (-1)^{|S|-|T|} w(T). \tag{78}$$

We define the extension operator χ by

$$\chi w(c) = \sum_{S \subset N} \alpha_S(w) \gamma_S(c). \tag{79}$$

Lemma 13.2. *The 2^n functions γ_S satisfy*

$$\gamma_S(C_T) = \begin{cases} 1 & \text{if } T \supset S \\ 0 & \text{if } S \cap \text{comp} \, T \neq \emptyset \end{cases} \tag{80}$$

and χw interpolates w in the sense that

$$\forall S \subset N, \quad \chi w(c_S) = w(S). \tag{81}$$

Proof. Formula (80) is self-evident. We calculate

$$\begin{aligned} \chi w(c_T) &= \sum_{S \subset N} \alpha_S(w) \gamma_S(c_T) \\ &= \sum_{S \subset T} \alpha_S(w) \\ &= \sum_{S \subset T} \sum_{R \subset S} (-1)^{|S|-|R|} w(R) \\ &= \sum_{R \subset T} \Big(\sum_{R \subset S \subset T} (-1)^{|S|-|R|} \Big) w(R). \end{aligned}$$

Since there are $\binom{|T|-|R|}{|S|-|R|}$ coalitions S between R and T with $|S|$ elements, it follows that

$$
\begin{aligned}
\sum_{R \subset S \subset T} (-1)^{|S|-|R|} &= \sum_{|R| \le |S| \le |T|} \binom{|T|-|R|}{|S|-|R|} (-1)^{|S|-|R|} \\
&= \sum_{0 \le k \le |T|-|R|} \binom{|T|-|R|}{k} (-1)^k \\
&= (1-1)^{|T|-|R|} \\
&= \begin{cases} 0 & \text{if } |R| < |T| \\ 1 & \text{if } |R| = |T|. \end{cases}
\end{aligned}
$$

Thus, we have shown that $\chi w(c_T) = w(T)$. $\qquad \square$

Since χw is continuously differentiable, it follows that $\phi = S \circ \chi$ is a linear mapping from the space $\mathcal{W} := R^{\mathcal{D}(N)}$ of standard games into $\mathbb{R}^n$.

Definition 13.7. *We shall say that the mapping ϕ defined by*

$$
\forall w \in \mathcal{W}, \quad \phi(w) := S(\chi w) \in \mathbb{R}^n \tag{82}
$$

is the **Shapley value.**

We saw that $S(\gamma_T)_i = 1/|T|$ if $i \in T$ and 0 otherwise and we thus deduce that the ith component of the Shapley value of a game w is defined by

$$
\phi(w)_i = \sum_{T \ni i} \frac{1}{|T|} \alpha_T(w). \tag{83}
$$

The following two propositions are independent.

Proposition 13.5. *The Shapley value is given by the following formulae:* $\forall i = 1, \ldots, n,$

$$
\begin{aligned}
\phi(w)_i &= \sum_{T \ni i} \frac{(|T|-1)!(n-|T|)!}{n!} (v(T) - v(T - \{i\})) \tag{84} \\
&= \frac{1}{n!} \sum_\theta (v(T_\theta(i)) - v(T_\theta^0(i))).
\end{aligned}
$$

where, when θ is one of the $n!$ permutations of the set of players N,

$$
T_\theta(i) = \{j | \theta(j) \le \theta(i)\} \text{ and } T_\theta^0(i) = \{j | \theta(j) < \theta(i)\}. \tag{85}
$$

Proof. Formulae (78) and (83) imply that

$$
\begin{aligned}
\phi(w)_i &= \sum_{T \ni i} \frac{1}{|T|} \sum_{S \subset T} (-1)^{|T|-|S|} w(S) \\
&= \sum_{S \subset N} w(S) \left(\sum_{T \supset S \cup \{i\}} \frac{(-1)^{|T|-|S|}}{|T|} \right).
\end{aligned}
$$

We set

$$\mu_i(S) := \sum_{T \supset S \cup \{i\}} \frac{(-1)^{|T|-|S|}}{|T|}. \tag{86}$$

If S does not contain i, then $R = S \cup \{i\}$ does contain it. Thus, it is easy to see that

$$
\begin{aligned}
\mu_i(S) &= \sum_{T \supset R} \frac{(-1)^{|T|-|R|+1}}{|T|} \\
&= - \sum_{T \supset R = R \cup \{i\}} \frac{(-1)^{|T|-|R|}}{|T|} \\
&= -\mu_i(R) \\
&= -\mu_i(S \cup \{i\}).
\end{aligned}
$$

Whence, we may write

$$
\begin{aligned}
\phi(w)_i &= \sum_{S \subset N} \mu_i(S) w(S) \\
&= \sum_{S \ni i} \mu_i(S) w(S) + \sum_{S \not\ni i} \mu_i(S) w(S) \\
&= \sum_{S \ni i} \mu_i(S) w(S) - \sum_{S \not\ni i} \mu_i(S \cup \{i\}) w(S) \\
&= \sum_{S \ni i} \mu_i(w(S) - w(S - \{i\}))
\end{aligned}
$$

since, for any coalition S which does not contain i, we may write $S \cup \{i\} = T - \{i\}$ where T does contain i, and vice versa.

Next we must calculate $\mu_i(S)$ when $S \ni i$. There are exactly $\binom{|N|-|S|}{|T|-|S|}$ coalitions T between S and N. Thus, we obtain

$$
\begin{aligned}
\mu_i(S) &= \sum_{S \subset T \subset N} \frac{(-1)^{|T|-|S|}}{|T|} \\
&= \sum_{|S| \le |T| \le |N|} (-1)^{|T|-|S|} \binom{|N|-|S|}{|T|-|S|} \int_0^1 x^{|T|-1} dx \\
&= \int_0^1 \sum_{s \le t \le n} (-1)^{t-s} \binom{n-s}{t-s} \int_0^1 x^{t-1} dx \\
&= \int_0^1 x^{s-1} \sum_{0 \le t-s \le n-s} (-1)^{t-s} \binom{n-s}{t-s} x^{t-s} dx \\
&= \int_0^1 x^{s-1} (1-x)^{n-s} dx.
\end{aligned}
$$

But we know how to calculate this integral:

$$\mu_i(S) = \int_0^1 x^{|S|-1} (1-x)^{n-|S|} dx = \frac{(|S|-1)!(n-|S|)!}{n!}.$$

This gives the first formula. To derive the second formula, we fix a coalition T and $i \in T$. The number of permutations $n!$ such that $T = T_\theta(i)$ is equal to $(|T| - 1)!(n - |T|)!$, since the members of $T - \{i\}$ and $N - T$ may be arbitrarily permuted. □

This formula may be interpreted by the following scenario. We fix one of the $n!$ permutations of the players at random (with probability $\frac{1}{n!}$). For a permutation θ, we consider the difference between the losses $v(T_\theta(i)) - v(T_\theta^0(i))$, which we interpret as the loss of the ith player in this ordering. Then, the ith component of the Shapley value is the mathematical expectation of these losses.

As in the case of generalised games, the Shapley value may be justified by a system of axioms, which we define below:

Efficiency Axiom

$$\sum_{i=1}^{n} \phi(w)_i = w(N). \tag{87}$$

Symmetry axiom If θ is a permutation, and if $\theta * w$ is defined by $(\theta * w)(S) = w(\theta(S))$, then

$$\phi(\theta * w) = \theta * \phi(w). \tag{88}$$

Redundant-players axiom A player i is said to be redundant if $\forall S \subset N, v(S) = v(S \cup \{i\})$. The axiom states that

$$\text{for any redundant player,} \quad \phi(w)_i = 0. \tag{89}$$

Proposition 13.6 *The Shapley value is the unique* **linear** *operator from $\mathcal{W}$ to $\mathbb{R}^n$ which satisfies the efficiency, symmetry and redundant-players axioms.*

Proof. Lemma 13.2 shows that χ is an isomorphism from $\mathcal{W}$ to the space generated by the 2^n functions γ_S. Thus, the functions $w_S = \chi^{-1}\gamma_S$ form a basis for $\mathcal{W}$. They are defined by

$$w_S(T) = \begin{cases} 1 & \text{if } T \supset S \\ 0 & \text{if } S \cap \operatorname{comp} T \neq \emptyset \end{cases} \tag{90}$$

and any function w may be written in the form

$$w = \sum_{S \subset N} \alpha_S(w) w_S. \tag{91}$$

Firstly, the Shapley value clearly satisfies the three axioms. Suppose that ψ is another linear operator from $\mathcal{W}$ to $\mathbb{R}^n$ which also satisfies these axioms. It is sufficient to show that ϕ and ψ coincide on the elements of the basis of $\mathcal{W}$ formed by the w_S. Let us consider w_S. If i does not belong to S, then i is a redundant player for the game w_S; in fact, if $T \supset S$, then $T \cup \{i\} \supset S$ and $w_S(T) = w_S(T \cup \{i\}) = 1$. If $S \cap \text{comp}\, T \neq \emptyset$, then $S \cap \text{comp}\, (T \cup \{i\}) \neq \emptyset$ since i does not belong to S. Thus, $w_S(T) = w_S(T \cup \{i\}) = 0$. Whence,

$$\text{if } i \notin S, \text{ then } \psi(w_S)_i = 0. \tag{92}$$

Suppose now that i and j belong to S and that θ is a permutation which interchanges i and j and leaves the other players unchanged. Then $\theta(S) = S$, whence $\theta * w_S = w_S$. The symmetry axiom implies that

$$\psi(w_S)_i = \psi(\theta * w_S)_i = \theta * \psi(w_S)_i = \psi(w_S)_j.$$

Lastly, the efficiency axiom implies that

$$\sum_{i=1}^{n} \psi(w_S)_i = \sum_{i \in S} \psi(w_S)_i = w_S(N) = 1.$$

It then follows that

$$\text{if } i \in S, \text{ then } \psi(w_S)_i = \frac{1}{|S|}. \tag{93}$$

Whence, we have obtained

$$\forall i = 1, \ldots, n, \quad \psi(w_S)_i = \phi(w_S)_i \tag{94}$$

This completes the proof of Proposition 13.6. □

Example: Simple Games. A game w is simple if for any coalition $S \subset N$ we have $w(S) = 1$ (winning coalition) or $w(S) = 0$ (losing coalition) and any coalition of winning coalitions is again a winning coalition. In this case, the terms $v(T) - v(T - \{i\})$ either have value 0 (if T and $T - \{i\}$ are both losing coalitions or if $T - i$ is a winning coalition) or value 1 if T is winning coalition and $T - \{i\}$ is a losing coalition.

We shall denote the set of winning coalitions S such that $S - \{i\}$ is a losing coalition by $G(i)$. Then, for a simple game we obtain

$$\phi(w)_i = \sum_{S \in G(i)} \frac{(|S| - 1)!(n - |S|)!}{|S|} \tag{95}$$

Remark. Many authors have suggested that the Shapley value of simple games should be interpreted as a *power index*. It is a matter of definition. Consider, for example, a game with three players. Each of these players is attributed side weights: 1, 48 and 49. The winning coalitions are those for which the sum of the weights is greater than 50 (electoral game).

The process which consists of attributing to each player the power index proportional to his weight would give $\frac{1}{100}$, $\frac{48}{100}$ and $\frac{49}{100}$. This partition is obtained by applying the fuzzy Shapley value to the weighting game defined by $v(c) := (c_1 c_2^{48} c_3^{49})^{1/100}$. But this fuzzy game v is not a good description, since it does not use the rule defining the winning coalitions. This shows that the winning coalitions are $\{1,3\}$, $\{2,3\}$ and $\{1,2,3\}$. If power means participating in a winning coalition then we see that player 3, who may participate in three winning coalitions, is more powerful than the other two players and that these two players have the same power of participation in two winning coalitions. Thus, we may think of attributing the power indices $\frac{2}{3}$, $\frac{2}{3}$ and $\frac{3}{3}$ to these players. (The Shapley value of the associated simple game attributes the players with indices $\frac{1}{6}$, $\frac{1}{6}$, $\frac{4}{6}$.)

Do we have enough information to define a power index?

The first two players have the same probability of participating in a winning coalition and are reliant on the choice of the third player. Thus, we see that this last player may use an optimisation mechanism; for example, the third player may pay the other players to participate.

This leads us to a very common paradox. Since he has a very low weight the first player may be less demanding as regards the compensation which he claims from the third player; thus, he may well participate in a winning coalition. We see that the definition of power indices depends on the information available to describe the game.

Part II

Nonlinear Analysis: Examples

14. Exercises

14.1 Exercises for Chapter 1 – Minimisation Problems: General Theorems

The proofs of the following results are left as an exercise:

- Proposition 1.1

- Proposition 1.2

- Proposition 1.5

- Proposition 1.6

- Proposition 1.7

- Proposition 1.8

Exercise 1.1

We consider the problem

$$\alpha = \inf_{x \in X} f(x).$$

A sequence x_n such that

$$\forall n, \quad f(x_n) \leq \alpha + \frac{1}{n}$$

is called a *minimising sequence*.

(a) Say why such a sequence exists.

(b) Give a proof of Theorem 1.1 using minimising sequences.

Exercise 1.2

Suppose $f : \mathbb{R}^n \to \mathbb{R} \cup \{+\infty\}$ is a nontrivial function satisfying

$$\lim_{\|x\| \to \infty} f(x) = +\infty.$$

Show that f is lower semi-compact. Prove that this is in particular the case for functions satisfying

$$\lim_{\|x\| \to \infty} \frac{f(x)}{\|x\|} > 0.$$

Exercise 1.3

Suppose that X and Y are two sets and that A is a mapping from X to Y. Suppose in addition that we have two functions $f : X \to \mathbb{R} \cup \{+\infty\}$ and $g : Y \to \mathbb{R} \cup \{+\infty\}$. We associate f with the function $Af : Y \to \overline{\mathbb{R}}$ defined by

$$Af(y) := \inf_{Ay=y} f(x)$$

where, by convention, we set $Af(y) = +\infty$ if $y \notin A(\mathrm{Dom}\, f)$. Thus,

$$\mathrm{Dom}\, Af = f(\mathrm{Dom}\, A).$$

(a) Note that

$$\mathrm{Dom}\,(gA) = A^{-1}\mathrm{Dom}\, g.$$

(b) Prove that if $M \subset Y$, then

$$\inf_{Ax \in M} (f(x) + g(Ax)) = \inf_{y \in M} (Af(y) + g(y)).$$

(c) Deduce that the following assertions are equivalent

  (i) $\bar{x}$ minimises $f(x) + g(Ax)$ on $A^{-1}(M)$.
  (ii) $\bar{y} = A(\bar{x})$ minimises $Af(y) + g(y)$ on M and $\bar{x}$ minimises $f(x)$ on $A^{-1}(\bar{y})$.

(d) If B is a mapping from Y to Z, deduce that $(BA)(f) = B(A(f))$.

Exercise 1.4

Uses Exercise 1.3.
Suppose that Y is a Hilbert space and that $f_1, \ldots, f_n$ are n nontrivial functions from Y to $\mathbb{R} \cup \{+\infty\}$.

We set

$$f_1 \square f_2 \square \ldots \square f_n(x) := \inf_{x_1 + \ldots + x_n = x} \sum_{i=1}^{n} f_i(x_i)$$

($f_1 \square \ldots \square f_n$ is said to be the *inf convolution* of the functions f_i).

We also define $B \in L(Y^n, Y)$ by

$$B(y_1, \ldots, y_n) = \sum_{i=1}^{n} y_i.$$

(a) Let $f_1 + \ldots f_n$ be the function defined on Y^n by

$$(f_1 + \ldots f_n)(y) := f_1(y_1) + \ldots + f_n(y_n).$$

Show that

$$f_1 \square \ldots \square f_n = B(f_1 + \ldots + f_n).$$

(b) Suppose that we have n Hilbert spaces X_i and n mappings A_i from X_i to Y. Show that

$$\inf_{\sum_{i=1}^{n} A_i(x_i) = y} f_i(x_i) = (A_1 f_1 \square A_2 f_2 \square \ldots \square A_n f_n)(y).$$

Deduce that $\bar{x} = (\bar{x}_1, \ldots, \bar{x}_n)$ minimises $\sum_{i=1}^{n} f_i(x_i)$ with the constraint that $\sum_{i=1}^{n} A_i(x_i) = y$ if and only if

(i) $\bar{y} = (\bar{y}_1, \ldots, \bar{y}_n)$ minimises $\sum_{i=1}^{n}(A_i f_i)(y_i)$ with the constraint that $\sum_{i=1}^{n} y_i = y$;

(ii) for all i, $\bar{x}_i$ minimises $f_i(x_i)$ with the constraint that $A_i(x_i) = \bar{y}_i$.

Remark. This a *decentralisation* principle. First we divide the resource y amongst the $\bar{y}_i$ and then, for each i, we solve independent (decentralised) minimisation problems.

Exercise 1.5

Suppose that f and g are two functions from the Hilbert space X to $\mathbb{R} \cup \{+\infty\}$. We define

$$f \square g(x) := \inf_{y + z = x} (f(y) + g(z))$$

(inf convolution of f and g).

(a) Show that

$$f \square g(x) = \inf_{y \in X} (f(x - y) + g(y)) = \inf_{z \in X} (f(z) + g(x - z)).$$

(b) Prove that

$$\text{Ep}(f) + \text{Ep}(g) \subset \text{Ep}(f \square g) \subset \text{closure}(\text{Ep}(f) + \text{Ep}(g)).$$

Exercise 1.6

Suppose X and Y are two Hilbert spaces. Consider a nontrivial function $g : X \times Y \to \mathbb{R} \cup \{+\infty\}$ and its *marginal* function f defined by

$$f(x) := \inf_{y \in Y} g(x, y).$$

We let π denote the mapping from $X \times Y \times \mathbb{R}$ to $X \times \mathbb{R}$ defined by

$$\pi(x, y, \lambda) = (x, \lambda).$$

Show that

$$\pi(\text{Ep } g) \subset \text{Ep}(f) \subset \text{closure}(\pi(\text{Ep } g)).$$

Exercise 1.7

Suppose that f is a nontrivial function from X to $\mathbb{R} \cup \{+\infty\}$. We associate f with the function $\bar{f}$ defined by

$$\bar{f}(x) := \liminf_{y \to x} f(y)$$

where we suppose that for all x, $\bar{f}(x) > -\infty$. Show that

$$\text{Ep}(\bar{f}) = \overline{\text{Ep}(f)}.$$

Exercise 1.8

Uses Exercise 1.3.
Suppose that X, Y and Z are three Hilbert spaces, where $A \in L(X, Z)$ and $B \in L(Y, Z)$ are two continuous linear operators and $f : X \to \mathbb{R} \cup \{+\infty\}$ and $g : Y \to \mathbb{R} \cup \{+\infty\}$ are two nontrivial functions.

(a) Show that

$$\inf_{Ax = By + z} (f(x) + g(y)) = \inf_y (Af(z + By) + g(y)).$$

(b) Deduce that the following two assertions are equivalent

(i) $(\bar{x}, \bar{y})$ minimises $f(x) + g(y)$ with the constraint $Ax = By + z$.

(ii) $\bar{y}$ minimises $Af(z + By) + g(y)$ and $\bar{x}$ minimises the function $f(x)$ with the constraint $Ax = B\bar{y} + z$.

(c) Suppose we have

- n Hilbert spaces X_i, m Hilbert spaces Y_j

- n operators $A_i \in L(X_i, Z)$, m operators $B_j \in L(Y_j, Z)$
- n functions $f_i : X_i \to \mathbb{R} \cup \{+\infty\}$, m functions $g_j : Y_j \to \mathbb{R} \cup \{+\infty\}$.

Show that

$$\inf_{\sum_{i=1}^n A_i x_i = z + \sum_{j=1}^m B_j y_j} \left(\sum_{i=1}^n f_i(x_i) + \sum_{j=1}^m g_j(y_j) \right)$$
$$= \inf_{y_j \in Y_j} \left((A_1 f_1 \square \ldots \square A_n f_n) \left(z + \sum_{j=1}^m B_j Y_j \right) + \sum_{j=1}^m g_j(y_j) \right)$$

and deduce a decomposition principle analogous to that of the previous question.

(d) Find an economic interpretation of this decomposition principle.

Exercise 1.9

Show that if K and L are compact, then $K + L$ is compact.
Show that if K is compact and L is closed, then $K + L$ is closed.
Give a counter-example to show that if K and L are closed, then $K + L$ is not necessarily closed.

Exercise 1.10

Show that

$$\psi_K + \psi_L = \psi_{K \cap L}$$

and that

$$\psi_L \circ A = \psi_{L^{-1}(A)}.$$

Exercise 1.11

Suppose that $f : X \to \mathbb{R} \cup \{+\infty\}$ is a nontrivial lower semi-continuous function. Show that

$$\sup_{x \in K} f(x) = \sup_{x \in \overline{K}} f(x).$$

Exercise 1.12

We associate a function $f : L \times M \to \mathbb{R}$ with the mapping F from L to the space $\mathbb{R}^M$ of all continuous real-valued functions on M by

$$F(x)(y) := f(x, y).$$

We denote the cone of non-negative functions on M by $\mathbb{R}_+^M$. We assume that the space $\mathbb{R}^M$ has the topology of simple convergence (or the product topology) which makes $\mathbb{R}^M$ a separate locally convex space.

Show that if L is compact and if

$$\forall y \in M, \quad x \to f(x, y) \text{ is lower semi-continuous} \tag{$*$}$$

then

$$F(L) + \mathbb{R}_+^M \text{ is closed in } \mathbb{R}^M \tag{$**$}$$

(see Proposition 1.8).

14.2 Exercises for Chapter 2 – Convex Functions and Proximation, Projection and Separation Theorems

The proofs of the following results are left as an exercise:

- Proposition 2.1

- Proposition 2.2

- Proposition 2.3

- Proposition 2.4

- Proposition 2.5

- Proposition 2.6

- Theorem 2.4

- Corollary 2.4

Exercise 2.1

Show that the image and the inverse image of a convex set under a linear operator are convex.

Exercise 2.2

Show that any intersection of convex sets is convex.

Exercise 2.3

Show that any product of convex sets is convex.

Exercise 2.4

Show that if K and L are convex, then $K + L$ is convex.

Exercise 2.5

Let K be a convex subset of a Hilbert space. Suppose that

$$x_0 \in \mathrm{Int}(K), \quad x_1 \in K \tag{$*$}$$

Show that

$$\forall \lambda \in]0, 1], \quad \lambda x_0 + (1 - \lambda)x_1 \in \mathrm{Int}(K). \tag{$**$}$$

Hint: show that if $B(x_0, \varepsilon) \subset K$ and $y \in B(x_1, \frac{\varepsilon\lambda}{1-\lambda})$, then $B(\lambda x_0 + (1-\lambda)y, \lambda\varepsilon) \subset K$ and contains $\lambda x_0 + (1 - \lambda)x_1$.

Exercise 2.6

Follows Exercise 2.5.
Let K be a convex subset of a Hilbert space. Show that $\overline{K}$ and $\mathrm{Int}\, K$ are convex, that the interiors of K and $\overline{K}$ coincide and that if $\mathrm{Int}\, K \neq \emptyset$, then $\overline{K} = \overline{\mathrm{Int}\, K}$.

Exercise 2.7

Show that $\mathrm{co}(K)$ (the smallest convex set containing K) is the set of convex combinations $\sum_{\text{finite}} \lambda_i x_i$ of elements x_i of K.

Exercise 2.8

Show that $\overline{\mathrm{co}}(K)$ (the smallest convex closed set containing K) is the closure of $\mathrm{co}(K)$.

Exercise 2.9

Show that the convex hull $\mathrm{co}(\cup_{i=1}^{n} K_i)$ of the union of n compact sets K_i is compact (whence, equal to its closure).

Deduce that the convex hull of a finite set is compact.

Exercise 2.10 (Carathéodory's Theorem)

Let K be a non-empty subset of $\mathbb{R}^n$. Show that the convex hull $\mathrm{co}(K)$ of K is the set of convex combinations of $(n + 1)$ elements of K

$$x = \sum_{i=0}^{n} \lambda_i x_i \text{ where } \lambda \in M^{n+1} := \left\{ \lambda \in \mathbb{R}_+^{n+1} \big| \sum_{i=0}^{n} \lambda_i = 1 \right\}.$$

Hint: show that if $x = \sum_{i=0}^{k} \lambda_i x_i$ for $k > n$, then $x = \sum_{j=0}^{k-1} \mu_j x_j$, using the fact that the $x_1 - x_0, \ldots, x_k - x_0$ are linearly dependent.

Exercise 2.11

Follows Exercise 2.10.
Show that the convex hull of a compact subset of $\mathbb{R}^n$ is compact.

Exercise 2.12

Let p be a linear form on X. Show that

$$\sup_{x \in K} \langle p, x \rangle = \sup_{x \in \mathrm{co}(K)} \langle p, x \rangle.$$

Exercise 2.13

Let K be a convex closed subset. Show that the functions $x \to d_K(x)$ and $\frac{1}{2} d_K(x)^2$ are convex functions.

Exercise 2.14

Let K be a convex subset of $X \times \mathbb{R}$ and suppose that f is the continuous function from X to $\mathbb{R} \cup \{+\infty\}$ defined by

$$f(x) := \inf\{\lambda | (x, \lambda) \in K\}.$$

Show that f is convex.

Exercise 2.15

Let g be a nontrivial function from X to $\mathbb{R}$ (not necessarily convex). We associate g with the function f defined by

$$f(x) := \inf\{\lambda | (x, \lambda) \in \mathrm{co}(\mathrm{Ep}(g))\}. \tag{$*$}$$

Show that

$$f(x) := \inf_{\sum_{\text{finite}} \lambda_i x_i = x} \left(\sum_{\text{finite}} \lambda_i g(x_i) \right). \tag{$**$}$$

Deduce that f is the largest convex function minorising g.

Exercise 2.16

Let f and g be two nontrivial convex functions from X to $\mathbb{R} \cup \{+\infty\}$. Show that the function $f\square g$ from X to $\mathbb{R} \cup \{+\infty\}$ defined by

$$f\square g(x) = \inf_{y \in X} (f(x - y) + g(y))$$

is convex.

Exercise 2.17

Suppose that f_i are n nontrivial convex functions from X to $\mathbb{R} \cup \{+\infty\}$ ($i = 1, \ldots, n$). Show that

$$f(x) := \inf_{\sum_{i=1}^{n} x_i = x} f_i(x_i)$$

is convex.

Exercise 2.18

Suppose that g and h are two convex functions and that f is the function defined by

$$f(x) := (g(x))^2 + h(x).$$

Let M be the set of x which minimise f. Show that g and h are constant functions on M (use the fact that $a \to a^2$ is strictly convex).

Exercise 2.19

Suppose that $f : X \to \mathbb{R} \cup \{+\infty\}$ is a nontrivial convex function and that $A : X \to Y$ is a linear operator. Show that the function $Af : Y \to \mathbb{R} \cup \{+\infty\}$ defined by

$$Af(y) := \inf_{Ax = y} f(x)$$

is also convex.

Exercise 2.20

With any function $f : L \times M \to \mathbb{R}$, we associate the mapping F from L to $\mathbb{R}^M$ (the vector space of all real-valued functions on M) defined by

$$F(x)(y) := f(x, y).$$

Show that if K is convex and if $\forall y \in M$, $x \to f(x, y)$ is convex, then $F(L) + \mathbb{R}^M_+$ is convex (see Proposition 2.6).

Exercise 2.21

Suppose that $f : X \to \mathbb{R} \cup \{+\infty\}$ is nontrivial and convex and that

$$\forall x \in \mathrm{Dom}(f), \quad \exists \xi \in \mathrm{Dom}(f) \text{ such that } f(\xi) < f(x).$$

Show that if $K \subset \mathrm{Dom}\, f$ and if $\bar{x} \in K$ minimises f over K, then $\bar{x}$ belongs to the boundary of K.

Exercise 2.22

Suppose that $f : X \to \mathbb{R} \cup \{+\infty\}$ is a nontrivial convex function and $x_0 \in \mathrm{Dom}(f)$. Show that f is lower semi-continuous at x_0 if and only if for all $\varepsilon > 0$, there exists $\eta > 0$ such that

(i) $f(x_0) \leq f(x) + \varepsilon$ if $\|x - x_0\| < \eta$
(ii) $f(x_0) \leq f(x) + \frac{\varepsilon}{\eta}\|x - x_0\|$ if $\|x - x_0\| \geq \eta$.

Exercise 2.23

Suppose that $f : X \to \mathbb{R} \cup \{+\infty\}$ is a nontrivial convex function and $x_0 \in \mathrm{Dom}(f)$. Show that f is continuous at x_0 if and only if there exist constants $c > 0$ and $\eta > 0$ such that

(i) $f(x_0) \leq f(x) + c\|x - x_0\|$ $\forall x \in X$
(ii) $f(x) \leq f(x_0) + c\|x - x_0\|$ if $\|x - x_0\| \leq \eta$.

Exercise 2.24

Suppose that $f : X \to \mathbb{R} \cup \{+\infty\}$ is convex and continuous at a point x_0. Suppose also that $A \in L(X, Y)$ is a continuous linear operator and that Af is the function defined on Y by

$$Af(y) := \inf_{Ax=y} f(x).$$

Show that Af is convex and continuous at Ax_0.

Exercise 2.25

Suppose that $f(x) := \frac{1}{2}\|x - u\|^2$ where $u \in X$ and K is a convex subset of X. We set

$$v := \inf_{x \in K} f(x)$$

and

$$S(f, \lambda) := \{x \in K \mid f(x) \leq \lambda\}.$$

Show that for all $\lambda > v$,

$$\text{Diam } S(f, \lambda) \leq \sqrt{8(\lambda - v)}.$$

Deduce that if K is closed, then f admits a unique minimum $\bar{x}$ on K (which is the projection of u onto K). Hint: prove that the $S(f, v + \frac{1}{n})$ form a decreasing sequence of closed sets the diameter of which tends to 0.

Exercise 2.26

Let $g : X \to \mathbb{R} \cup \{+\infty\}$ be a nontrivial, convex, lower semi-continuous function. Adapt Exercise 2.25 for the functions

$$f_\lambda(x) := g(x) + \frac{1}{2\lambda} \|x - u\|^2.$$

Exercise 2.27

Show that if K is a convex subset of the finite-dimensional vector space $\mathbb{R}^n$ which is dense in $\mathbb{R}^n$, then $K = \mathbb{R}^n$ (use the Large Separation Theorem, Theorem 2.25).

Exercise 2.28

Let P be a closed cone in a Hilbert space X. Show that if $X = P + \varepsilon B$ ($\varepsilon > 0$, B is the unit ball), then $X = P$ (apply the separation theorem for a closed set and a point separated by a neighbourhood).

14.3 Exercises for Chapter 3 – Conjugate Functions and Convex Minimisation Problems

The proofs of the following results are left as an exercise:

- Proposition 3.1

- Proposition 3.2

- Proposition 3.3

- Proposition 3.4

- Proposition 3.5

- Proposition 3.6

- Corollary 3.1

- Proposition 3.7

- Corollary 3.2

- Corollary 3.3

- Proposition 3.8

- Theorem 3.3

- Theorem 3.4

- Proposition 3.9

Derive the formulae for support functions (formulae (62) to (71) of Chapter 3).

Exercise 3.1

Let $\{f_i\}_{i \in I}$ be a family of nontrivial functions from X to $\mathbb{R} \cup \{+\infty\}$. Show that

$$\left(\inf_{i \in I} f_i\right)^* = \sup_{i \in I} f_i^*$$

and that

$$\left(\sup_{i \in I} f_i\right)^* \leq \inf_{i \in I} f_i^*.$$

Exercise 3.2

Suppose that $\varphi : \mathbb{R} \to \mathbb{R} \cup \{+\infty\}$ is a nontrivial, convex, lower-semi-continuous function with conjugate φ^*. Show that

$$\varphi(\|\cdot\|)^* = \varphi^*(\|\cdot\|_*)$$

where $\|\cdot\|_*$ is the dual norm of $\|\cdot\|$.

Exercise 3.3

If $\alpha > 1$, calculate the conjugate of the function φ defined by

$$\varphi(x) := \frac{1}{\alpha}|x|^\alpha.$$

Hint: use Hölder's inequality $ab \leq \frac{1}{\alpha}a^\alpha + \frac{1}{\alpha^*}b^{\alpha^*}$ where $\frac{1}{\alpha} + \frac{1}{\alpha^*} = 1$.

Exercise 3.4

Uses Exercises 3.2 and 3.3.
Suppose that X is a Hilbert space with norm $\|\cdot\|$ and dual norm $\|\cdot\|_*$ on X^*. Show that

$$\left(\frac{1}{\alpha}\|\cdot\|\right)^*(p) = \frac{1}{\alpha^*}\|p\|_*^{\alpha^*} \quad \text{where} \quad \frac{1}{\alpha} + \frac{1}{\alpha^*} = 1.$$

Exercise 3.5

Let $f : X \to \mathbb{R} \cup \{+\infty\}$ be a nontrivial function. Show that

$$f^*(p) = \sigma_{\mathrm{Ep}(f)}(p, -1).$$

Exercise 3.6

Show that if $f : X \to \mathbb{R} \cup \{+\infty\}$ is a convex lower semi-continuous function satisfying

$$f^*(p) \le \alpha(\|p\|_*)$$

where $\alpha : \mathbb{R} \to \mathbb{R} \cup \{+\infty\}$ is nontrivial, then f satisfies the inequality

$$\forall x \in X, \quad f(x) \ge \alpha^*(\|x\|).$$

Exercise 3.7

Show that if the conjugate f^* of a nontrivial function $f : X \to \mathbb{R} \cup \{+\infty\}$ is continuous at 0, then

$$\forall \lambda \in \mathbb{R}, \quad \{x | f(x) \le \lambda\} \text{ is bounded.}$$

Exercise 3.8

Suppose that f is a nontrivial, convex, lower semi-continuous function from X to $\mathbb{R} \cup \{+\infty\}$. Show that

$$f^*(p) \le a, \quad \forall p \in p_0 + \eta B$$

if and only if

$$f(x) \ge \eta\|x\| + \langle p_0, x \rangle - a.$$

Exercise 3.9

If $f : X \to \mathbb{R} \cup \{+\infty\}$ is nontrivial, convex and lower semi-continuous, show that there exist $p \in X^*$ and $a \in \mathbb{R}$ such that

$$\forall x \in X, \quad f(x) \geq \langle p, x \rangle + a.$$

Exercise 3.10

Calculate the conjugate functions of the functions $x \to f(x) + \frac{1}{2}\|x\|^2$ and $x \to f(x) + \frac{1}{2\lambda}\|x - u\|^2$.

Exercise 3.11

Calculate the conjugate functions for $x \to f(x) + \|x\|$ and $x \to f(x) + \|x - u\|$.

Exercise 3.12

Suppose that $f : X \to \mathbb{R} \cup \{+\infty\}$ is a nontrivial, convex, lower semi-continuous function. Calculate

$$f_\lambda(x) + (f^*)_\mu(q).$$

For what values of λ and μ is this sum zero?

Exercise 3.13

Consider n functions $f_i : X \to \mathbb{R} \cup \{+\infty\}$ and the function $f : X \to \mathbb{R} \cup \{+\infty\}$ defined by

$$f(x) := \inf_{\sum_{i=1}^n x_i = x} f_i(x_i).$$

Show that

$$f^*(q) = \sum_{i=1}^n f_i^*(q).$$

Exercise 3.14

We consider n Hilbert spaces X_i, n functions $f_i : X_i \to \mathbb{R} \cup \{+\infty\}$ and n continuous linear operators $A_i \in L(X_i, Y)$ from X_i to the same Hilbert space Y. We define

$$f(x) := \inf_{\sum_{i=1}^n A_i x_i = x} f_i(x_i).$$

Show that

$$f^*(q) = \sum_{i=1}^{n} f_i^*(A_i^* q).$$

Exercise 3.15

Suppose that f and g are two nontrivial functions from X to $\mathbb{R} \cup \{+\infty\}$ and that

$$h(x) := \inf_{y \in X} (f(x - y) + g(y)).$$

Show that $h^*(q) = f^*(q) + g^*(q)$.

Exercise 3.16

Suppose $g(y) := \inf_{x \in X} f(x, y)$ where f is a nontrivial function from $X \times Y$ to $\mathbb{R} \cup \{+\infty\}$. Show that $g^*(q) = f^*(0, q)$.

Exercise 3.17

Let K be a convex closed subset of a Hilbert space X. Show that

$$\forall \lambda \in \mathbb{R}, \quad \{x \in X | \langle p_0, x \rangle \leq \lambda\} \text{ is bounded}$$

if and only if the support function $p \to \sigma_K(p)$ is continuous at x_0 (use Exercise 3.7).

Deduce that the support function of a bounded set is continuous.

Exercise 3.18

Let $A \in L(X, Y)$, $L \subset X$ and $M \subset Y$. Show that if L and M are cones, the following conditions are equivalent

(i) $0 \in \text{Int}(A(L) - M)$
(ii) $A(L) - M = Y$.

Exercise 3.19

Prove that the *Slater condition*

$$\exists x_0 \in L \text{ such that } A(x_0) \in \text{Int}(M)$$

implies the condition

$$0 \in \text{Int}(A(L) - M).$$

Exercise 3.20

Let M be a subset of $\mathbb{R}^l$. Show that

$$\text{Int}(M + \mathbb{R}^l_+) = M + \text{Int}\,\mathbb{R}^l_+$$

(use balls of the form $[-\varepsilon, +\varepsilon]^n$).

Exercise 3.21

Suppose that $f : X \to \mathbb{R} \cup \{+\infty\}$ is a nontrivial, convex, lower semi-continuous function, that $A \in L(X, Y)$ and $B \in L(X, Z)$ are two continuous linear operators and that $M \subset Z$ is a non-empty, convex, closed subset. Consider the problem

$$v := \inf\{f(x) | Ax = 0 \text{ and } Bx \in M\}.$$

(a) Show that the condition

$$0 \in \text{Int}\,(\text{Im}\,A^* + B^*b(M) + \text{Dom}\,f^*)$$

implies that there exists a solution of the problem v.

(b) Write down the dual problem.

(c) Show that the condition

$$\forall y, z \in \eta B, \quad \exists x \text{ such that } Ax = y \text{ and } Bx \in M + z \qquad (*)$$

implies that there exists a solution of the dual problem.

Exercise 3.22

Suppose that $g : X \times Y \to \mathbb{R} \cup \{+\infty\}$ is a nontrivial, convex, lower semi-continuous function. Consider the minimisation problem

$$v = \inf_{y \in Y} g(0, y).$$

Characterise the associated dual problem and give sufficient conditions which imply the existence of solutions of the primal and dual problems.

Exercise 3.23

Suppose that $g : X \times X \to \mathbb{R} \cup \{+\infty\}$ is a nontrivial, convex, lower semi-continuous function. Characterise the dual problem of the minimisation problem

$$v := \inf_{x \in X} f(x, x)$$

and give sufficient conditions for these problems to have solutions.

Exercise 3.24

Let K be a subset. We set

$$K^0 := \{p \in X^* | \sigma_K(p) \le 1\} \tag{1}$$

(K^0 is called the (negative) polar set associated with K).

(a) Show that K^0 is a convex closed subset with $0 \in K^0 \subset b(K)$.

(b) Show that $K^{00} = \overline{\text{co}}(K \cup \{0\})$.

(c) If K is a cone, show that $K^0 = K^-$.

Exercise 3.25

Suppose that $P \subset X$ is a cone and that $K \subset X$ is a subset. Show that

$$(P + K)^0 = P^- \cap K^0.$$

Exercise 3.26

Let X and Y be two Hilbert spaces, $A \in L(X, Y)$ a continuous linear operator from X to Y and $p \in X^*$ fixed. We suppose that

$$\text{Im } A^* \text{ is closed} \tag{*}$$

and that

$$\langle p, x \rangle = 0, \quad \forall x \in X \text{ satisfying } Ax = 0. \tag{**}$$

(a) Show that there exists $q \in Y^*$ with $p = A^*q$. Deduce that if A is surjective, then q is unique.

(b) Show that if $\text{Im}(A)$ is not dense in Y, then (a) implies that there exists $q \in Y^*$, $q \ne 0$, such that $A^*q = 0$.

Exercise 3.27

Suppose that X is a Hilbert space, that Y is a finite-dimensional space, that $A \in L(X, Y)$ and that $K \subset Y^*$ is a closed subset. We also suppose that

$$\text{Im } A + b(K) = Y. \tag{*}$$

Consider a sequence of elements $q_n \in Y^*$ satisfying

$$A^*q_n \text{ converges to } p \text{ in } X^*. \tag{**}$$

(a) Show that there exists a subsequence of q_n which converges to an element $q \in K$. Hint: use (*) to show that for all $y \in Y$, $\langle q_n, y \rangle$ is bounded.

(b) Deduce that $A^*(K)$ is closed in X^*.

(c) Deduce that the sets $\{q \in K | A^*q = p\}$ are compact.

Exercise 3.28

Suppose that K is a closed cone in $X \times Y$.

(a) Show that $A(0) := \{y \in Y | (0, y) \in K\}$ is a convex closed cone.

(b) We set $A(x) := \{y \in Y | (x, y) \in K\}$. Show that for all x, $A(x)$ is a convex closed set with asymptotic cone $A(0)$.

Exercise 3.29

Uses Exercise 2.17.
Let X be a Hilbert space, which is identified with its dual. Suppose that P is a convex closed cone and that K is a convex closed subset. We denote the orthogonal projector onto the convex closed cone P by π_P.

(a) Show that $\pi_P(K) \subset K - P^-$.

(b) If A is a convex closed subset, we set

$$m(A) := \pi_A(0) \tag{2}$$

(the projection of best approximation of 0 by the elements of A).
Show that $m(\pi_P(K)) = m(K - P^-)$.

(c) Show that the element $\bar{x} = \pi_P(K)$ is characterised by

(i) $\qquad\qquad\qquad\qquad \bar{x} \in P$
(ii) $\qquad\qquad\qquad\quad \sigma_K(-\bar{x}) + \|\bar{x}\|^2 \leq 0.$

Exercise 3.30

Suppose that X and Y are finite-dimensional spaces, that $L \subset X$ and $M \subset Y$ are non-empty, convex, closed subsets and that A is a continuous, *injective*, linear operator from X to Y.

(a) Show that the operator $A^- \in L(Y, X)$ defined by $A^- y := (A^* A)^{-1} A^*$ satisfies $A^- A = 1$ and prove that

$$\|y\|_Y := \|A^- y\|_X \tag{*}$$

is a norm.

(b) Show that for all $y \in Y$, the solution $x_y \in L$ which minimises $\|Ax - y\|$ over L is given by

$$x_y = \pi_L(A^- y)$$

where π_L is the projection of best approximation by the elements of L when Y has the norm of $(*)$.

(c) We denote the projection of 0 at K by $m(K)$. Show that

$$m(A(L) - M) = m((A\pi_L(A^-) - 1)(M)).$$

Exercise 3.31

Let K be a convex closed subset of a Hilbert space. Show that $x_0 \in \text{Int}(K)$ if and only if

$$\{p \,|\, \sigma_K(p) \leq \langle p, x_0 \rangle + 1\}$$

is bounded.

Exercise 3.32

Suppose that K is a convex compact subset. Consider the family $\mathcal{M}_K$ of non-empty closed subsets M of K satisfying

$$\forall y, z \in K, \quad \text{if} \quad \lambda y + (1 - \lambda)z \in \quad \text{then} \quad y, z \in M \qquad (*)$$

(a) Show that K belongs to $\mathcal{M}_K$.

(b) If $f : K \to \mathbb{R}$ is convex and upper semi-continuous and if $M \in \mathcal{M}_K$, show that

$$\{x \in M \,|\, f(x) = \sup_{y \in M} f(y)\} \in \mathcal{M}_K. \qquad (**)$$

(c) Consider a decreasing sequence of non-empty closed subsets $M_i \in \mathcal{M}_K$. Show that

$$M := \cap_{i \in I} M_i \quad \text{belongs to} \quad \mathcal{M}_K.$$

(d) Show that any minimal set $M_0 \in \mathcal{M}_K$ reduces to a point. Hint: if not, there would exist two distinct points x_0 and x_1 of M_0 and a continuous linear form $p \in X^*$ separating them; whence $\{x \in M_0 \,|\, \langle p, x \rangle = \sigma_{M_0}(p)\} \neq M_0$.

(e) A point $x_0 \in K$ such that $\{x_0\} \in \mathcal{M}_K$ is called an *extremal point* of K. Characterise such points.

(f) Use Zorn's lemma to deduce that any set $M \in \mathcal{M}_K$ contains an extremal point and in particular, that if f is convex and upper semi-continuous, then $\{x \in K | f(x) = \sup_{y \in K} f(y)\}$ contains an extremal point.

(g) Use the Separation Theorem to deduce that any convex compact set is the convex hull of its extremal points.

14.4 Exercises for Chapter 4 – Subdifferentials of Convex Functions

The proofs of the following results are left as an exercise:

- Proposition 4.2

- Corollary 4.1

- Theorem 4.2

- Corollary 4.2

- Theorem 4.4

- Proposition 4.3

- Corollary 4.4

Derive the formulae for calculating tangent and normal cones (formulae (45) to (51) of Chapter 4).

Exercise 4.1

Let X be a Hilbert space. We denote its duality operator by $L \in L(X, X^*)$. Suppose $\varphi(x) := \|x\|$. Show that

$$
D\varphi(x)(v) = \begin{cases} \langle L\frac{x}{\|x\|}, v \rangle & \text{if } x \neq 0 \\ \|v\| & \text{if } x = 0. \end{cases}
$$

Deduce that φ is not differentiable at 0, but that it has a right derivative and that

$$
\partial\varphi(x) = \begin{cases} L\frac{x}{\|x\|} & \text{if } x \neq 0 \\ B_* & \text{if } x = 0 \end{cases}
$$

where B_* is the unit ball of the dual.

Exercise 4.2

Let X be a Hilbert space. We denote its duality operator by $L \in L(X, X^*)$. We set

$$\varphi^2(x) := \|x\|^2.$$

Prove that

$$D\varphi^2(x)(v) := 2\langle Lx, v \rangle$$

and show that φ^2 is Fréchet differentiable and that its derivative is the duality operator $2L$.

Exercise 4.3

Suppose that K is a convex closed subset and consider the function $d_K^2(x) = \inf_{y \in K} \|x - y\|^2$.

(a) Show that

$$d_K^2(x + v) - d_K^2(x) - 2\langle x - \pi_K(x), v \rangle \leq \|v\|^2$$

and that

$$d_K^2(x) - d_K^2(x + v) + 2\langle x - \pi_K(x), v \rangle$$
$$\leq (\|\pi_K(x + v) - \pi_K(x)\| + \|v\|)\|v\|.$$

(b) Deduce that $x \to d_K^2(x)$ is Fréchet differentiable and that

$$\nabla d_K^2(x) = 2(x - \pi_K(x))$$

where π_K is the projector of best approximation onto K.

(c) Show also that this result is a consequence of Theorem 5.2.

Exercise 4.4

Suppose that K is a convex closed subset and consider the function d_K defined by $d_K(x) := \inf_{y \in K} \|x - y\|$. Suppose that $T_K(x)$ is the tangent cone to K at x.

(a) If $y \in K$, show that for all $w \in T_K(y)$, we have

$$Dd_K(y)(v) \leq \|v - w\|.$$

(b) If $z \notin K$, deduce that for all $w \in T_K(\pi_K(z))$, we have

$$Dd_K(z)(v) \leq \|v - w\|,$$

whence also

$$Dd_K(z)(v) \leq d(v, T_K(\pi_K(z))).$$

(c) If $x \in K$ and $v \in X$, show that the convex function $f(t) := d_K(x + tv)$ has a right derivative $f'_+(t)$ satisfying

$$f'_+(t) \le d(v, T_K(\pi_K(x + tv))).$$

Exercise 4.5

Let $f : X \to \mathbb{R} \cup \{+\infty\}$ be a nontrivial (not necessarily convex) function. Show that the following conditions are equivalent:

$$x_0 \text{ minimises } x \to f(x) - \langle p_0, x \rangle \qquad (*)$$

and

(i) $x_0 \in \partial f^*(p_0)$
(ii) $f(x_0) = f^{**}(x_0).$ \qquad (**)

Deduce that the set of x which minimise $x \to f(x) - \langle p_0, x \rangle$ is

$$M := \{x \in \partial f^*(p) | f(x) = f^{**}(x)\}.$$

Exercise 4.6

Consider a continuous strictly increasing function $\gamma : \mathbb{R}_+ \to \mathbb{R}_+$ such that $\gamma(0) = 0$ and $\lim_{t \to \infty} \gamma(t) = \infty$. We associate this with a convex function φ defined by

$$\varphi(t) := \begin{cases} \int_0^t \gamma(\tau) d\tau & \text{if } t \ge 0 \\ +\infty & \text{if } t < 0. \end{cases}$$

Let f be the function defined on X by

$$f(x) := \varphi(\|x\|).$$

(a) Show that

$$\partial f(x) = \{p \in X^* | \langle p, x \rangle = \|p\|_* \|x\| \text{ and } \|p\|_* = \varphi(\|x\|)\}.$$

Hint: Use the fact that $f^*(p) = \varphi^*(\|p\|_*).$

(b) Show that if $p \in \partial f(x)$ and $q \in \partial f(y)$, then

$$\langle p - q, x - y \rangle \ge (\gamma(\|x\|) - \gamma(\|y\|))(\|x\| - \|y\|).$$

Exercise 4.7

Suppose that the function f is subdifferentiable at x_0. Show that

$$\forall x \in X, \quad f(x_0) \le f(x) + \sigma(\partial f(x_0), x_0 - x).$$

Exercise 4.8

Suppose that f is differentiable in the neighbourhood of a convex closed set K.

(a) Show that if $\bar{x} \in K$ minimises f over K, then $\bar{x}$ is a solution of the variational inequality

(i)
(ii)
$$\bar{x} \in K$$
$$\langle \nabla f(\bar{x}), \bar{x} - x \rangle \leq 0 \ \forall x \in K. \qquad (*)$$

(b) Prove that if a minimum $\bar{x}$ belongs to the interior of K, the variational inequality becomes the equation

$$\bar{x} \in \text{Int}(K) \text{ and } \nabla f(\bar{x}) = 0 \qquad (**)$$

(Fermat's rule).

Exercise 4.9

Show that if a nontrivial function $f : X \to \mathbb{R} \cup \{+\infty\}$ is both differentiable (Gâteaux differentiable) and subdifferentiable at a point $x_0 \in \text{Dom } f$, then

$$\partial f(x_0) = \{\nabla f(x_0)\}. \qquad (*)$$

In this case, any solution x_0 of the variational inequality

(i)
(ii)
$$x_0 \in K$$
$$\langle \nabla f(x_0), x_0 - x \rangle \leq 0 \ \forall x \in K$$

minimises f over K.

Exercise 4.10

Let g and h be two functions such that

(i) g is convex and differentiable in the neighbourhood of K;

(ii) h is convex;

(iii) K is convex and closed.

Then $\bar{x} \in K$ minimises $f := g + h$ over K if and only of $\bar{x}$ is a solution of

(i)
(ii)
$$\bar{x} \in K$$
$$\forall x \in K, \ \langle \nabla g(\bar{x}), \bar{x} - x \rangle + h(\bar{x}) - h(x) \leq 0. \qquad (*)$$

Exercise 4.11

If the function f is nontrivial, convex and subdifferentiable at x_0, show that f is right differentiable and that

$$\sigma(\partial f(x_0), v) \leq Df(x_0)(v).$$

Under what conditions is there equality?

Exercise 4.12

Consider n convex functions f_i all continuous at a point x_0. Set

$$f(x) := \max_{i=1,\ldots,n} f_i(x)$$

and

$$I(x) := \{i = 1, \ldots, n | f(x) = f_i(x)\}.$$

Show that

$$Df(x_0)(u) = \max_{i \in I(x_0)} Df_i(x_0)(u)$$

and deduce that

$$\partial f(x_0) = \mathrm{co}\left(\bigcup_{i \in I(x_0)} \partial f_i(x_0)\right).$$

Exercise 4.13

Let K be a convex closed subset. Show that

$$\forall x \in K, \quad N_K(x) \subset b(K)$$

and deduce that

$$\forall x \in K, \quad T_K(x) \text{ contains the asymptotic cone.}$$

Exercise 4.14

Suppose that the function $f : X \to \mathbb{R} \cup \{+\infty\}$ is nontrivial and convex. Show that for any finite sequence of points $x_1, \ldots, x_n$ of $\mathrm{Dom}\, f$ and any sequence $p_i \in \partial f(x_i)$, $(i = 1, \ldots, n)$, we have

$$\langle p_1, x_1 - x_2 \rangle + \langle p_2, x_2 - x_3 \rangle + \ldots + \langle p_n, x_n - x_1 \rangle \geq 0.$$

The set-valued map F from X to X^* is said to be *cyclically monotone* if

$$\forall (x_i, f_i) \in \mathrm{Graph}(F), \ 1 \leq i \leq n, \quad \langle p_1, x_1 - x_2 \rangle + \ldots + \langle p_n, x_n - x_1 \rangle \geq 0. \quad (1)$$

Exercise 4.15

Consider a cyclically monotone set-valued map F from X to X^* (satisfying equation (1) of Exercise 4.14). We fix $(x_0, p_0) \in \text{Graph}(F)$ and construct the function $f : X \to \mathbb{R} \cup \{+\infty\}$ defined by

$$f(x) := \sup\{\langle p_n, x - x_n \rangle + \langle p_{n-1}, x_n - x_{n-1} \rangle + \ldots + \langle p_0, x_1 - x_0 \rangle\}$$

where the supremum is taken over all *finite sequences* of points (x_i, p_i) of the graph of F $(i = 1, \ldots, n)$.

(a) Show that f is convex and lower semi-continuous.

(b) Show that $f(x_0) = 0$ (whence, that f is nontrivial).

(c) Show that

$$\forall x \in \text{Dom}(F), \quad F(x) \subset \partial f(x).$$

Hint: take $p \in F(x)$ and show that

$$\forall y \in X, \quad f(x) - f(y) \leq \langle p, x - y \rangle$$

by associating any $\varepsilon > 0$ with a finite sequence of $(x_i, p_i) \in \text{Graph}(F)$ such that $f(x) \leq \langle p_n, x - x_n \rangle + \ldots + \langle p_0, x_1 - x_0 \rangle + \varepsilon$.

(d) We say that a cyclically monotone set-valued map F from X to X^* is *maximal* if there is no other cyclically monotone set-valued map $G \neq F$ with $\text{Graph}(G) \supset \text{Graph}(F)$.

Deduce that any maximal, cyclically monotone, set-valued map is the sub-differential of a nontrivial, convex, lower semi-continuous function (Rockafellar's Theorem).

Exercise 4.16

Consider the Sobolev space

$$H^1(0, T; \mathbb{R}^n) = \{x(\cdot) \in L^2(0, T; X) | x'(t) \in L^2(0, T; X)\}$$

and the Hilbert subspace

$$H = \{x \in H^1(0, T; X) | x(0) = x_0\}.$$

Suppose that $f : X \to \mathbb{R} \cup \{+\infty\}$ is a nontrivial, convex, lower semi-continuous function. Let ϕ be the functional defined on H by

$$\phi(x) = \int_0^T (f(x(t)) + f^*(-x'(t))) dt + \frac{1}{2}\|x(T)\|^2 - \frac{1}{2}\|x_0\|^2.$$

(a) Show that

$$\forall x \in H, \quad \phi(x) \geq 0.$$

(b) Show that the following conditions are equivalent

$$\phi(x) = 0 \qquad (*)$$

and

$$-x'(t) \in \partial f(x(t)) \text{ and } x(0) = x_0. \qquad (**)$$

Deduce that any solution $x(\cdot)$ of the differential inclusion $(**)$ minimises ϕ and note that the converse is not necessarily true.

Exercise 4.17

Suppose that the function $f : X \to \mathbb{R} \cup \{+\infty\}$ is nontrivial, convex and lower semi-continuous. Let $x_0 \in \mathrm{Dom}(\partial f)$. Consider a solution of the differential inclusion

$$-x'(t) \in \partial f(x(t)), \quad x(0) = x_0. \qquad (*)$$

(a) Show that $t \to f(x(t))$ is a decreasing function satisfying

$$\frac{d}{dt} f(x(t)) + \|x'(t)\|^2 = 0. \qquad (**)$$

Hint: find upper bounds for $f(x(t)) - f(x(t+h))$ and $f(x(t+h)) - f(x(t))$, divide by $h > 0$ and let h tend to 0.

(b) Deduce that there exists at most one solution of the differential inclusion $(*)$. Hint: if x_1 and x_2 are two solutions, differentiate $t \to \frac{1}{2}\|x_1(t) - x_2(t)\|^2$.

(c) Deduce from $(**)$ that the solution $x(\cdot)$ of $(*)$ is in fact the solution of the differential equation

$$-x'(t) = \pi_{\partial f(x(t))}(0), \quad x(0) = x_0. \qquad (***)$$

Remark. This involves showing that there exists a (unique) solution of the differential inclusion $(*)$ (Crandall–Rabinowitz Theorem) even though the second term of the differential equation is not continuous.

(d) By integrating $(**)$ from s to t, show that

$$\lim_{s,t \to \infty} \int_s^t \|x'(\tau)\|^2 = 0$$

and deduce from the Cauchy criterion that

$$\int_0^\infty \|x'(\tau)\|^2 d\tau < +\infty. \qquad (****)$$

(e) Let A_ε be the set of $t > 0$ such that $\|x'(t)\| < \varepsilon$. Deduce from (****) that $\mathrm{meas}(A_\varepsilon) = \infty$. Hint: suppose that $\mathrm{meas}(A_\varepsilon)$ is finite and obtain a contradiction to (****).

(f) Deduce that for all $y \in X$

$$\inf_{t>0} f(x(t)) \leq f(y).$$

Hint: show that $f(x(t)) \leq f(y) - \langle x'(t), x(t) - y \rangle$, use question (e) and let ε tend to 0.

(g) Deduce that any limit point of $x(t)$ as $t \to \infty$ minimises f.

14.5 Exercises for Chapter 5 – Marginal Properties of Solutions of Convex Minimisation Problems

The proofs of the following results are left as an exercise:

- Corollary 5.1

- Corollary 5.2

- Corollary 5.3

- Corollary 5.4

Exercise 5.1

Let X and Y be two Hilbert spaces and $A \in L(X, Y)$ a continuous linear operator. Suppose that $u \in X$ and $w \in \mathrm{Im}\, A$ are given. We also suppose that we have a semi-scalar product $\lambda(x, y)$ associated with a self-transpose positive semi-definite operator $L \in L(X, Y)$ by the formula

$$\lambda(x, y) = \langle Lx, y \rangle \tag{1}$$

(see the section of Chapter 1 entitled 'Examples of Convex Functions'). We suppose that

$$\mathrm{Im}\,(L) \text{ is closed.} \tag{2}$$

We set

(i) $$\lambda^2(x) := \lambda(x, x),$$
(ii) $$\text{if } p \in \mathrm{Im}(L), \ \lambda_*^2(p) = \langle p, x \rangle \text{ where } Lx = p. \tag{3}$$

We recall that

$$
\left(\frac{1}{2}\lambda^2\right)^*(p) = \begin{cases} \frac{1}{2}\lambda_*^2(p) & \text{if } p \in \mathrm{Im}(L) \\ +\infty & \text{if } p \notin \mathrm{Im}(L) \end{cases}
\tag{4}
$$

and that

$$
\partial\lambda^2(x) = 2Lx.
\tag{5}
$$

We consider the minimisation problem

(i)
(ii)
$$
\begin{aligned}
A\bar{x} &= w \\
\lambda^2(\bar{x} - u) &= \inf_{Ax=y}\lambda^2(x - u).
\end{aligned}
\tag{6}
$$

Show that if

$$
\mathrm{Im}\,A^* + \mathrm{Im}\,L = X^*
$$

then there exists a solution of the problem (6).

Exercise 5.2

We assume the framework of Exercise 5.1 and that

$$
A \text{ is surjective and } \mathrm{Im}(L) = X^*.
\tag{7}
$$

(a) Write down the dual problem associated with (6).

(b) Show that $(\bar{x}, \bar{q})$ is a solution of the Hamiltonian system if and only if

(i)
(ii)
$$
\begin{aligned}
L\bar{x} + A^*\bar{q} &= Lu \\
A\bar{x} &= w
\end{aligned}
$$

(c) Show that $\bar{q}$ is a solution of the equation

$$
(AL^{-1}A^*)\bar{q} = Au - w
\tag{8}
$$

and deduce that this problem has a unique solution equal to

$$
\bar{q} = (AL^{-1}A^*)^{-1}(Au - w).
$$

(d) Show that $\bar{x}$ is equal to

$$
\bar{x} = u - L^{-1}A^*(AL^{-1}A^*)^{-1}(Au - w).
\tag{9}
$$

Exercise 5.3

We assume the framework of Exercise 5.2.

(a) Show that

$$\xi(q) := u - L^{-1}A^*q$$

minimises $x \to \frac{1}{2}\lambda^2(x - u) + \langle q, Ax \rangle$ over X.

(b) If A is surjective, show that the solution $\bar{q}$ of the dual problem minimises

$$q \to \frac{1}{2}\lambda^2(\xi(q)) + \langle q, w \rangle$$

over Y^* and that the solution $\bar{x}$ of the problem (6) is equal to $\bar{x} = \xi(\bar{q})$.

Exercise 5.4

We assume the framework of Exercise 5.2. Prove that $\bar{x} = \xi(\bar{q}) := u - L^{-1}A^*\bar{q}$ is a solution of the problem (6) if and only if $A\xi(\bar{q}) = w$.

Exercise 5.5

We assume the framework of Exercise 5.2. Show that the solution $\bar{x}$ of the problem (6) is the solution of the problem

(i) $\langle A^*\bar{q}, \bar{x} \rangle = \langle \bar{q}, w \rangle$

(ii) $\lambda^2(\bar{x} - u) = \displaystyle\min_{\langle A^*\bar{q},x \rangle = \langle \bar{q},w \rangle} \lambda^2(x - u).$

Exercise 5.6

We assume the framework of Exercise 5.2. We set

$$\varphi(w) = \inf_{Ax=w} \lambda^2(x - u).$$

Show that φ is differentiable (Fréchet differentiable) and that

$$\nabla\varphi(w) = \bar{q}$$

where $\bar{q} = (AL^{-1}A^*)^{-1}(Au - w)$.

Exercise 5.7

Suppose that $A \in L(X,Y)$ is *surjective*, that $B \in L(X,Z)$ is injective, that $u \in Z$ and $w \in Y$ are given and that we have a scalar product λ on Z. We consider the minimisation problem

$$\begin{aligned}
\text{(i)} && A\bar{x} &= w \\
\text{(ii)} && \lambda^2(B\bar{x} - u) &= \min_{Ax=w} \lambda^2(Bx - u).
\end{aligned}$$

Show that there exists a unique solution $\bar{x}$ of this problem, which is given by

$$\begin{aligned}
\text{(i)} && \bar{x} &= w - M^{-1}A^*\bar{p}, \text{ where } M := B^*LB, \text{ where} \\
\text{(ii)} && \bar{p} &= (AM^{-1}A^*)^{-1}(Av - w), \text{ where} \\
\text{(iii)} && v &:= M^{-1}B^*Lu.
\end{aligned}$$

Exercise 5.8

We assume the framework of Exercise 5.2. We set

$$\pi(A) := (AL^{-1}A^*)^{-1}(Au - w).$$

(a) Show that if $B \in L(X,Y)$, then

$$\lim_{h \to 0+} \frac{\pi(A + hB) - \pi(A)}{h}$$
$$= (AL^{-1}A^*)^{-1}(B(uL^{-1}A^*\pi(A)) - AL^{-1}B^*\pi(A)).$$

(b) We set

$$\begin{aligned}
A^+ &:= L^{-1}A^*(AL^{-1}A^*)^{-1} \in L(Y, X) \\
S(A) &:= (1 - A^+A)L^{-1} \in L(X, X)
\end{aligned}$$

and we denote by $\Delta(A)$ the solution

$$\Delta(A) := u - L^{-1}A^*\pi(A) = u - L^{-1}A^*(AL^{-1}A^*)^{-1}(Au - w)$$

of the problem

$$\psi(A) = \inf_{Ax=w} \lambda^2(x - u).$$

Deduce from (a) that

$$\lim_{h \to 0+} \frac{\Delta(A + hB) - \Delta(A)}{h} = -A^+B\Delta(A) - S(A)B^*\pi(A).$$

Exercise 5.9

We assume the framework of Exercise 5.2.

(a) Show that

$$A^+ := L^{-1}A^*(AL^{-1}A^*)^{-1} \in L(Y, X)$$

satisfies

$$AA^+y = y$$

(A^+ is a right inverse of A) and

$$\lambda^2(A^+y) = \min_{Ax=y} \lambda^2(x).$$

(b) Prove that

$$(AL^{-1}A^*)^{-1} = (A^+)^*LA^+$$

is the duality operator associated with the scalar product $\mu(y_1, y_2)$ defined on Y by

$$\mu(y_1, y_2) := \lambda(A^+y_1, A^+y_2).$$

(c) Show that the scalar product μ_*, the dual of μ on Y^*, is defined by

$$\mu_*(q_1, q_2) = \lambda_*(A^*q_1, A^*q_2).$$

(d) Prove that the solution $\bar{x}$ of the problem (6) is defined by

$$\bar{x} = A^+w + (1 - A^+A)u.$$

Exercise 5.10

We assume the framework of Exercise 5.9. Suppose that $A_1 \in L(X, Y_1)$ and $A_2 \in L(Y_1, Y_2)$ are two continuous, surjective, linear operators. Show that

$$(A_2A_1)^+ = A_1^+A_2^+$$

where $A_2^+ = L_1^{-1}A_2^*(A_2L_1^{-1}A_2^*)^{-1}$ and $L_1 := (A_1L^{-1}A_1^*)^{-1}$.

Exercise 5.11

Suppose that $B \in L(X, Z)$ is a continuous linear operator satisfying

$$B \text{ is injective and its image is closed.} \qquad (*)$$

Suppose that $\mu(z_1, z_2)$ is the scalar product on Z and that $M \in L(Z, Z^*)$ is its duality operator. We set

$$B^- := (B^*MB)^{-1}B^*M \in L(Z, X).$$

(a) Show that

$$B^- Bx = x$$

(B^- is a left inverse of B) and that

$$\mu^2(BB^{-1}u - u) = \inf_{x \in X} \mu^2(Bx - u).$$

(b) Show that

(i)	$(B^-)^* =$	$(B^*)^+$
(ii)	$(B^*MB)^{-1} =$	$(B^-)M^{-1}(B^-)^*$
(iii)	$B =$	$(B^-)^+.$

(c) Show that

$$\mu^2(Bx - u)^2 = \mu^2(B(x - B^-u)) + \mu^2(u) - \mu^2(BB^-u)$$

and deduce that minimising the distance from Bx to u is equivalent to minimising the distance from Bx to BB^-u.

Exercise 5.12

We assume the framework of Exercises 5.7, 5.9 and 5.11. Show that $\bar{x}$ is a solution of the problem

(i)	$A\bar{x} =$	w
(ii)	$\lambda^2(B\bar{x} - u) =$	$\min_{Ax=w} \lambda^2(Bx - u)$

if and only if $\bar{x} = B^-\bar{z}$ where $\bar{z}$ is the unique solution of the problem

$$\lambda^2(\bar{z} - u) = \min_{AB^- z = w} \lambda^2(z - u).$$

Exercise 5.13

We consider

- 4 Hilbert spaces X, U, Y and Z;

- a nontrivial, convex, lower semi-continuous function
  $f : X \times U \to \mathbb{R} \cup \{+\infty\}$;

- 3 continuous linear operators $A \in L(X, Y)$, $B \in L(U, Y)$ and
  $C \in L(X, Z)$;

- non-empty, convex, closed subsets $L \subset Y$ and $M \subset Y$.

We consider the minimisation problem

$$v := \inf\{f(x, u) | Ax + Bu \in L \text{ and } Cx \in Z\}.$$

(a) Write down the dual problem of this problem.

(b) Show that the surjectivity condition: $\exists \gamma > 0$ such that

$$\forall y, z \in \gamma B, \quad \exists (x, u) \in \mathrm{Dom}(f) \text{ satisfying}$$
$$Ax + Bu \in L + y \text{ and } Cx \in M + z, \tag{$*$}$$

implies the existence of a solution of the dual problem.

(c) Show that $(*)$ is satisfied if we assume that

(i) $\forall x \in X$, $\exists u$ such that $f(x, u) < +\infty$;
(ii) $A \times C \in L(X, Y \times Z)$ is surjective.

(d) We suppose that $(*)$ is satisfied and that there exists a solution $(\bar{x}, \bar{u})$ in
    $\mathrm{Dom} f$ of the problem v. Show that there exists a solution $(\bar{p}, \bar{q})$ satisfying

(i) $\bar{p} \in N_L(A\bar{x} + B\bar{u}), \quad \bar{q} \in N_M(C\bar{x})$
(ii) $(-A^*\bar{p} - C^*\bar{q}, -B^*\bar{p}) \in \partial f(\bar{x}, \bar{u})$. \hfill $(+)$

(e) We now assume that

$$f(x, u) = g(x) + h(u)$$

where $g : X \to \mathbb{R} \cup \{+\infty\}$ and $h : U \to \mathbb{R} \cup \{+\infty\}$ are nontrivial, convex
and lower semi-continuous.

Show that (+) may be written the form: there exists a solution $\bar{p}$ of

(i) $$\bar{p} \in N_L(A\bar{x} + B\bar{u})$$
(ii) $$-A^*\bar{p} \in \partial g(\bar{x}) + C^* N_M(C\bar{x}) \qquad (++)$$

(adjoint equation) satisfying

$$\bar{u} \text{ minimises } u \to H(u, \bar{p}) \qquad (+++)$$

where $H(u, \bar{p})$ is defined by

$$H(u, \bar{p}) = h(u) + \langle \bar{p}, Bu \rangle$$

(maximum principle).

(f) Show that if there exists $\delta > 0$ such that

$$\forall \pi \in \delta B, \quad \forall \omega \in \delta B, \quad \exists p \in b(L) \text{ and } q \in b(M) \text{ such that}$$
$$f^*(\pi - A^*p - C^*q, \omega - B^*p) < +\infty \qquad (**)$$

then there exists a solution $(\bar{x}, \bar{u})$ of the problem v.

(g) Deduce in particular that this solution exists if we suppose that there exist $p_0 \in b(L)$ and $q_0 \in b(M)$ such that

$$(-A^*p_0 - C^*q_0, B^*p_0) \in \text{Int}(\text{Dom} f_*).$$

14.6 Exercises for Chapter 6 – Generalised Gradients of Locally Lipschitz Functions

The proofs of the following results are left as an exercise:

- Proposition 6.3

- Proposition 6.4

- Proposition 6.5

- Proposition 6.6

- Proposition 6.7

- Corollary 6.1

- Proposition 6.8

Exercise 6.1

Let $f : X \to \mathbb{R} \cup \{+\infty\}$ be a nontrivial function. If $x \in \text{Dom}(f)$ and $v \in X$, we set

$$D_+ f(x)(v) := \liminf_{\substack{h \to 0+ \\ v' \to v}} \frac{f(x + hv') - f(x)}{h} \in \overline{\mathbb{R}} \tag{1}$$

$(D_+ f(x)(v)$ is called the *contingent epiderivative* of f at x in the direction v).

(a) Show that $v \to D_+ f(x)(v)$ is positively homogeneous and that if for all v, $D_+ f(x)(v) > -\infty$, it is lower semi-continuous.

(b) Show that if f is Lipschitz (with constant L) in the neighbourhood of $x \in \text{Int}\,(\text{Dom}\,f)$, then

$$D_+ f(x)(v) = \liminf_{h \to 0+} \frac{f(x + hv) - f(x)}{h}$$

and

$$-L\|v\| \le D_+ f(x)(v) \le D_c f(x)(v) \le L\|v\|.$$

(c) Show that if f is convex, then

$$D_+ f(x)(v) = \liminf_{v' \to v} D f(x)(v).$$

Exercise 6.2

Suppose that $f : X \to \mathbb{R} \cup \{+\infty\}$ is a nontrivial function and that $\bar{x} \in \text{Dom}\,f$. Show that if $\bar{x}$ is a local minimum of f, then

$$\forall v \in (X), \;\; 0 \le D_+ f(x)(v).$$

Compare with Proposition 6.5.

Exercise 6.3

Suppose that f and g are two nontrivial functions from X to $\mathbb{R} \cup \{+\infty\}$. Show that if $x \in \text{Dom}\,f \cap \text{Dom}\,g$, then

$$D_+ f(x)(v) + D_+ g(x)(v) \le D_+ (f + g)(x)(v).$$

Compare with Proposition 6.3.

Exercise 6.4

Suppose that $A \in L(X, Y)$ and that $g : X \to \mathbb{R} \cup \{+\infty\}$ is a nontrivial function. Suppose we have $x \in X$ such that $Ax \in \mathrm{Dom}\, g$. Show that

$$D_+ f(Ax)(Av) \leq D_+(f \circ A)(x)(v).$$

Compare with Corollary 6.1.

Exercise 6.5

We consider n nontrivial functions $f_i : X \to \mathbb{R} \cup \{+\infty\}$ and $x \in \cap_{i=1}^{n} \mathrm{Dom}\, f_i$. We set

$$I(x) := \left\{ i = 1, \ldots, n \,|\, f_i(x) = \max_{j=1,\ldots,n} f_j(x) \right\}.$$

Show that

$$\max_{i \in I(x)} D_+ f_i(x) \leq D_+ \left(\max_{j=1,\ldots,n} f_j \right)(x)(v).$$

Compare with Proposition 6.9.

Exercise 6.6

Suppose that $A \in L(X, Y)$ and that $f : X \to \mathbb{R} \cup \{+\infty\}$ is a nontrivial function. We set

$$g(y) := \inf_{Ax=y} f(x)$$

and we suppose that there exists a solution $\bar{x}$ of the problem $A\bar{x} = y$ and that $g(y) = f(\bar{x})$. Show that

$$\forall x \in X, \quad D_+ g(y)(Av) \leq D_+ f(\bar{x})(v).$$

Deduce that if g is differentiable (Gâteaux differentiable) at y and f is differentiable at $\bar{x}$, then

$$\nabla f(\bar{x}) - A^* \nabla g(y) = 0. \tag{$*$}$$

In other words, $\nabla g(y)$, the gradient of the marginal function, is a Lagrange multiplier. Compare with Proposition 6.8. Hint: use the inequality

$$g(y + hAv) - g(y) \leq f(\bar{x} + hv) - f(\bar{x}).$$

Exercise 6.7

Consider a function $f : X \to \mathbb{R} \cup \{+\infty\}$ and $x \in \mathrm{Dom}\, f$. We define

$$\hat{D}f(x)(v) := \limsup_{\substack{h \to 0+ \\ v' \to v \\ (y,\lambda) \to (x,f(x)) \\ (y,\lambda) \in \mathrm{Ep}(f)}} \frac{f(y + hv') - \lambda}{h}.$$

(a) Show that

$$D_c f(x)(v) \le \hat{D}f(x)(v).$$

(b) Show that $(x, v) \to \hat{D}f(x)(v)$ is upper semi-continuous and that

$$\{v | \hat{D}(x)(v) \in \mathbb{R}\}$$

is open.

(c) Show that

$$\hat{D}f(x)(-v) = \hat{D}(-f)(x)(v).$$

Exercise 6.8

Let K be a non-empty subset of X. Consider the normal cone to K at x, $N_K(x)$ (Proposition 6.14).

(a) Show that if $p \in X^*$ satisfies

$$\langle p, x_0 \rangle = \max_{y \in K} \langle p, y \rangle$$

then $p \in N_K(x_0)$.

(b) Show that if $y \notin K$ and if $\bar{x} \in \overline{K}$ satisfies

$$\|y - \bar{x}\| = \inf_{x \in K} \|y - x\| = d_K(y),$$

then $y - \bar{x}$ belongs to $N_K(\bar{x})$. Hint: use Proposition 6.15 applied to the functions $x \to \langle -p, x \rangle$ and $x \to \|y - x\|$.

Exercise 6.9

Let K be a closed subset of a finite-dimensional vector space X. Consider the tangent cone to K at x (Definition 6.4). Show that the following assertions are equivalent.

(i) $v \in T_K(x)$.

(ii) $\lim_{\substack{h \to 0+ \\ y \to x \\ y \in K}} \frac{d_K(y+hv)}{h} = 0$.

(iii) $\forall \varepsilon > 0$, $\exists \eta > 0$ such that $\forall h \in]0, \eta[$ and $\forall y \in B_K(x, \eta)$, $\exists u \in v + \varepsilon B$ such that $x + hu \in K$.

(iv) For any sequences $h_n \to 0+$ and $x_n \in K$ converging to x, there exists a sequence v_n converging to v such that $x_n + h_n v_n \in K$ for all n.

Hint: derive the inequalities

$$\sup_{\substack{\|y-x\| \le \alpha \\ y \in K}} \frac{d_K(y+hv)}{h} \quad \le \quad \sup_{\|y-x\| \le \alpha} \frac{d_K(y+hv) - d_K(y)}{h}$$

$$\le \quad \sup_{\substack{\|z-x\| \le 2\alpha \\ z \in K}} \frac{d_K(z+hv)}{h}$$

by taking $z \in K$ such that $\|y - z\| = d_K(y) \le \|y - x\|$ when $y \notin K$ and deduce the equivalence of (i) and (ii).

Exercise 6.10

Let K be a subset of X and $x \in K$. Show that the following assertions are equivalent.

(i) $D_+ d_K(x)(v) = 0$.

(ii) $\forall \varepsilon > 0$, $\forall \alpha > 0$, $\exists h \in]0, \alpha]$, $\exists u \in v + \varepsilon B$ such that $x + hu \in K$.

(iii) For any sequence $h_n \to 0+$, there exists a sequence v_n converging to v such that $x + h_n v_n \in K$ for all n.

The closed cone $T_K^b(x)$ of the elements v of X satisfying one of these equivalent conditions is called the *contingent*   cone to K at x.

Exercise 6.11

Show that for all $x \in K$, we have

$$T_K(x) \subset T_K^b(x).$$

If $A \in L(X, Y)$, show that

$$AT_K^b(x) \subset T_{A(K)}(Ax).$$

Exercise 6.12

Let X be a finite-dimensional space. Let K be a closed subset of X.

(a) Suppose that $y \in K$. Show that for all $w \in T_K^b(x)$ (see Exercise 6.10),

$$D_+ d_K(y)(v) \le \|v - w\|.$$

(b) Suppose that $z \notin K$ and take $y \in K$ such that $\|y - z\| = d_K(z)$. Show that for all $w \in T_K^b(y)$

$$D d_K(z)(v) \le \|v - w\|$$

and thus, that

$$D d_K(z)(v) \le \inf_{\substack{\|y-z\|=d_K(z) \\ y \in K}} d(v, T_K^b(y)).$$

(c) We now suppose that for all $v \in T_K^b(x)$

the function $y \to d(v, T_K^b(y))$ is upper semi-continuous at x.   (∗)

Show that for all $\varepsilon > 0$, there exists $\eta > 0$ such that, for all $y \in B_K(x, \eta)$, $h \in]0, \eta]$, we have

$$\frac{d_K(y + hv)}{h} \le \varepsilon$$

Hint: set $f(t) := d_K(y + tv)$, deduce from the previous inequalities that for almost all t, in a neighbourhood of 0, $f'(t) \le d(v, T_K(z)) \le \varepsilon$ when $z \in K$ satisfies $d_K(y + tv) = \|y + tv - z\|$ and integrate from 0 to h.

(d) Deduce that if the regularity condition (∗) is satisfied at x, then

$$T_K(x) = T_K^b(x).$$

Exercise 6.13

Let ψ_K be the indicator function of K. Show that

$$D_+ \psi_K(x) = \psi_{T_K^b(x)}.$$

Exercise 6.14

Suppose that $f : X \to \mathbb{R} \cup \{+\infty\}$ is a nontrivial function. Show that

$$\mathrm{Ep}\, D_+ f(x) = T_{\mathrm{Ep}(f)}^b(x, f(x)).$$

Exercise 6.15

Suppose that $f : X \to \mathbb{R} \cup \{+\infty\}$ is a nontrivial function, Lipschitz in the neighbourhood of a point $x \in \mathrm{Int}(\mathrm{Dom}\, f)$. Show that

$$\mathrm{Ep}\, D_c f(x) = T_{\mathrm{Ep}(f)}(x, f(x))$$

and that

$$p \in \partial f(x) \Leftrightarrow (p, -1) \in N_{\mathrm{Ep}(f)}(x, f(x)).$$

Note that these properties may be used to extend $D_c f(x)$ and $\partial f(x)$ to any nontrivial function $f : X \to \mathbb{R} \cup \{+\infty\}$ and deduce that in this case we have the formula

$$D_c \psi_K(x) = \psi_{T_K(x)}.$$

Exercise 6.16

A set K is said to be *pseudo convex* at x if

$$T_K^b(x) = \text{closure}\left(\bigcup_{h>0} \tfrac{1}{h}(K - x)\right). \qquad (*)$$

(a) Prove that any convex set is pseudo convex at any point $x \in K$.

(b) If K is pseudo convex at x and if $A \in L(X, Y)$, show that

$$AT_K^b(x) = T_{A(K)}^b(Ax).$$

Compare with Exercise 6.11.

Exercise 6.17

A nontrivial function $f : X \to \mathbb{R} \cup \{+\infty\}$ is said to be pseudo convex at $x \in \mathrm{Dom} f$ if and only if

$$\forall y \in X, \quad D_+ f(x)(y - x) \leq f(y) - f(x). \qquad (*)$$

a Show that any convex function is pseudo convex at all $x \in \mathrm{Dom} f$.

(b) Show that f is pseudo convex at x if and only if

$$\mathrm{Ep}(f) \text{ is pseudo convex at } (x, f(x)). \qquad (**)$$

See Exercise 6.16.

Exercise 6.18

Suppose that X and Y are two Hilbert spaces. We consider a nontrivial function $g : X \times Y \to \mathbb{R} \cup \{+\infty\}$ and its marginal function f defined by

$$f(x) := \inf_{y \in Y} g(x, y).$$

Let $\bar{y}$ be a point at which the minimum is attained: $f(x) = g(x, \bar{y})$.

(a) Show that we have

$$D_+ f(x)(u) \leq \inf_{v \in Y} D_+ g(x, \bar{y})(u, v).$$

(b) Show that if g is pseudo convex at $(x, \bar{y})$ and that if the problem $f(x')$ has a solution for all x' in a neighbourhood of x, then

$$D_+ f(x)(u) = \inf_{v \in Y} D_+ g(x, \bar{y})(u, v).$$

Hint: use Exercise 6.17.

Exercise 6.19

We say that K is star shaped around $x \in K$ if and only if

$$\forall y \in K, \quad x + \theta(y - x) \in K \text{ for } \theta \in [0, 1].$$

Show that K is pseudo convex at x.

14.7 Exercises for Chapter 8 – Two-person Zero-sum Games: Theorems of Von Neumann and Ky Fan

The proofs of the following results are left as an exercise:

- Lemma 8.1

- Lemma 8.3

- Corollary 8.1

Exercise 8.1

Show that in Proposition 8.2 and its consequences (Theorems 8.1, 8.2 and 8.5) the assumption

$$E \text{ is compact} \tag{$*$}$$

may be replaced by

$$\exists y_1, \ldots, y_n \in F \text{ such that } x \to \max_{i=1,\ldots,n} f(x, y_i) \text{ is lower semi-compact.} \tag{$**$}$$

Exercise 8.2

We now consider n Hilbert spaces X_i, n subsets $E_i \subset X_i$ and n functions $f_i : E_i \times F \to \mathbb{R}$. We set

$$f(x, y) := \sum_{i=1}^{n} f_i(x_i, y).$$

We assume that

$$\forall i = 1, \ldots, n, \ \ \exists y_i \in F \text{ such that } x_i \to f_i(x_i, y_i) \text{ is lower semi-compact}$$
$$\text{and such that } \inf_{x_j \in E_j} f_j(x_j, y_i) > -\infty \ \forall j \neq i \tag{$*$}$$

and that

$$\forall y \in F, \ \forall i = 1, \ldots, n, \ \ x_i \to f_i(x_i, y) \text{ is lower semi-continuous.} \tag{$**$}$$

Show that there exist $\bar{x}_i \in F_i$ such that

$$\sup_{y \in F} \sum_{i=1}^{n} f_i(\bar{x}_i, y) = v^\natural$$

where

$$v^\natural = \sup_{K \in S} \inf_{x_i \in E_i} \max_{y \in K} \sum_{i=1}^{n} f_i(x_i, y).$$

Hint: use Exercise 8.1.

Exercise 8.3

We consider a *metric space* E, a subset F of a Hilbert space Y and a function $f : E \times F \to \mathbb{R}$. We use $\overline{D}$ to denote the canonical set-valued map from F to E defined by

$$\overline{D}(y) := \{\bar{x} \in E | f(\bar{x}, y) = \inf_{x \in E} f(x, y)\}.$$

We suppose that

(i) F is convex and compact
(ii) $\forall x \in E, \ y \to f(x, y)$ is concave and upper semi-continuous $\tag{$*$}$

and that

(i) E is compact
(ii) $\forall y \in F, \ x \to f(x, y)$ is lower semi-continuous. $\tag{$**$}$

(a) Show that there exists $\bar{y} \in F$ such that

$$f^\flat(\bar{y}) := \inf_{x \in E} f(x, \bar{y}) = v^\flat := \sup_y f^\flat(y).$$

(b) We take any $\bar{x}_\lambda$ in $D^b((1-\lambda)\bar{y}+\lambda y)$. Use the fact that f is concave with respect to y to deduce that

$$\forall \lambda \in [0,1], \quad f(\bar{x}_\lambda, y) \le f^b(\bar{y}).$$

(c) Suppose that, in addition, $\overline{D}$ is one-to-one and continuous. Deduce that there exists $\bar{x} \in E$ such that:

$$\forall y, \quad f(\bar{x}, y) \le f^b(\bar{y}) = v^b$$

and thus that $(\bar{x}, \bar{y})$ is a saddle point of f.

(d) Show that we obtain the same conclusion if we simply assume that $y \to D^b(y)$ is lower semi-continuous with non-empty values.

Note that we have proved another minimax theorem where the convexity assumptions on E and on $x \to f(x,y)$ are replaced by the assumption that $\overline{D}$ is lower-semi continuous.

Exercise 8.4

Suppose that f is a nontrivial convex function from X to $\mathbb{R} \cup \{+\infty\}$ and that K is a convex, closed, bounded subset of X^* such that

$$\forall x \in X, \quad f(x_0) \le f(x) + \sigma(K, x_0 - x).$$

Show that $\partial f(x_0)$ is non-zero.

Hint: consider the function $\varphi(x,p) := f(x) + \langle p, x_0 - x \rangle$ on $\mathrm{Dom} f \times K$ and the fact that when X is not finite dimensional, convex, closed bounded sets are weakly compact.

Compare with Exercise 4.7

Exercise 8.5

Let X be a Hilbert space. Consider the family $\mathcal{K}$ of non-empty, convex, closed subsets of X. We define the Hausdorff semi-distance by

$$d\!L(K,L) := \sup_{x \in K} \inf_{y \in L} \|x - y\| = \sup_{x \in K} d(x, L).$$

(a) Show that

$$d\!L(K,L) \le d\!L(K,M) + d\!L(M,L).$$

(b) Show that

$$d\!L(K,L) = 0 \Leftrightarrow K \subset L.$$

(c) Prove that

$$\sup_{\|p\|_* \le 1} (\sigma_K(p) - \sigma_L(p)) = d\!L(K,L).$$

Hint: if X is infinite dimensional, use the weak topology.

Exercise 8.6

Consider two convex compact subsets $E \subset X$ and $F \subset Y$ and a function $f : E \times F \to \mathbb{R}$. Suppose that $p \in X^*$ and $q \in Y^*$ are two continuous linear forms on X and Y, respectively. We set

$$v^\#(p, q) := \inf_{x \in E} \sup_{y \in F}(f(x, y) - \langle p, x \rangle + \langle q, y \rangle)$$

and

$$v^\flat(p, q) := \sup_{y \in F} \inf_{x \in E}(f(x, y) - \langle p, x \rangle + \langle q, y \rangle).$$

(a) Suppose that y_0 is a max inf:

$$\inf_{x \in E}(f(x, y_0) - \langle p_0, x \rangle + \langle q_0, y_0 \rangle) = v^\#(p_0, q_0).$$

   Show that

$$v^\#(p_0, q_0) - v^\#(p_0, q) \leq \langle q_0 - q, y_0 \rangle. \qquad (*)$$

   (in other words that $y_0 \in \partial v^\#(p_0, \cdot)(q_0)$) and that

$$\inf_{x \in E}(f(x, y_0) - \langle p_0, x \rangle) = \inf_{q \in Y^*}(v^\#(p_0, q) - \langle q, y_0 \rangle). \qquad (**)$$

(b) Show that if y_0 satisfies $(*)$ and $(**)$, then y_0 is a max inf.

(c) Suppose that $\forall x \in E$, $y \to f(x, y)$ is concave and upper semi-continuous. Show that the condition $(**)$ is satisfied and thus deduce from (b) that if y_0 satisfies $(*)$ ($y_0 \in \partial v^\#(p_0, \cdot)(q_0)$) then y_0 is a max inf.

(d) Similarly, show that y_0 is a conservative strategy for Frances if and only if

   (i) $y_0 \in \partial v^\flat(p_0, \cdot)(q_0)$
   (ii) $\inf_{x \in E}(f(x, y_0) - \langle p_0, y \rangle) = \inf_q(v^\flat(p_0, q) - \langle q, y_0 \rangle).$

Exercise 8.7

Suppose that K is a subset of a Hilbert space X defined by a family of constraints

$$K := \{x \in X | \forall p \in P, \; \gamma(x, p) \leq 0\}.$$

Suppose also that

(i) P is a convex closed cone in a Hilbert space Z,
(ii) $\gamma : X \times P \to \mathbb{R}$ is positively homogeneous in p. $\qquad (*)$

We set

$$f_K(x) := \begin{cases} f(x) & \text{if } x \in K \\ +\infty & \text{otherwise} \end{cases}$$

and

$$\ell(x, p) := f(x) + \gamma(x, p)$$

(the Lagrangian for the problem).

(a) Show that

$$f_K(x) := \sup_{p \in P} \ell(x, p)$$

and that

$$\inf_{x \in K} f(x) = \inf_{x \in X} \sup_{p \in P} \ell(x, p).$$

(b) $\bar{p} \in P$ is said to be a Lagrange multiplier if and only if

$$\inf_{x \in X} \sup_{p \in P} \ell(x, p) = \inf_{x \in X} \ell(x, \bar{p})$$

($\bar{p}$ is a max inf of the Lagrangian).

Show that $\bar{p} \in P$ is a Lagrange multiplier and that $\bar{x} \in X$ minimises f over K if and only if

(i) $\forall x \in X, \quad f(\bar{x}) + \gamma(\bar{x}, \bar{p}) \leq f(x) + \gamma(x, \bar{p})$
   ($\bar{x}$ minimises $x \to f(x) + \gamma(x, \bar{p})$ over X).
(ii) $\forall p \in P, \quad \gamma(\bar{x}, p) \leq 0$ ($\bar{x}$ belongs to K), and
(iii) $\gamma(\bar{x}, \bar{p}) = 0$.

Exercise 8.8

Suppose that X and Y are two Hilbert spaces and $A \in L(X, Y)$ and that $f : X \to \mathbb{R} \cup \{+\infty\}$ and $g : Y \to \mathbb{R} \cup \{+\infty\}$ are two nontrivial, convex, lower semi-continuous functions such that $0 \in A\text{Dom } f - \text{Dom } g$.

(a) Show that

$$\inf_{x \in X} (f(x) + g(Ax)) = \inf_{x \in X} \sup_{q \in Y^*} \ell(x, q)$$

where

$$\ell(x, q) := f(x) + \langle q, y \rangle - g^*(q).$$

(b) Deduce that the following conditions are equivalent

$$\inf_{x \in X} (f(x) + g(Ax)) = \inf_{x \in X} \ell(x, \bar{q})$$

and

$$\inf_{x \in X} (f(x) + g(Ax)) + f^*(-A^*\bar{q}) + g^*(\bar{q}) = 0.$$

14.8 Exercises for Chapter 9 – Solution of Nonlinear Equations and Inclusions

The proofs of the following results are left as an exercise:

- Proposition 9.1

- Proposition 9.2

- Proposition 9.3

- Proposition 9.4

- Proposition 9.5

- Corollary 9.1

- Theorem 9.5

- Theorem 9.6

- Theorem 9.7

Exercise 9.1

Let C be a set-valued map which is upper hemi-continuous at x_0 and bounded in a neighbourhood of x_0 in the sense that

$$\sup_{x \in B(x_0, \eta)} \sup_{v \in C(x)} \|v\| < +\infty.$$

Show that

$$(x, p) \to \sigma(C(x), p)$$

is upper semi-continuous at (x_0, p_0) for all p_0.

Exercise 9.2

Suppose that C is a set-valued map from K to $\mathbb{R}^n$ which is upper semi-continuous at x_0. Show that if $C(x_0)$ is bounded, then $(x, p) \to \sigma(C(x), p)$ is upper semi-continuous at (x_0, p_0) for all p_0.

Exercise 9.3

Consider three metric spaces X, Y and U and a mapping $f : X \times U \to Y$ with an associated set-valued map F from X to Y defined by

$$F(x) := \{f(x, u)\}_{u \in U}.$$

(a) Show that if

$$\forall u \in U, \quad x \to f(x, u) \text{ is continuous}$$

then F is lower semi-continuous.

(b) Show that if

    (i) U is compact

    (ii) f is continuous from $X \times U$ to Y

then F is upper semi-continuous.

Exercise 9.4

Consider a metric space K, two Hilbert spaces X and Y, set-valued maps $x \to L(x) \subset X$ and $x \to M(x) \subset Y$ with convex closed values and a mapping $x \to A(x) \in L(X, Y)$. We define the set-valued map

$$R(x) := \{u \in L(x) | A(x)u \in M(x)\}.$$

(a) Show that if

$$\forall x \in K, \quad 0 \in \text{Int}(A(x)L(x) - M(x)),$$

then

$$\sigma(R(x), p) = \inf_{q \in Y^*} (\sigma(L(x), p - A(x)^* q) + \sigma(M(x), q)).$$

(b) Deduce that if $x \to A(x) \in L(X, Y)$ is continuous, if the set-valued maps are upper hemi-continuous and if L is bounded in the neighbourhood of each point, then R is upper hemi-continuous.

Exercise 9.5

Suppose that X is a finite-dimensional space and that C is a set-valued map from the unit ball $B \subset X$ to X with non-empty, convex, closed values. We suppose that

$$\forall x \in X, \quad \|x\| = 1, \quad \sigma(F(x), x) \geq 0.$$

Show that there exists a solution $\bar{x} \in B$ of $0 \in F(\bar{x})$.

    Hint: apply Theorem 8.5 to the function φ defined on $B \times Y$ by $\varphi(x, y) := -\sigma(C(x), y)$ and for the continuous mapping $r : Y \to B$ defined by

$$r(y) := y \text{ if } y \in B, \qquad r(y) := y/\|y\| \text{ if } y \notin B.$$

Exercise 9.6

Suppose that K is a convex compact subset of a Hilbert space X and that C is a set-valued map from K to X which is upper hemi-continuous with convex closed values. Suppose that π_K is the projector of best approximation onto K. Show that if

$$\forall x \in X, \ \sigma(C(\pi_K(x)), x) \geq 0$$

then there exists a unique solution $\bar{x} \in K$ of the inclusion $0 \in F(\bar{x})$. Hint: the proof is similar to that of Exercise 9.5.

Exercise 9.7

We define the *zone of support* of a convex closed subset $K \subset X$ at $p \in X^*$ to be the set

$$\partial \sigma_K(p) := \{x \in K | \langle p, x \rangle = \sigma_K(p)\}.$$

Suppose that C is a set-valued map from K to X which is upper hemi-continuous with convex closed values. Show that the condition

$$\forall p \in X^*, \ \forall x \in \partial \sigma_K(p), \ \sigma(C(x), -p) \geq 0 \qquad (*)$$

is equivalent to

$$\forall x \in K, \ \forall p \in N_K(x), \ \sigma(C(x), -p) \geq 0. \qquad (**)$$

Exercise 9.8

Suppose that X and Y are metric spaces. Let $F : X \to Y$ be a lower semi-continuous set-valued map. Show that $x \to \overline{F(x)}$ is also lower semi-continuous.

Exercise 9.9

Suppose that X is a metric space and that Y is a normed space. Let $F : X \to Y$ be a lower semi-continuous set-valued map with convex values. Show that $\forall \varepsilon > 0$, there exists a continuous function f from X to Y such that

$$\forall x \in X, \ d(f(x), F(x)) \leq \varepsilon.$$

Hint: use continuous partitions of unity.

Exercise 9.10

Show that the graph of an upper semi-continuous set-valued map with closed values is closed.

Exercise 9.11

Show that if F is upper semi-continuous with compact values, then the image of any compact set under F is compact.

Exercise 9.12

Suppose that K is a convex compact subset of a Hilbert space X and that C is a set-valued map from K to X with non-empty, convex, closed values.

(a) Show that a necessary and sufficient condition for the existence for all $x \in K$ of an 'explicit' solution of the dynamical system

(i) $$x_0 = x$$
(ii) $$x_{n+1} - x_n \in C(x_n), \quad n = 0, 1, \dots$$

which is 'viable' in the sense that

$$\forall n \geq 0, \quad x_n \in K$$

is that

$$\forall x \in K, \quad \sigma(C(x), -p) \geq \langle p, x \rangle - \sigma_K(p). \tag{$*$}$$

Hint: use the Separation Theorem.

(b) Note that the condition

$$\forall x \in K, \quad \forall p \in b(K), \quad \sigma(C(x), -p) \geq 0$$

implies both the condition $(*)$ and the condition

$$\forall x \in K, \quad \forall p \in N_K(x), \quad \sigma(C(x), -p) \geq 0.$$

Exercise 9.13

Let K be a convex compact subset of a Hilbert space X and C an upper hemi-continuous set-valued map from K to X with non-empty, convex, closed values. We suppose that

$$\forall x \in K, \quad \forall p \in N_K(x), \quad \sigma(C(x), -p) \geq 0.$$

Show that for all $x \in K$ it is possible to construct an 'implicit' solution of the dynamical system

(i) $$x_0 = x$$
(ii) $$x_{n+1} - x_n \in C(x_n), \quad n = 0, 1, \dots$$

which is 'viable' in the sense that

$$\forall n \geq 0, \quad x_n \in K.$$

Compare with Exercise 9.12. Hint: use Theorem 9.4.

Exercise 9.14

Consider a convex compact subset K of X and an upper hemi-continuous set-valued map from K to Y with convex closed values. Consider also a continuous mapping $x \in K \to A(x) \in L(X, Y)$. We suppose that

$$\forall x \in K, \quad C(x) \cap \text{closure}(A(x)T_K(x)) \neq \emptyset.$$

Modify the proof of Theorem 9.3 to show that

(i) $\exists \bar{x} \in K, \quad 0 \in F(\bar{x})$

(ii) $\forall y \in K, \quad \exists \hat{x} \in K$ such that $A(\hat{x})(\hat{x} - y) \in F(\hat{x})$.

Exercise 9.15

Let K be a convex closed subset of $\mathbb{R}^n$. Let C be an upper hemi-continuous set-valued map from K to $\mathbb{R}^n$ with non-empty, convex, closed values. We suppose that

$$C(K) := \cup_{x \in K} C(x)$$

is contained in a compact subset M of K. Show that C has a fixed point. Hint: consider the restriction of C to the convex closure of M.

Exercise 9.16

Suppose that C is a set-valued map from the unit ball B to itself satisfying

$$\forall x \in B, \quad \|x\| = 1, \quad -x \in C(x).$$

Show that

(i) $\exists \bar{x} \in B, \quad \bar{x} \in C(\bar{x})$

(ii) $B \subset C(B).$

Hint: apply Theorem 9.4 to $C(x) - x$ and $C(x) - y$.
    If C is one-to-one, we obtain Borsuk's Antipodal Theorem.

Exercise 9.17

Suppose that K is a convex compact subset of X. Consider an upper hemi-continuous set-valued map C from K to X with convex closed values satisfying

$$\forall p \in X^*, \quad \forall x \in \partial \sigma_K(p), \quad C(x) \cap \partial \sigma_K(p) \neq \emptyset.$$

Show that

(i) $\exists \bar{x} \in K$ such that $\bar{x} \in C(\bar{x})$

(ii) $\forall y \in K, \quad \exists \hat{x} \in K$ such that $y \in C(\bar{x}).$

Hint: apply Theorem 9.4 to the set-valued map $x \to x - C(x)$.

Exercise 9.18

Let X be a Hilbert space, $A \in L(X, X^*)$ an X-elliptical operator satisfying

$$\exists c > 0 \text{ such that } \forall x \in X, \quad \langle Ax, x \rangle \geq c\|x\|^2 \qquad (*)$$

and K a non-empty, convex, closed subset of X. We consider the following problem: if $p \in X^*$, find $\bar{x}$ such that

(i) $\qquad\qquad\qquad\qquad \bar{x} \in K$
(ii) $\qquad\qquad \forall y \in K, \quad \langle A\bar{x} - p, \bar{x} - y \rangle \leq 0. \qquad (**)$

(a) Show that there is at most one solution $\bar{x}$ of $(**)$.

(b) If π_K denotes the projector of best approximation onto K, show that $(**)$ is equivalent to

$$\bar{x} = \pi_K(\bar{x} + \lambda L^{-1}(A\bar{x} - p)) \qquad (***)$$

where L is the duality operator and λ is an arbitrary positive scalar.

(c) Show that

$$\|(1 - \lambda L^{-1}A)x\|^2 \leq (1 + \lambda^2\|A\|^2 - 2\lambda c)\|x\|^2.$$

(d) For a judicious choice of $\lambda > 0$, deduce from the Banach–Picard Theorem (Theorem 1.5) that $(***)$ has a fixed point and thus that there exists a unique solution of the variational inequality $(***)$.

(e) If $G(p)$ denotes this unique solution, show that

$$\|G(p) - G(q)\| \leq c^{-1}\|p - q\|_*.$$

(f) Deduce that $A \in L(X, X^*)$ is an isomorphism.

14.9 Exercises for Chapter 10 – Introduction to the Theory of Economic Equilibrium

The proof of the following result is left as an exercise:

- Corollary 10.1

Exercise 10.1

Consider n companies j described by production sets Z_j. We suppose that each consumer has a vector w_i of initial resources and 'shares' θ_{ij} of the yield for the company j such that $\sum_{i=1}^{n} \theta_{ij} = 1$ for all j. Show that the set M of available commodities is

$$M := \sum_{i=1}^{n} w_i + \sum_{j=1}^{m} Z_j - \mathbb{R}_+^{\ell}$$

and that

$$r_i(p) = \langle p, w_i \rangle + \sum_{j=1}^{m} \theta_{ij} \sigma(Z_j, p).$$

Exercise 10.2

We assume the framework of Exercise 10.1. We consider a Walrasian equilibrium $\bar{p} \in M^{\ell}$ and the vector $\bar{y} := \sum_{i=1}^{n} \bar{x}_i$ of associated demands, which may be written as

$$\bar{y} = \sum_{i=1}^{n} w_i + \sum_{j=1}^{m} \bar{z}_j \text{ where } \bar{z}_j \in Z_j.$$

Show that for each company $j = 1, \ldots, m$,

$$\langle \bar{p}, \bar{z}_j \rangle = \max_{z \in Z_j} \langle \bar{p}, z \rangle$$

or, in other words, that each production $\bar{z}_j$ associated with the Walrasian equilibrium price maximises the profit $\langle \bar{p}, z \rangle$ over the net production set Z_j.

Exercise 10.3

The space of commodities $\mathbb{R}^{\ell}$ has a scalar product $\lambda(x, y)$ and we consider the associated quadratic function $\lambda^2(x) := \lambda(x, x)$ and its duality operator L. We suppose that each consumer i tries his utmost to approximate to an ideal consumption u_i subject to the financial constraint $\langle p, x \rangle = r$. Show that the demand is then equal to

$$D_i(p, x) := u_i - \frac{\langle p, u_i \rangle - r}{\lambda_*(p)^2} L^{-1} p.$$

Exercise 10.4

We assume the framework of Exercise 10.3. We denote the initial resources of each consumer by $w_i \in \mathbb{R}^{\ell}$ and their ideal consumptions by $u_i \in \mathbb{R}^{\ell}$. Show that there exists a Walrasian equilibrium price, which is unique up to a non-zero scalar α, given by

$$\bar{p} = \alpha L \left(\sum_{j=1}^{n} u_j - w_j \right)$$

and that the associated demands are equal to

$$\bar{x}_i = u_i - \frac{\lambda(u_i - w_i, \sum_{j=1}^{m} u_j - w_j)}{\lambda^2(\sum_{j=1}^{m}(u_j - w_j))} \sum_{j=1}^{m}(u_j - w_j).$$

Exercise 10.5

We assume the framework of Exercise 10.4, where an appropriate institution wishes to redistribute the yields so that the Walrasian equilibrium mechanism provides consumptions as close as possible to $(v_1, \ldots, v_n)$ chosen in advance. In other words, if $w := \sum_{i=1}^{n} w_i$ and $r_i w$ is the new yield attributed to consumer i, find $\bar{r} \in \mathbb{R}^n$ such that

(i)
$$\sum_{i=1}^{n} \bar{r}_i = 1$$

(ii)
$$\sum_{i=1}^{n} \lambda^2 (D_i(\bar{p}, \bar{r}_i \langle \bar{p}, w \rangle) - v_i) = \inf_{\sum_{i=1}^{n} r_i = 1} \lambda^2 (D_i(\bar{p}, r_i \langle \bar{p}, w \rangle) - v_i)$$

where $\bar{p} = L(\sum_{i=1}^{n}(u_j - w_j))$. Show that

$$\bar{r}_i = \frac{\langle \bar{p}, v^i \rangle - \frac{1}{n} \langle \bar{p}, \sum_{i=1}^{n}(v_i - w_i) \rangle}{\langle \bar{p}, \sum_{i=1}^{n} w_i \rangle}.$$

Exercise 10.6

Suppose we have a consumption set L and a set of resources M such that

$$M = M_0 - \mathbb{R}_+^{\ell} \quad \text{where } M_0 \text{ is compact.}$$

(a) Show that $p \to r(p) = \sup_{y \in M} \langle p, y \rangle$ is continuous on $\mathbb{R}_+^{\ell}$. Deduce that the graph of the budget set-valued map $(p, x) \to B(p, x)$ defined by

$$B(p, r) := \{x \in L | \langle p, x \rangle \leq r(p)\}$$

is closed. Deduce from Problem 1 (Chapter 15) that if

$$L \text{ is compact} \qquad\qquad (*)$$

then B is upper semi-continuous.

(b) We suppose that

(i) L is convex
(ii) $\forall p \in \mathbb{R}_+^\ell, \ \exists x \in L$ such that $\langle p, x \rangle < r(p)$. $(**)$

Show that the set-valued map $(p, x) \to B(p, r)$ is upper semi-continuous.

(c) We consider a continuous function f from L to $\mathbb{R}$ and we construct the following set-valued demand map

$$D(p, r) := \{\bar{x} \in B(p, r) | f(\bar{x}) = \inf_{x \in B(p,r)} f(x)\}.$$

Deduce from Problem 6 (Chapter 15) that the set-valued demand map $D(p, r)$ is upper semi-continuous when the assumptions $(*)$ and $(**)$ are satisfied.

Exercise 10.7

We suppose that each consumer is described by

(i) a convex compact consumption set L_i;

(ii) a convex continuous loss function $f_i : L_i \to \mathbb{R}$;

(iii) a set of initial resources M_i which is convex, compact, has a non-empty interior and satisfies $L_i \cap \operatorname{Int} M_i \neq \emptyset$.

We consider the yield functions $r_i(p) := \sigma(M_i, p)$ and the set-valued demand maps

$$D_i(p, r) := \{\bar{x} \in B_i(p, x) | f_i(\bar{x}) = \inf_{x \in B_i(p,r)} f_i(x)\}$$

where

$$B_i(p, r) := \{x \in L_i | \langle p, x \rangle \leq r\}.$$

Let $M := \sum_{i=1}^n M_i - \mathbb{R}_+^\ell$. Deduce from Exercise 10.6 and Theorem 10.1 that there exist $\bar{p} \in M^\ell$ and $\bar{x}_i \in D_i(\bar{p}, r_i(\bar{p}))$ such that

$$\sum_{i=1}^n \bar{x}_i \in M.$$

Exercise 10.8

We assume the framework of Theorem 10.2. Let $h > 0$. Show that we may associate any initial allocation x_0 with a sequence of allocations $(x_1^t, \ldots x_n^t)$ and prices $p^t \in M^\ell$ such that, for all $i = 1, \ldots, n$,

$$x_i^t - x_i^{t-1} = hc_i(x_i^t, p^t), \quad t = 1, \ldots, \tag{1}$$

which are 'viable' in the sense that

$$\forall t = 1, \ldots, \quad x_i^t \in L_i \text{ and } \sum_{i=1}^n x_i^t \in M. \tag{2}$$

Exercise 10.9

We assume the framework of Exercise 10.8. Show that if $(x_1^t \ldots, x_n^t, p^t)$ is a solution of the problem (1), (2), then

$$\sum_{i=1}^n c_i(x_i^t, p^t) \in T_M \left(\sum_{i=1}^n x_i^{t-1} \right), \quad t = 1, \ldots,$$

Exercise 10.10

We assume the framework of Theorem 10.2, with the assumption (28)(ii) replaced by

$$\forall q \in b(L_i), \quad \forall p \in M, \quad \langle q, c_i(x,p) \rangle \le 0.$$

Let $h > 0$. Show that we may associate any initial allocation x_0 with a sequence of allocations $(x_1^t \ldots, x_n^t)$ and prices $p^t \in M^\ell$ such that, for all $i = 1, \ldots, n$,

$$x_i^t - x_i^{t-1} = hc_i(x_i^{t-1}, p^t), \quad t = 1, \ldots, n \tag{3}$$

which are viable in the sense that

$$\forall t = 1, \ldots, n, \quad x_i^t \in L_i \text{ and } \sum_{i=1}^n x_i^t \in M. \tag{4}$$

Hint: use Exercise 9.12 or prove this exercise directly from the Separation Theorem. Compare with Exercise 10.8.

Exercise 10.11

We assume the framework of Exercise 10.10. Show that if $(x_1^t, \ldots, x_n^t, p^t)$ is a solution of the problem (3), (4), then

$$\sum_{i=1}^n c_i(x_i^{t-1}, p^t) \in T_M \left(\sum_{i=1}^n x_i^{t-1} \right). \tag{5}$$

14.10 Exercises for Chapter 11 – The Von Neumann Growth Model

The proofs of the following results are left as an exercise:

- Corollary 11.1

- Theorem 11.5

- Theorem 11.6

See also Problem 44.

14.11 Exercises for Chapter 12 – n-person Games

The proofs of the following results are left as an exercise:

- Proposition 12.1

- Theorem 12.2

- Proposition 12.2

- Theorem 12.3

- Proposition 12.3

- Proposition 12.4

Exercise 12.1

Show that Proposition 12.1 remains true if we replace the function φ defined by (8) by the function

$$\psi(x, y) := \sup_{i=1,\dots,n} \left(f_i(x_i, x_i) - f_i(y^i, x_i) \right).$$

Find other examples of functions φ for which Proposition 12.1 remains true.

Exercise 12.2

We consider n nontrivial functions $f_i : \prod_{i=1}^{n} X_i \to \mathbb{R} \cup \{+\infty\}$ such that

$$\forall i, \quad y_i \to f_i(y_i, x_i) \text{ is convex and lower semi-continuous.}$$

Show that $x = (x_1, \dots, x_n)$ is a non-cooperative equilibrium if and only if

$$\forall i = 1, \dots, n, \quad 0 \in \partial f_i(\cdot, x_i)(x_i)$$

or, if and only if

$$0 \in \prod_{i=1}^{n} \partial f_i(\cdot, x_{\hat{i}})(x_i)$$

(Fermat's rule). Here, $\partial f_i(\cdot, x_{\hat{i}})(x_i)$ denotes the subdifferential of $y_i \to f_i(\cdot, x_{\hat{i}})$ at x_i.

Deduce that the non-cooperative equilibria are the fixed points of the set-valued map

$$x \to \prod_{i=1}^{n} \partial f_i^*(\cdot, x_{\hat{i}})(0).$$

Exercise 12.3

We assume the framework of Exercise 12.2. We consider n convex closed subsets $E_i \subset X_i$ such that

$$\forall i, \quad \forall x_{\hat{i}}, \quad 0 \in \mathrm{Int}(K_i - \mathrm{Dom}\, f_i(\cdot, x_{\hat{i}})).$$

Show that $x = (x_1, \ldots, x_n) \in \prod_{i=1}^{n} E_i$ is a non-cooperative equilibrium if and only if

$$\forall i = 1, \ldots, n, \quad 0 \in \partial f_i(\cdot, x_{\hat{i}})(x_i) + N_{E_i}(x_i).$$

Deduce that if $y_i \to f_i(y_i, x_{\hat{i}})$ is differentiable (Gâteaux differentiable) at all $x_i \in E_i$ for all $x_{\hat{i}} \in E_{\hat{i}}$, then the non-cooperative equilibria are the solutions of the variational inequalities

(i) $$\forall i = 1, \ldots, n, \quad x_i \in E_i$$

(ii) $$\sum_{i=1}^{n} \langle \nabla_i f_i(x_i, x_{\hat{i}}), x_i - y_i \rangle \leq 0$$

where $\nabla_i f_i(x_i, x_{\hat{i}})$ denotes the gradient of $y_i \to f_i(y_i, x_{\hat{i}})$ at x_i.

Exercise 12.4

We assume we are in the framework of Exercise 12.2. We introduce n other Hilbert spaces Y_i, n continuous linear operators $A_i \in L(X_i, Y_i)$ and n nontrivial, convex, lower semi-continuous functions $g_i : Y_i \to \mathbb{R} \cup \{+\infty\}$. We consider the dual problem

$$v_{i*}(x_{\hat{i}}) := \inf_{q \in Y_i^*} \left(\sum_{i=1}^{n} f_i^*(-A_i^* q, x_{\hat{i}}) + g_i^*(q) \right). \qquad (*)$$

(a) Show that

$$v_{i*}(x_{\hat{i}}) + f_i(y_i, x_{\hat{i}}) + g_i(A_i y_i) \geq 0.$$

(b) Show that if $\bar{q}_i$ is a solution of the dual problem $v_{i*}(x_i)$ for all i and if

$$\forall i = 1, \ldots, n, \quad v_{i*}(x_i) + f_i(x_i, x_i) + g_i(A_i x_i) = 0 \qquad (**)$$

then $(x_1, \ldots, x_n)$ is a non-cooperative equilibrium for the loss functions $x \to f_i(x) + g_i(A_i x)$.

(c) Show that in this case, we have the relations

$$\forall i = 1, \ldots, n, \quad 0 \in \partial f_i(\cdot, x_i)(x_i) + A_i^* \partial g_i(A_i x_i)$$

and that $(x_1, \ldots, x_n, \bar{p}_1, \ldots, \bar{p}_n)$ is a solution of the system

(i) $0 \in \partial f_i(\cdot, x_i)(x_i) + A_i^* p_i$
(ii) $0 \in -A_i x_i + \partial g_i^*(p_i).$

Exercise 12.5

We consider n finite-dimensional spaces X_i, their product $X = \prod_{i=1}^{n} X_i$, an operator $M \in L(X, X)$ defined by $(Mx)_j = \sum_{k=1}^{m} M_{jk} x_k$ where $M_{jk} \in L(X_k, X_j)$, $M_{ii} = \mathrm{Id}$, n scalar products $\lambda_i(x, y)$ and their duality operators $L_i \in L(X_i, X_i^*)$, an operator $A \in L(X, Y)$ defined by $Ax = \sum_{j=1}^{n} A_j x$ where $A_j \in L(X_j, Y)$ and $u = (u_1, \ldots, u_n) \in X$. We make the following assumption

$$\forall x \neq 0, \quad \langle Lx, Mx \rangle = \sum_{j=1}^{n} \sum_{k=1}^{n} \langle L_j x_j, M_{jk} x_k \rangle > 0. \qquad (*)$$

(a) Show that $(*)$ implies that M is invertible.

(b) Show that if A is surjective then $AM^{-1}L^{-1}A^*$ is invertible.

(c) Prove that $\bar{x} = (\bar{x}_1, \ldots, \bar{x}_n) \in X$ defined by

$$\bar{x} = M^{-1}(u - L^{-1}A^*\bar{p}),$$

where

$$\bar{p} = (AM^{-1}L^{-1}A^*)^{-1}(AM^{-1}u - w),$$

is the solution of the following problem

(i)
$$\sum_{i=1}^{n} A_i \bar{x}_i = w$$

(ii)
$$\sum_{j=1}^{n} \lambda_j^2 \left(\bar{x}_j + \sum_{k \neq j} M_{jk} \bar{x}_k - u_j \right)$$

$$= \inf_{\sum_{i=1}^{n} A_i x_i = w} \sum_{j=1}^{n} \lambda_j^2 \left(x_j + \sum_{k \neq j} M_{jk} \bar{x}_k - u_j \right).$$

Exercise 12.6

We assume we are in the framework of Exercise 12.5. We consider n convex closed subsets $K_i \subset X_i$. Show that $(*)$ implies that there exists a unique non-cooperative equilibrium $x \in \prod K_i$ for the functions

$$x \to \lambda_j^2 \left(x_j + \sum_{k \neq j} M_{jk} x_k - u_i \right).$$

Hint: use Exercise 9.18.

Exercise 12.7

We consider

(i) n convex compact subsets $L_i \subset \mathbb{R}^\ell$

(ii) n convex compact subsets $M_i \subset \mathbb{R}^\ell$

such that

$$\forall i = 1, \ldots, n, \ \ 0 \in L_i - M_i + \mathring{\mathbb{R}}_+^\ell.$$

We consider n continuous functions

$$f_i : \prod X_i \times \mathbb{R}_+ \to \mathbb{R}$$

such that

$$\forall p, \ \forall x_{\hat{i}}, \ \ x_i \to f_i(x_i, x_{\hat{i}}, p) \ \text{ is convex.}$$

We set $r_i(p) := \sigma(M_i, p)$ and $r(p) := \sum_{i=1}^n r_i(p)$. Show that there exist $\bar{x}_1, \ldots, \bar{x}_n, \bar{p}$ such that

(i) $\displaystyle\sum_{i=1}^n \bar{x}_i \in \sum_{i=1}^n M_i$

(ii) $\forall i = 1, \ldots, n, \ \ \bar{x}_i \in B_i(\bar{p}, r_i(\bar{p}))$

(iii) $\forall i = 1, \ldots, n, \ \ f_i(\bar{x}_i, \bar{x}_{\hat{i}}, \bar{p}) = \min\{f_i(y_i, \bar{x}_{\hat{i}}, \bar{p}) | y_i \in B_i(\bar{p}, r_i(\bar{p}))\}.$

Hint: use Theorem 12.3 applied to an $(n+1)$-person game where the $(n+1)$th player is the 'market' which has loss function

$$f_0(x, p) = r(p) - \left\langle p, \sum_{i=1}^n x_i \right\rangle.$$

Exercise 12.8

We consider

(i) n subsets $L_i \subset X_i$;

(ii) a convex subset $M \subset Y$, where Y is finite dimensional;

(iii) n continuous linear operators $A_i \in L(X_i, Y)$;

(iv) n functions $f_i : L_i \to \mathbb{R}$.

Suppose we have $\bar{\lambda} \in M^n$, $\bar{p} \in b(M)$ (barrier cone of M) and $\bar{x} \in \prod_{i=1}^n L_i$ satisfying

(i) $\forall i = 1, \ldots, n$, $\bar{\lambda}_i f_i(\bar{x}_i) + \langle \bar{p}, A_i \bar{x}_i \rangle = \inf\{\bar{\lambda}_i f_i(x_i) + \langle \bar{p}, A_i x_i \rangle | x_i \in L_i\}$

(ii) $\displaystyle\sum_{i=1}^n \langle \bar{p}, A_i \bar{x}_i \rangle = \sigma_M(p).$

We set

$$K := \left\{ x \in \prod_{i=1}^n L_i | \sum_{i=1}^n A_i x_i \in M \right\}.$$

(a) Show that $(*)$ implies that $\bar{x} \in K$ is a Pareto optimum for the f_i on K.

(b) We set

$$\phi(x, y) := \left(f_1(x), \ldots, f_n(x), \sum_{i=1}^n A_i x_i - y \right)$$

where $\phi : L_i \times M \to \mathbb{R}^n \times Y$.

Show that $\bar{x}$ is a Pareto optimum on K if and only if

$$(\mathbf{f}(\bar{x}), 0) \notin \phi \left(\prod_{i=1}^n L_i \times M \right) + \mathring{\mathbb{R}}_+^n \times \{0\}.$$

(c) We now suppose that

(i) the sets L_i and M are convex;

(ii) the functions f_i are convex.

Show that there exist $\bar{\lambda} \in \mathbb{R}_+^n$ and $\bar{p} \in Y^*$ such that

(i) $(\bar{\lambda}, \bar{p}) \neq 0$

(ii) $\displaystyle\sum_{i=1}^n \bar{\lambda}_i f_i(\bar{x}_i) = \inf \left(\sum_{i=1}^n \bar{\lambda}_i f_i(x_i) + \langle \bar{p}, A_i x_i \rangle - \sigma_M(p) \right).$

(d) Show that the condition

$$0 \in \mathrm{Int} \left(\sum_{i=1}^n A_i(L_i) - M \right)$$

enables us to take $\bar{\lambda} \in M^n$ and thus to show that $(*)$ holds.

Exercise 12.9

We consider a cooperative game without side payments (see Definition 12.2) defined by subsets $V(S)$. For all $\lambda \in \mathbb{R}^n_+$, we set

$$w(S, \lambda) := \inf_{v \in V(S)} \langle \lambda, v \rangle.$$

We also set

$$M^T := \{\lambda \in M^n | \lambda_i = 1 \ \forall i \notin T\}$$

and

$$\alpha(r) := \sup_{S \in N} \inf_{\lambda \in M^S} (\langle \lambda, c_S \cdot r \rangle - w(S, \lambda)).$$

(a) Show that r is in the core of the game if and only if

$$r \in V(N) \ \text{ and } \ \alpha(r) \leq 0.$$

(b) Consider the set

$$\mathcal{O} := \left\{ \bar{r} \in V(N) | \alpha(\bar{r}) = \inf_r \{\alpha(r), r \in V(N)\} \right\}.$$

Show that if

$$V(N) \text{ is closed and bounded below,} \qquad (*)$$

then $\mathcal{O}$ is non-empty and compact. Show that $\mathcal{O}$ is always contained in the set of Pareto optima and in the core when the latter is non-empty.

Exercise 12.10

We assume we are in the framework of Exercise 12.9. We set

$$\beta(r) := \inf_{\lambda \in M^n} \sup_{S \subset N} (\langle \lambda, c_S \cdot r \rangle - w(S, \lambda)).$$

We say that $\bar{r} \in V(N)$ is an 'equilibrium' of the game if there exists $\bar{\lambda} \in M^n$ such that

$$\forall S, \ \langle \bar{\lambda}, c_S \cdot \bar{r} \rangle \leq w(S, \bar{\lambda}) \qquad (*)$$

(a) Show that any equilibrium $\bar{r}$ belongs to the core of the game.

(b) Show that we may write

$$\beta(r) := \inf_{\lambda \in M^n} \sup_{c \in [0,1]^n} (\langle \lambda, c \cdot r \rangle - \hat{w}(c, \lambda))$$

where

$$\hat{w}(c, \lambda) := \inf_{\substack{\sum m(S)c_S = c \\ m(S) \geq 0}} \left(\sum m(S)w(S, \lambda) \right)$$

Hint: write

$$\sup_{S} (\langle \lambda, c_S \cdot r \rangle - w(S, r))$$

$$= \sup_{\alpha(S) \geq 0, \sum \alpha(S) = 1} (\langle \lambda, (\sum \alpha(S)c_S) \cdot r - \sum \alpha(S)w(S, \lambda))$$

$$= \sup_{c \in [0,1]^n} \sup_{\sum \alpha(S)c_S = c} (\langle \lambda, c \cdot r \rangle - \sum \alpha(s)w(S, \lambda))$$

(c) We set

$$\hat{V} := \overline{\text{co}} \bigcup_{\substack{\sum m(S)c_S = c \\ m(S) \geq 0}} \left(\sum_S m(S)V(S) \right).$$

Show that

$$\hat{w}(c, \lambda) := \inf_{\lambda \in \hat{V}(c)} \langle \lambda, r \rangle.$$

(d) Deduce that any equilibrium $\bar{r}$ associated with a $\bar{\lambda} \in \mathring{\mathbb{R}}_+^n$ is contained in the 'fuzzy core' of the game, namely the set of multilosses $r \in V(N)$ such that

$$\forall c \in [0,1]^n, \quad c \cdot r \notin \hat{V}(c) + c \cdot \mathbb{R}_+^n.$$

(e) Conversely, deduce from Theorem 8.1 that any element of the fuzzy core is an equilibrium.

Hint: show that if r belongs to the fuzzy core then

$$\hat{\alpha}(r) := \sup_{c \in [0,1]^n} \inf_{\lambda \in M^n} (\langle \lambda, c \cdot r \rangle - \hat{w}(c, \lambda)) \leq 0.$$

Exercise 12.11

We assume we are in the framework of Exercise 12.10. We suppose that the game is strongly balanced, in the sense that for any balancing m (see Definition 9.9), we have

$$\sum_S m(S)V(S) \subset V(N).$$

(a) Show that if the game is strongly balanced then it is balanced.

(b) We set

$$Q(\lambda) := \{r \in V(N) | \langle \lambda, c \cdot r \rangle \leq \hat{w}(c, \lambda) \ \forall c \in [0, 1]^n\}.$$

Show that $Q(\lambda)$ has convex compact values and that

$$\sigma(Q(\lambda), \lambda) \geq \hat{w}(c_N, \lambda) = v(N, \lambda) = -\sigma(-V(N), \lambda).$$

(c) Deduce from Theorem 9.2 (Gale–Nikaïdo–Debreu) that if $\lambda \to Q(\lambda)$ is upper semi-continuous, then there exists $\bar{\lambda} \in M^n$ such that

$$C(\bar{\lambda}) \cap V(N) \neq \emptyset.$$

Show that the elements of this intersection are equilibria.

14.12 Exercises for Chapter 13 – Cooperative Games and Fuzzy Games

The proofs of the following results are left as an exercise:

- Theorem 13.1

- Theorem 13.2

- Proposition 13.2

- Theorem 13.4

- Proposition 13.4

Exercise 13.1

We assume we are in the framework of Theorem 13.1. Show that we may associate any $x^0 \in L$ with a sequence of $x^t \in L$ and fuzzy coalitions $c^t \in [0, 1]^n$ such that

$$x^{t+1} - x^t = \sum_{i=1}^{n} c_i^{t+1} f_i(x^{t+1}).$$

Deduce a law of evolution for fuzzy coalitions.

Exercise 13.2

We consider a 'continuum of players' identified with an interval Ω of the straight line. The 'fuzzy coalitions' of players are identified with the measurable functions from Ω to $[0, 1]$.

We associate each player with its action $f(\omega, \cdot)$ where $f : \Omega \times L \to \mathbb{R}^n$ and each fuzzy coalition $c(\omega)$ with its action $\int_\Omega f(\omega, x)c(\omega)d\omega$. We suppose that

(i) $\forall x \in L, \; \omega \to f(\omega, x) \in L^1$;

(ii) $x \to f(w, x)$ is continuous for almost all ω;

(iii) $\sup_{x \in L} \sup_{i=1,\dots,n} |f_i(\omega, x)| \leq g_0(\omega)$ where $g_0 \in L^1$.

If $L \subset \mathbb{R}^n$ is convex and compact, show that there exist $\bar{x} \in L$ and $\bar{c} \in L^\infty(\Omega, [0, 1])$ such that

$$\int_\Omega f(\omega, x)\bar{c}(\omega)d\omega = 0.$$

Exercise 13.3

We consider a compact set $L \subset \mathbb{R}^\ell$, n operators $A_i \in L(\mathbb{R}^\ell, \mathbb{R}^k)$, n closed sets $M_i \subset \mathbb{R}^k$ and n continuous mappings f_i from L to $\mathbb{R}^\ell$. We suppose that $\forall x \in L$, $\exists c \in [0, 1]^n$ such that

(i) $$\sum_{i=1}^n c_i A_i x \in \sum_{i=1}^n c_i M_i$$

(ii) $$\sum_{i=1}^n c_i f_i(x) \in T_L(x).$$

(a) Show that there exists $\bar{x} \in L$ and $\bar{c} \in [0, 1]^n$ such that

(i) $$\sum_{i=1}^n \bar{c}_i A_i \bar{x} \in \sum_{i=1}^n \bar{c}_i M_i$$

(ii) $$\sum_{i=1}^n \bar{c}_i f_i(\bar{x}) = 0.$$

(b) Show that for any $x^0 \in L$ and c^0 such that $\sum c_i^0 A_i x^0 \in \sum c_i^0 M_i$, there exists a sequence of x^t and c^t such that

(i) $$\sum_{i=1}^n c_i^t A_i x^t \in \sum_{i=1}^n c_i^t M_i$$

(ii) $$x^{t+1} - x^t = \sum_{i=1}^n c_i^{t+1} f_i(x^{t+1}).$$

Hint: use the set-valued map C defined by

$$C(x) = \left\{ \sum_{i=1}^n c_i f_i(x) \middle| \sum_{i=1}^n c_i A_i x \in \sum_{i=1}^n c_i M_i \right\}.$$

Exercise 13.4

A share-out game v is said to be *inessential* if

$$\forall S, \ w(S) = \sum_{i \in S} a_i.$$

Show that the core of such a game consists of a single vector $a = (a_1, \ldots, a_n)$.

Exercise 13.5

Suppose that the fuzzy game v may be written as

$$\forall c \in \mathbb{R}_+^n, \ v(c) := \sup_{p \in P} w(c, p)$$

where

(i) P is compact;

(ii) $\forall c \in \mathbb{R}_+^n$, $p \to w(c, p)$ is upper semi-continuous;

(iii) $\forall p \in P$, $c \to w(c, p)$ is convex, positively homogeneous and finite.

We consider the core C of the game v and the cores $C(p)$ of the games $w(\cdot, p)$. Let

$$P(N) := \{p \in P | v(c_N) = w(c_N, p)\}.$$

Show that

$$C = \overline{co}\left(\bigcup_{p \in P(N)} C(p)\right).$$

Hint: use Theorem 4.4.

Exercise 13.6

We consider the space V^n of functions $v : [0, 1]^n \to \mathbb{R}$ which are zero at 0 and continuously differentiable at all points of the diagonal. $\{tc_N\}_{t \in [0,1]}$. We let $\sigma(v)$ denote the vector with components

$$\sigma(v)_i = \int_0^1 \frac{\partial}{\partial c_i} v(t, t, \ldots, t) dt.$$

(a) Show that for any operator $A \in L(\mathbb{R}^m, \mathbb{R}^n)$ such that $Ac_N = c_N$, we have

$$\sigma(v \circ A) = A^* \sigma(v).$$

(b) Show that if v is positively homogeneous

$$\sigma(v) = S(v) := \nabla v(c^N).$$

(c) Show that the efficiency axiom

$$\sum_{i=1}^{n} \sigma(v)_i = v(c_N)$$

is satisfied (calculate $\frac{d}{dv}v(tc_N)$ and integrate from 0 to 1).

(d) Show that the symmetry axiom

$$\sigma(\theta * v) = \theta * \sigma(v) \text{ for any permutation } \theta$$

is satisfied (for A use the operator defined by formula (35) of Chapter 13).

(e) Show that the atomicity axiom is satisfied (for A use the operator defined by formula (38) of Chapter 13).

(f) Show that the redundant-players axiom is satisfied (for A use the projection operator from $\mathbb{R}^m$ onto $\mathbb{R}^n$).

(g) The space V^n is assigned the scalar product

$$((v, w)) = \sum_{i=1}^{n} \int_0^1 \frac{\partial v}{\partial c_i}(t, \dots, t) \frac{\partial w}{\partial c_i}(t, \dots, t) dt.$$

Show that $\sigma(v)$ is the projection of v onto the space of linear functions $\mathbb{R}^n \subset V^n$.

15. Statements of Problems

15.1 Problem 1 – Set-valued Maps with a Closed Graph

(a) Show that the graph of any upper semi-continuous set-valued map F from X to Y with closed values is closed.

(b) Show that the converse is true if the codomain Y is compact.

15.2 Problem 2 – Upper Semi-continuous Set-valued Maps

We consider two set-valued maps F and G from X to Y such that

$$F \text{ is upper semi-continuous at } x_0 \tag{1}$$
$$F(x_0) \text{ is compact} \tag{2}$$
$$\text{the graph of } G \text{ is closed} \tag{3}$$
$$\forall x \in X, \quad F(x) \cap G(x) \neq \emptyset. \tag{4}$$

Let N be an open neighbourhood of $F(x_0) \cap G(x_0)$.

(a) If $F(x_0) \subset N$, deduce that $x \to F(x) \cap G(x)$ is upper semi-continuous at x_0.

(b) Otherwise, set $K := F(x_0) \cap N$ and show that we may associate any $y \in K$ with neighbourhoods $U_y(x_0)$ of x_0 and $W(y)$ of y such that

$$\forall x \in U_y(x_0), \quad G(x) \cap W(y) = \emptyset. \tag{5}$$

(c) Deduce that there exists a neighbourhood $U(x_0)$ of x_0 and an open subset M of Y such that

$$\forall x \in U(x_0), \quad F(x) \subset M \cup N \quad \text{and} \quad G(x) \cap M = \emptyset. \tag{6}$$

(d) Now show that $x \to F(x) \cap G(x)$ is upper semi-continuous at x_0 under assumptions (1)–(4).

(e) Deduce that if the graph of a set-valued map G from X to a *compact* set Y is closed, then G is upper semi-continuous.

15.3 Problem 3 – Image of a Set-valued Map

Consider three metric spaces X, Y and U and

- a set-valued map G from X to U
- a *continuous* mapping from $\mathrm{Graph}(G)$ to Y.

We define the (set-valued) image map F from X to Y by

$$F(x) := \{f(x, u)\}_{u \in G(x)}. \tag{1}$$

(a) Show that if

$$G \text{ is lower semi-continuous} \tag{2}$$

the same is true of F.

(b) Show that if

$$G \text{ is upper semi-continuous with compact values} \tag{3}$$

the same is true of F.

15.4 Problem 4 – Inverse Image of a Set-valued Map

We consider a metric space K, two set-valued maps T and F from K to vector spaces Y and U and a mapping f from $\mathrm{Graph}(F)$ to Y. We define the associated set-valued map R by

$$R(x) := \{u \in F(x) | f(x, u) \in T(x)\}. \tag{1}$$

(a) Suppose that f is continuous, that the graph of T is closed and that F is upper semi-continuous with convex compact values. Show that R is upper semi-continuous. Hint: use Problem 2.

(b) Suppose that

$$F \text{ is lower semi-continuous with convex closed values} \tag{2}$$
$$\forall x \in K, \ T(x) \text{ is convex and closed, } \mathrm{Int}\, T(x) \neq \emptyset \text{ and the graph of}$$
$$x \to \mathrm{Int}\, T(x) \text{ is open in } K \times Y \tag{3}$$
$$f \text{ is continuous and affine.} \tag{4}$$

Show successively that

(i) The set-valued map S defined by

$$S(x) := \{u \in F(x) | f(x, u) \in \mathrm{Int}\, T(x)\} \tag{5}$$

is lower semi-continuous.

(ii) $\forall x \in K$, $\overline{S(x)} = R(x)$.

(iii) The set-valued map R is lower semi-continuous.

(c) Show that if assumptions (2), (4) together with the following

T is lower semi-continuous with convex values,	(6)
$\exists \gamma > 0$ such that $\forall x \in K, \gamma B \subset f(x, F(x)) - T(x)$,	(7)
the images $F(x)$ are contained in a bounded set,	(8)

are satisfied, then R is lower semi-continuous.

15.5 Problem 5 – Polars of a Set-valued Map

Let Y be a finite-dimensional space. Consider a set-valued map T from a metric space X to Y.

(a) Show that if T is lower semi-continuous, then the set-valued map $x \to T(x)^-$ has a closed graph.

(b) Show that if the images $T(x)$ are convex closed cones, then the converse is true. Hint: use the theorem relating to projections onto convex closed cones (Problem 17, below).

15.6 Problem 6 – Marginal Functions

Consider two metric spaces X and Y, a function $g : X \times Y \to \mathbb{R}$ and a set-valued map F from X to Y with the marginal function f defined by

$$f(x) = \sup_{y \in F(x)} g(x, y). \tag{1}$$

(a) Suppose that g and F are lower semi-continuous and prove that f is then lower semi-continuous.

(b) Suppose that g is upper semi-continuous and that F is upper semi-continuous with compact values; prove that f is upper semi-continuous.

(c) Suppose that g is continuous and that F is continuous with compact values (then the marginal function is continuous). Using Problem 2, prove that the marginal set-valued map M defined by

$$M(x) := \{y \in F(x) | f(x) = g(x, y)\} \tag{2}$$

is upper semi-continuous.

(d) Suppose that g is Lipschitz and that F is Lipschitz in the sense that

$$\forall x_1, x_2, \quad \sup_{y_1 \in F(x_1)} d(y_1, F(x_2)) \le c \|x_1 - x_2\|. \tag{3}$$

Prove that the marginal function is also Lipschitz.

15.7 Problem 7 – Generic Continuity of a Set-valued Map with a Closed Graph

Consider a set-valued map F from X to a compact set Y.

(a) Consider a countable family of open sets U_n such that for any open set U and for any $x \in U$, there exists U_n with $x \in \overline{U}_n \subset U$ (this is possible, because, since Y is compact, there exists a countable dense sequence in Y). Set

$$K_n := \{x \in X | F(x) \cap \overline{U}_n \neq \emptyset\}. \tag{1}$$

Show that if F is lower semi-continuous at x, then there exists K_n such that $x \in \operatorname{Int} K_n$, the interior of K_n.

(b) Deduce that the set of points of discontinuity of F is contained in $\cup_{n=1}^{\infty} \partial K_n$.

(c) Deduce from Baire's theorem that F is lower semi-continuous (whence continuous) on a dense subset of X.

15.8 Problem 8 – Approximate Selection of an Upper Semi-continuous Set-valued Map

(a) Consider the set-valued map F from $\mathbb{R}$ to $\mathbb{R}$ defined by

$$\begin{aligned}
F(x) &= \{-1\} &&\text{if } x < 0 \\
F(0) &= [-1, +1] && \\
F(x) &= \{+1\} &&\text{if } x > 0
\end{aligned} \tag{1}$$

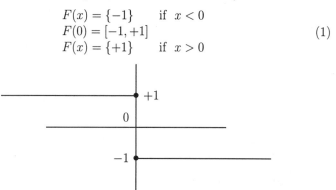

Verify that this set-valued map is upper semi-continuous and has no continuous selection, in other words there is no continuous mapping f such that $f(x) \in F(x)$, $\forall x$.

(b) Suppose that F is a convex-valued upper semi-continuous function from a compact metric space X to a Hilbert space Y. Show that given any $\varepsilon > 0$, there exists a continuous mapping f_ε from X to Y such that

$$\forall x \in X, \quad f_\varepsilon(x) \in F(x) + \varepsilon B. \tag{2}$$

Hint: use continuous partitions of unity to construct f_ε.

15.9 Problem 9 – Continuous Selection of a Lower Semi-continuous Set-valued Map

Consider a *lower semi-continuous* set-valued map F from a compact metric space X to a Hilbert space Y with *convex closed* values.

(a) Show that given any $\varepsilon > 0$, there exists a continuous mapping f_ε from X to Y such that

$$\forall x \in X, \quad d(f_\varepsilon(x), F(x)) \leq \varepsilon. \tag{1}$$

Hint: use continuous partitions of unity to construct f_ε.

(b) Prove inductively that there exist continuous mappings $f_n : X \to Y$ satisfying

(i) $$\forall x \in X, \quad d(f_n(x), F(x)) \leq \frac{1}{2^n}$$

(ii) $$\forall x \in X, \quad \|f_n(x) - f_{n-1}(x)\| \leq \frac{1}{2^{n-2}} \tag{2}$$

(c) Deduce that the f_n converge uniformly to a continuous mapping f from X to Y satisfying

$$\forall x \in X, \quad f(x) \in F(x) \tag{3}$$

(continuous selection of F).

15.10 Problem 10 – Interior of the Image of a Convex Closed Cone

Suppose that X and Y denote Hilbert (or Banach) spaces and that B denotes the unit ball of X or Y.

(a) We consider a convex closed subset $K \subset X$ containing 0 and a continuous, linear operator $A \in L(X, Y)$ from X to Y. We suppose that

$$0 \in \mathrm{Int}(A(K)). \tag{1}$$

Using Baire's theorem (if the interior of the union of a countable family of closed subsets of a complete metric space is non-empty then the interior of one of the closed sets is non-empty), show that

$$0 \in \mathrm{Int}\left(\mathrm{closure}\left(A(K \cap B)\right)\right). \tag{2}$$

(b) We set

$$T := A(K \cap B).$$

Show that this set has the property

$$\frac{1}{2} \sum_{k=0}^{\infty} 2^{-k} T \subset T. \tag{3}$$

Hint: use the fact that $K \cap B$ is bounded, convex and closed and the fact that X is complete.

(c) Show that if T is a set satisfying property (3) and that if

$$0 \in \text{Int} \left(\text{closure} \left(T \right) \right) \tag{4}$$

then

$$0 \in \text{Int} \left(T \right). \tag{5}$$

Deduce that assumption (1) implies

$$\exists \gamma > 0 \text{ such that } \gamma B \subset A(K \cap B). \tag{6}$$

What is the corresponding result when $K = X$?

(d) We now suppose that

(i) K is a convex closed cone
(ii) $A(K) = Y$. $\tag{7}$

Deduce that there exists a constant $c > 0$ such that for all $x_1 \in K$ and $y_2 \in Y$, there exists a solution $x_2 \in K$ of the equation $Ax_2 = y_2$ satisfying the inequality

$$\|x_2 - x_1\| \le c\|Ax_1 - y_2\| \tag{8}$$

or, more concisely, for all $y_1, y_2 \in Y$

$$K \cap A^{-1}(y_1) \subset K \cap A^{-1}(y_2) + c\|y_1 - y_2\|B \tag{9}$$

(the set-valued map $y \to K \cap A^{-1}(y)$ is Lipschitz).

(e) We take convex closed cones $P \subset X$ and $Q \subset Y$ and a continuous linear operator $L \in L(X,Y)$ from X to Y. With any $y \in Y$, we associate the set

$$M(y) := \{x \in P | Lx \in Q + y\}. \tag{10}$$

Deduce from (d) that there exists $c > 0$ such that for all $x_0 \in P$ and $y \in Y$,

$$d(x_0, M(y)) \le cd_Q(Lx_0 - y). \tag{11}$$

Hint: set $K := P \times Q$ and $A(x,y) := Lx - y$.

(f) We consider a set-valued map F from X to Y the graph of which is a convex closed cone (F is said to be a convex closed process). We suppose that F is surjective:

$$\text{Im}\,(F) = Y. \tag{12}$$

Deduce from (d) that there exists a constant $c > 0$ such that

$$\forall y_1 \in Y, \ \ \forall x_1 \in F^{-1}(y_1), \ \forall y_2 \in Y, \ \ \exists x_2 \in F^{-1}(y_2)$$

such that

$$\|x_1 - x_2\| \le c\|y_1 - y_2\| \tag{13}$$

or, such that

$$F^{-1}(y_1) \subset F^{-1}(y_2) + c\|y_1 - y_2\|. \tag{14}$$

Hint: set $K := \text{Graph}(F)$ and $A := \pi_Y$, the canonical projection from $X \times Y$ onto Y; note, moreover, that the properties (d), (e) and (f) are equivalent.

(g) We suppose that

(i) K is a convex closed set containing 0
(ii) $0 \in \text{Int}(A(K))$. \hfill (15)

Show that there exists a constant $\gamma > 0$ such that for all $x_0 \in K$ and all $y \in \gamma B$, there exists a solution $x \in K$ of the equation $y = Ax$ such that

$$\|x - x_0\| \le \frac{1}{\gamma}(1 + \|x_0\|)\|y - Ax_0\|. \tag{16}$$

(h) We consider

(i)   convex closed subsets $P \subset X$ and $Q \subset Y$
(ii)   a continuous linear operator $L \in L(X, Y)$ \hfill (17)

satisfying

$$0 \in \text{Int}(L(P) - Q). \tag{18}$$

We set

$$M(y) := \{x \in P | Lx \in Q + y\} \tag{19}$$

and we choose $\bar{x} \in M(0)$.

Deduce from (g) that there exist $\gamma > 0$ and $c > 0$ such that, for all $x_0 \in K$ and $z \in \gamma B$,

$$d(x_0, M(z)) \le c(\|x_0 - \bar{x}\| + 1)d_Q(Lx_0 - z). \tag{20}$$

Hint: Set $K := P \times Q - (\bar{x}, L\bar{x})$ and $A(x, y) := Lx - y$.

(i) We consider a set-valued map F from X to Y the graph of which is a convex closed cone. We suppose that

$$\bar{y} \in \text{Int} (\text{Im} (F)) \tag{21}$$

and we take $\bar{x} \in F^{-1}(\bar{y})$. Deduce from (g) that there exists a constant γ such that for all $(x_0, y_0) \in \text{Graph}(F)$ and for all $y \in \bar{y} + \gamma B$ we have

$$d(x_0, F^{-1}(y)) \le \frac{1}{\gamma}(1 + \max(\|x_0 - \bar{x}\|, \|y_0 - \bar{y}\|)\|y - y_0\|. \tag{22}$$

Note that the inequality

$$d(\bar{x}, F^{-1}(y)) \le \frac{1}{\gamma}\|y - \bar{y}\| \tag{23}$$

is a direct consequence. Hint: take $K := \text{Graph}(F) - (\bar{x}, \bar{y})$ and for A take the canonical projection of $X \times Y$ onto Y.

15.11 Problem 11 – Discrete Dynamical Systems

Suppose that X is a complete metric space and that g is a *continuous mapping* from X to X.

We suppose that there exists $f : X \to \mathbb{R}_+ \cup \{+\infty\}$ with the property that

$$\forall x \in X, \quad f(g(x)) + d(x, g(x)) \le f(x). \tag{1}$$

We consider a solution (x^t), $x^t \in X$, $t = 0, 1, 2, \ldots$ of the discrete dynamical system

$$x^{t+1} = g(x^t), \quad x^0 = x_0 \tag{2}$$

the equilibria of which are the fixed points of g.

(a) Show that assumption (1) implies that if $f(x_0)$ is finite, then the solution of (2) converges to an equilibrium.

(b) Show that if there exists a function f satisfying (1), then the function f_g defined by

$$f_g(x) := \sum_{t=0}^{\infty} d(g^{t+1}(x), g^t(x)) \in [0, \infty] \tag{3}$$

is the smallest function $f : X \to \mathbb{R}_+ \cup \{+\infty\}$ satisfying (1). Deduce that the domain of f_g contains an equilibrium point.

(c) Deduce the Banach–Picard Fixed-point Theorem for contractions.

(d) Let G be a set-valued map from X to X. We suppose that there exists a function $f : X \to \mathbb{R}_+ \cup \{+\infty\}$ satisfying

$$\forall x \in X, \; \exists y \in G(x) \text{ such that } f(y) + d(x,y) \leq f(x). \tag{4}$$

Show that if the graph of G is closed, then for all $x_0 \in \text{Dom} f$, there exists a solution (in X) of the dynamical system

$$x^{t+1} \in G(x^t), \quad x^0 = x_0 \tag{5}$$

which converges to an equilibrium point of G, in other words a fixed point $\bar{x} \in G(\bar{x})$.

(e) We consider the set $T(x)$ of solutions (x^t) of (5) based on x and the set $H(x)$, the *orbit* based on x defined by

$$H(x) := \bigcup_{(x^t) \in T(x)} \bigcup_{t \geq 0} \{x^t\}. \tag{6}$$

Show that the relation '$y \geq x$ if and only if $y \in H(x)$' is a preorder.

(f) With any solution (x^t) of (5) we associate the set

$$K(x^t) := \bigcap_{t \geq 0} \overline{H(x^t)} \tag{7}$$

(i) Show that the accumulation points (limit points) of the sequence x^t belong to $K(x^t)$.

(ii) Show that if G is lower semi-continuous, then $G(K(x^t)) \subset K(x^t)$.

(g) We now suppose that there exists a function $f : X \to \mathbb{R}_+ \cup \{+\infty\}$ satisfying

$$\forall x \in X, \; \forall y \in G(x), \; f(y) + d(x,y) \leq f(x). \tag{8}$$

For a given $x \in \text{Dom}(f)$, construct a solution (x^t) of (5) based on x, satisfying

$$x_{n+1} \in H(x_n) \text{ and } f(x_{n+1}) \leq v(x_n) + 2^{-n} \tag{9}$$

where

$$v(x) := \inf\{f(y) | y \in H(x)\}. \tag{10}$$

(i) Show, by adapting the proof of Theorem 1.2, that the sequence x^t converges to a limit $\bar{x}$.

(ii) Deduce that if G is also lower semi-continuous, then $G(\bar{x}) = \{\bar{x}\}$.

15.12 Problem 12 – Fixed Points of Contractive Set-valued Maps

We consider a nontrivial set-valued map G from a metric space E to itself. We denote by $T(x)$ the set of sequences x_n such that $x_{n+1} \in G(x_n)$, $x_0 = x$ (the set of solutions of the discrete dynamical system defined by G beginning at x). We set

$$f_G(x) = \inf_{T(x)} \sum_{n=0}^{\infty} d(x_n, x_{n+1}) \in \mathbb{R}_+ \cup \{+\infty\} \tag{1}$$

(minimum length of trajectories based on x).

(a) Show that if $f : E \to \mathbb{R}_+ \cup \{+\infty\}$ satisfies

$$\forall x \in E, \quad \exists y \in G(x) \text{ such that } f(y) + d(x, y) \le f(x), \tag{2}$$

then

$$\forall x \in E, \quad f_G(x) \le f(x). \tag{3}$$

(b) We associate any $\varepsilon > 0$ with the function f_ε defined by

$$f_\varepsilon(x) = \inf \left\{ \sum_{n=0}^{\infty} d(x_n, x_{n+1}) \,|\, x_0 = x \text{ and } x_{n+1} \in B(G(x_n), \varepsilon) \right\}. \tag{4}$$

Show that if $\varepsilon_2 \le \varepsilon_1$, then

$$f_{\varepsilon_1}(x) \le f_{\varepsilon_2}(x) \le f_G(x) \tag{5}$$

and that

$$\forall x \in E, \exists x_\varepsilon \in B(G(x), \varepsilon) \text{ such that } f_\varepsilon(x_\varepsilon) + d(x, x_\varepsilon) \le f_\varepsilon(x) + \varepsilon. \tag{6}$$

(c) We now suppose that

$$G \text{ is upper semi-continuous with compact values.} \tag{7}$$

Show that there exists a subsequence x_{ε_k} converging to an element $\bar{x} \in G(x)$ such that, for all $\delta > 0$, $\exists k_\delta$ such that $\forall k \ge k_\delta$,

$$f_\delta(\bar{x}) + d(x, x_{\varepsilon_k}) - d(\bar{x}, x_{\varepsilon_k}) \le f_{\varepsilon_k}(x) + \varepsilon_k. \tag{8}$$

(d) We set

$$f_0(x) = \lim_{\varepsilon \to 0} f_\varepsilon(x). \tag{9}$$

Show that $f_0 = f_G$ satisfies the property (2).

(e) We now suppose that the set-valued map G is a *contraction* from E to E, in the sense that there exists $\lambda \in]0, 1[$ such that

$$\forall x, y \in E, \quad G(y) \subset B(G(x), \lambda). \tag{10}$$

If G has compact values, show that G has a fixed point. Hint: use Theorem 1.4.

15.13 Problem 13 – Approximate Variational Principle

Suppose that X is a Hilbert space and that $f : X \to \mathbb{R}$ is a nontrivial, positive, lower semi-continuous, Gâteaux-differentiable function.

(a) Fix $\varepsilon > 0$, $\lambda > 0$ and $x_0 \in \text{Dom } f$ such that

$$f(x_0) \leq \inf_{x \in X} f(x) + \varepsilon\lambda. \tag{1}$$

Show that there exists $\bar{x}$ such that

$$f(\bar{x}) \leq f(x_0), \quad \|x_0 - \bar{x}\| \leq \lambda \quad \text{and} \quad \|Df(\bar{x})\| \leq \varepsilon. \tag{2}$$

(b) Deduce that there exists a sequence $\bar{x}_n \in X$ satisfying

(i) $\qquad\qquad\qquad\qquad f(\bar{x}_n) \to \inf_{x \in X} f(x)$
(ii) $\qquad\qquad\qquad\qquad Df(\bar{x}_n) \to 0 \text{ in } X^* \qquad\qquad$ (3)

(c) We also suppose that

$$\lim_{\|x\| \to \infty} \frac{f(x)}{\|x\|} = +\infty. \tag{4}$$

Show that $Df(X)$ is dense in X^*.

15.14 Problem 14 – Open Image Theorem

Consider a Hilbert (or Banach) space X, a finite-dimensional space Y, a non-empty *closed* subset K of X and a continuous linear operator A from X to Y.

We consider the tangent cone $T_K(x_0)$ and the normal cone $N_K(x_0)$ to K at x_0 (in the sense of Definition 6.4). Alternatively, we may suppose that K is convex and closed and take the tangent and normal cones of convex analysis (Definition 4.3).

The aim of the problem is to show that the condition

$$AT_K(x_0) = Y \tag{1}$$

implies that

$$Ax_0 \in \text{Int } A(K), \tag{2}$$

in other words, that there exists $\gamma > 0$ such that

$$\forall y \in A(x_0) + \gamma B, \quad \exists x \in K \text{ solution of, } \quad A(x) = y \tag{3}$$

(Frankowska's theorem).

(a) Suppose that $A(x_0) \notin \text{Int } A(K)$. Using Ekeland's Theorem (Theorem 1.2), show that there exist sequences $y_n \notin A(K)$ and $x_n \in K$ such that x_n converges to x_0 and

$$\forall x \in K, \quad \|A(x_n) - y_n\| \le \|A(x) - y_n\| + \frac{1}{n}\|x_n - x\|. \tag{4}$$

(b) Calculate the Fréchet derivative of the function $z \to \|A(x_n + z) - y_n\|$ and show that it may be written in the form $A^* p_n$, where p_n is in the unit sphere of Y.

(c) Deduce from (a) and (b), that there exists ρ in the unit sphere of S such that

$$\forall u \in T_K(x_0), \quad 0 \le \langle \rho, Au \rangle. \tag{5}$$

(d) Deduce that (1) implies (2).

(e) More generally, suppose that A is a continuously differentiable mapping from a neighbourhood of K to Y. Show that the condition

$$A'(x_0)T_K(x_0) = Y \tag{6}$$

implies that

$$A(x_0) \in \text{Int } A(K). \tag{7}$$

(f) Show that if $L \subset X$ is a closed subset of a Hilbert space, M a closed subset of a finite-dimensional space, $B \in L(X,Y)$ and $x_0 \in L \cap B^{-1}(M)$, then the condition

$$BT_L(x_0) - T_M(Bx_0) = Y \tag{8}$$

implies that

$$0 \in \text{Int}(B(L) - M). \tag{9}$$

Verify that this extends to the case in which B is continuously differentiable and (8) is replaced by

$$B'(x_0)T_L(x_0) - T_M(B(x_0)) = Y. \tag{10}$$

(g) Suppose that

$$K \text{ is a closed subset of } X \tag{11}$$

and consider a continuously differentiable mapping F from a neighbourhood of L to $\mathbb{R}^n$ with components f_i. Show that if x_0 is a Pareto optimum (see Definition 12.5), then there exists $\lambda \in \mathbb{R}^n_+$, $\lambda \ne 0$ such that

$$\sum_{i=1}^n \lambda_i f_i'(x_0) \in N_K(x_0). \tag{12}$$

15.15 Problem 15 – Asymptotic Centres

Let E be a finite-dimensional vector space (the problem also holds for Hilbert spaces, provided we use weak convergence). We consider a *bounded* sequence of elements x_t of E with which we associate:

1. The function v defined by

$$v(y) := \limsup_{t \to \infty} \|x_t - y\|^2 \qquad (= \inf_{s \geq 0} \sup_{t \geq s} \|x_t - y\|^2).$$

2. The set N (attractor for the sequence) defined by

$$N := \{y \in E | v(y) = \lim_{t \to \infty} \|x_t - y\|^2\}.$$

3. The set C which is the convex closed hull of the limit points of the sequence.

(a) Show that v is a finite strictly convex function satisfying

$$\lim_{\|y\| \to \infty} v(y) = +\infty.$$

Deduce that the function v has a unique minimum $\bar{x}$, which is called the *asymptotic centre* of the bounded sequence x_t.

(b) Let $\bar{y} \in C$ be the best approximation of the asymptotic centre $\bar{x}$ by the elements of C. Show that

$$v(\bar{y}) + \|\bar{x} - \bar{y}\|^2 \leq v(\bar{x}).$$

Hint: expand $\|x_t - \bar{y} + \bar{y} - \bar{x}\|^2$, pass to the limit of a suitable subsequence.

Deduce that the asymptotic centre $\bar{x}$ of a bounded sequence x_t always belongs to the set C and that if a sequence converges, its asymptotic centre is its limit (the asymptotic centre may be viewed as a virtual limit).

(c) We take $y \in N \cap C$, $z \in E$ and w a limit point of the sequence x_t. Show that

$$v(y) + \|y - z\|^2 + 2\langle w - y, y - z \rangle \leq v(z).$$

Deduce that

$$v(y) + \|y - z\|^2 \leq v(z) \quad \forall z \in E.$$

Thus, show that either $N \cap C = \emptyset$ or $N \cap C = \{\bar{x}\}$ (reduces to the asymptotic centre).

15.16 Problem 16 – Fixed Points of Non-expansive Mappings

Let K be a *convex, closed, bounded* subset of a Hilbert space E and f a mapping from K to K, which is non-expansive in the sense that

$$\forall x, y \in K, \ \ \|f(x) - f(y)\| \leq \|x - y\|. \tag{1}$$

Let $x_0 \in K$.

(a) Show that the mappings f_t defined by

$$f_t(x) := \frac{1}{t}x_0 + \left(1 - \frac{1}{t}\right)f(x) \tag{2}$$

are contractions from K to K.

(b) Deduce that there exists a sequence of points $x_t \in K$ such that

$$\lim_{t \to \infty} \|x_t - f(x_t)\| = 0. \tag{3}$$

(c) We now suppose that E is finite dimensional. If not, use weak convergence.

Let w be a limit point of the sequence x_t. Set $x_\lambda := (1-\lambda)w + \lambda f(w)$ and show that

$$\langle x_\lambda - f(x_\lambda), x_\lambda - w \rangle \geq 0. \tag{4}$$

Deduce that any limit point w of the sequence x_t is a fixed point of f.

(d) Show that the set of fixed points of f is convex and closed.

(e) We assume the solution of Problem 15 on asymptotic centres. We now consider the sequence $x_t := f^t(x_0)$.

   (i) Show that any fixed point $\bar{x}$ of f belongs the attractor set N for this sequence x_t.

   (ii) Deduce from question (c) of Problem 15 that if a limit point w of this sequence x_t is a fixed point then this limit point is the asymptotic centre of $f^t(x_0)$.

   (iii) Deduce from all the above that if

$$\lim_{t \to \infty} \|f^t(x_0) - f^{t+1}(x_0)\| = 0 \tag{5}$$

   then $f^t(x_0)$ converges to a fixed point of f.

15.17 Problem 17 – Orthogonal Projectors onto Convex Closed Cones

Suppose that K is a *convex closed cone* of X and consider the projector of best approximation defined in Theorem 2.3, which we shall denote by π_K.

(a) Show that the variational characterisation of π_K in the case of convex closed cones becomes

(i) $$\langle x - \pi_K(x), \pi_K(x) \rangle = 0$$
(ii) $$\langle x - \pi_K(x), z \rangle \leq 0 \quad \forall z \in K. \tag{1}$$

(b) Use (1) to show that π_K is positively homogeneous (in the sense that $\pi_K(\lambda x) = \lambda \pi_K(x)$ for all $\lambda > 0$).

(c) Show that

$$\|x\|^2 = \|\pi_K(x)\|^2 + \|x - \pi_K(x)\|^2 \tag{2}$$

(Pythagoras's equation) and deduce that

$$\|\pi_K(x)\| \leq \|x\|, \quad \|x - \pi_K(x)\| \leq \|x\|. \tag{3}$$

(d) If $K^- = \{y \in X | \forall x \in K, \langle y, x \rangle \leq 0\}$ denotes the *negative polar cone* of K, show that

$$1 - \pi_K \text{ is the projector of best approximation onto } K^-. \tag{4}$$

(e) Show that

$$K^- = \{x \in X | \pi_K(x) = 0\} \tag{5}$$

and

$$K = \{x \in X | x - \pi_K(x) = 0\}.$$

(f) Deduce that any element $x \in X$ may be uniquely written as

$$x = y + z, \quad y \in K, \quad z \in K^- \text{ and } \langle y, z \rangle = 0 \tag{6}$$

and that in this case, $y = \pi_K(x)$ and $z = (1 - \pi_K)(x)$.

(g) Show that if K is a closed vector subspace, then π_K is the orthogonal projector onto K, which is linear, has norm 1 and is self-adjoint

$$\langle \pi_K x, y \rangle = \langle x, \pi_K y \rangle \quad \forall x, y \in X. \tag{7}$$

When K is a convex closed cone, by convention, the projector of best approximation π_K is also called the *orthogonal projector* onto K.

15.18 Problem 18 – Gamma-convex functions

Consider a set E and denote the set of finite sequences $m := \{(\alpha_i, x_i)\}_{i=1,\dots,n}$ where $x_i \in E$, $\alpha_i \geq 0$ and $\sum_{i=1}^{n} \alpha_i = 1$ by $\mathcal{M}(E)$.

If f is a nontrivial function from E to $\mathbb{R} \cup \{+\infty\}$ we set

$$f^s(m) := \sum_{i=1}^{n} \alpha_i f(x_i) \quad \text{where} \quad m := \{(\alpha_i, x_i)\}_{i=1,\dots,n}. \tag{1}$$

(a) We suppose that E is a subset of a vector space X and define the mapping β from $\mathcal{M}(E)$ to X by

$$\beta(m) := \sum_{i=1}^{n} \alpha_i x_i. \tag{2}$$

(i) Characterise the sets E such that $\beta \mathcal{M}(E) \subset E$.

(ii) Characterise the functions $f : E \to \mathbb{R} \cup \{+\infty\}$ such that

$$f(\beta(m)) \leq f^s(m) \quad \forall m \in \mathcal{M}(E). \tag{3}$$

(b) We let E denote the family of convex compact subsets of a Hilbert space X and define the mapping α from $\mathcal{M}(E)$ to E by

$$\alpha(m) := \sum_{i=1}^{n} \alpha_i K_i \quad \text{where} \quad m := \{(\alpha_i, K_i)\}_{i=1,\dots,n}. \tag{4}$$

Let g be a nontrivial, convex, lower semi-continuous function from X to $\mathbb{R} \cup \{+\infty\}$. We associate this with a function f from E to $\mathbb{R}$ defined by

$$\forall K \in E, \quad f(K) := \inf_{x \in K} f(x). \tag{5}$$

Show that

$$\forall m \in \mathcal{M}(E), \quad f(\alpha(m)) \leq f^s(m). \tag{6}$$

More generally, we have

a set-valued map γ from $\mathcal{M}(E)$ to E with non-empty values. $\tag{7}$

We say that a nontrivial function f from E to $\mathbb{R} \cup \{+\infty\}$ is γ-convex if

$$\forall m \in \mathcal{M}(E), \quad \forall x \in \gamma(m), \quad f(x) \leq f^s(m). \tag{8}$$

(c) Show that if n functions f_i from E to $\mathbb{R}$ are γ-convex, then the set

$$F(E) + \mathbb{R}_+^n \quad \text{is convex} \tag{9}$$

where F is the mapping from E to $\mathbb{R}^n$ defined by

$$F(x) := (f_1(x), \dots, f_n(x)). \tag{10}$$

(d) Conversely, show that if n functions f_i from E to $\mathbb{R}$ have the property (9), then there exists a set-valued map γ from $\mathcal{M}(E)$ to E for which the n functions f_i are γ-convex.

(e) Consider n functions f_i from E to $\mathbb{R}$. We set

(i)
$$v := \inf_{x \in E} \sup_{i=1,\dots,n} f_i(x)$$

(ii)
$$v^b := \sup_{\lambda \in S^n} \inf_{x \in E} \sum_{i=1}^n \lambda_i f_i(x). \tag{11}$$

1. Show that $v^b \le v$.

2. Conversely, suppose that the n functions f_i are γ-convex. Show that for all $\varepsilon > 0$,

$$v \le v^b + \varepsilon. \tag{12}$$

Hint: set $\mathbf{1} = (1, \dots, 1)$ and show that

$$(v^b + \varepsilon)\mathbf{1} \in F(E) + \mathbb{R}^n_+. \tag{13}$$

(f) Suppose that the n functions f_i from E to $\mathbb{R}$ are γ-convex. Consider a Pareto optimum $\bar{x}$ (such that there does not exist $y \in E$ such that $f_i(y) < f_i(\bar{x})$ for all $i = 1, \dots, n$). Show that there exists $\bar{\lambda} \in S^n$ such that

$$\sum_{i=1}^n \bar{\lambda}_i f_i(\bar{x}) = \min_{x \in E} \sum_{i=1}^n \bar{\lambda}_i f_i(x). \tag{14}$$

Remark. Questions (a), (b) and (d) are independent. Questions (e) and (f) are independent of each other and depend only on (c).

15.19 Problem 19 – Proper Mappings

(a) We consider two metric spaces X and Y and a *continuous* mapping f from X to Y. Show that the following two conditions are equivalent:

Given any sequence $x_n \in X$ such that $f(x_n)$ converges in Y we may extract a convergent subsequence of x_n. $\hfill (1)$
f maps closed sets to closed subsets and for all $y \in Y$, $f^{-1}(y)$ is compact. $\hfill (2)$

We shall say that a *continuous* mapping which satisfies one of these equivalent properties is a *proper mapping*.

(b) Let f be a proper mapping. Show that:

(i) for any compact subset K of Y, $f^{-1}(K)$ is compact;

(ii) if X is compact, then any continuous mapping is proper;

(iii) the composition of two proper mappings is proper.

(c) Let X and Y be Hilbert spaces, $A \in L(X, Y)$ a continuous linear operator and K a convex closed subset of X. Suppose that

$$b(K) := \{p \in X^* \,|\, \sup_{x \in K} \langle p, x \rangle < +\infty\} \tag{3}$$

is the barrier cone of K (Definition 3.2). Show that if

$$\operatorname{Im} A^* + b(K) = X^* \tag{4}$$

then the mapping $A : K \to Y$ is proper and consequently that $A(K)$ is closed and $K \cap A^{-1}(y)$ is compact for all $y \in Y$ (suppose that X is finite dimensional if you do not wish to use the weak compactness of weakly closed bounded sets).

(d) Suppose X and Y are two Hilbert spaces, $L \in L(X, Y)$ is a continuous linear operator from X to Y and $P \subset X$ and $Q \subset Y$ are convex closed cones. Deduce from the previous question that if

$$L^* Q^- + P^- = X^* \tag{5}$$

then the mapping $(x, y) \in P \times Q \to Lx - y \in Y$ is proper. Deduce in particular that

(i) $L(P) - Q$ is a convex closed cone

(ii) $\forall y \in Y$, $\{x \in P | Lx \in Q + y\}$ is compact. $\tag{6}$

(e) Consider n subsets $L_i \subset \mathbb{R}^\ell$ satisfying

$$\forall i = 1, \dots, n \;\; L_i \text{ is closed and bounded below} \tag{7}$$

(in the sense that $L_i \subset u_i + \mathbb{R}^\ell_+$ where $u_i \in \mathbb{R}^\ell$).

Show that the mapping

$$x = (x_1, \dots, x_n) \in \prod_{i=1}^n L_i \to \sum_{i=1}^n x_i \in \mathbb{R}^\ell \tag{8}$$

is proper.

(f) Suppose that the subset L of $\mathbb{R}^\ell$ is closed and bounded below and that M is a subset of $\mathbb{R}^\ell$ satisfying

(i) M is convex and closed

(ii) $\exists w \in M$ such that $(M - w) \cap \mathbb{R}^\ell_+ = \{0\}$. $\tag{9}$

Show that

$$\{x, y\} \in L \times M \to x - y \in \mathbb{R}^l \tag{10}$$

is proper. Hint: first show that if $x_n - y_n \in L - M$ converges, the sequence y_n is bounded, by eliminating the case in which $\|y_n\| \to \infty$ using (9).

Deduce that the assumptions (7) and (9) imply that the mapping

$$\{x_1, \ldots, x_n, y\} \in \prod_{i=1}^n L_i \times M \to \sum_{i=1}^n x_i - y \in \mathbb{R}^\ell \tag{11}$$

is proper and that, in particular, the set

$$K := \left\{ x \in \prod_{i=1}^n L_i \mid \sum_{i=1}^n x_i \in M + y \right\} \tag{12}$$

is compact.

15.20 Problem 20 – Fenchel's Theorem for the Functions $L(x, Ax)$

Suppose that

(i) X is a Hilbert space and Y is a finite-dimensional space.

(ii) A is a continuous linear operator $A \in L(X, Y)$.

(iii) L is a nontrivial, convex, lower semi-continuous function from $X \times Y$ to $\mathbb{R} \cup \{+\infty\}$.

We set

$$v \quad := \quad \inf_{x \in X} L(x, Ax) \tag{1}$$

$$v_* \quad := \quad \inf_{q \in Y^*} L^*(-A^*q, q). \tag{2}$$

(a) Show that $v < +\infty$ if and only if

$$0 \in ((A \oplus -1)\operatorname{Dom} L) \tag{3}$$

where $A \oplus -1 \in L(X, Y)$ is defined by $(A \oplus -1)(x, y) = Ax - y$. Show also that $v_* < +\infty$ if and only if

$$0 \in (1 \oplus A^*)\operatorname{Dom} L^*. \tag{4}$$

Show that

$$v + v_* \geq 0 \tag{5}$$

and deduce that conditions (3) and (4) imply that v and v_* are finite.

(b) We define $\phi : X \times Y \to Y \times \mathbb{R}$ by

$$\phi(x, y) := (Ax - y, L(x, y)). \qquad (6)$$

Show that

(i) $\phi(\mathrm{Dom}\, L) + \{0\} \times]0, \infty[$ is convex
(ii) $(0, v) \notin (\mathrm{Dom}\, L) + \{0\} \times]0, \infty[. \qquad (7)$

(c) Deduce that there exist $p \in Y^*$ and $a \geq 0$ such that $(p, a) \neq (0, 0)$ and

$$(aL)^*(-A^*q, q) \leq -av. \qquad (8)$$

(d) Show that the assumption

$$0 \in \mathrm{Int}((A \oplus -1)\mathrm{Dom}L) \qquad (9)$$

implies that a is positive and deduce that there exists $\bar{q} \in Y^*$ such that

$$L^*(-A^*\bar{q}, \bar{q}) = v_* = -v. \qquad (10)$$

(e) We set

$$f(x) := L(x, Ax). \qquad (11)$$

Deduce from (d) that assumption (10) implies that if $p \in (1 \oplus A^*)\mathrm{Dom}L^*$, there exists $\bar{q} \in Y^*$ such that

$$f^*(p) = L^*(p - A^*\bar{q}, \bar{q}) = \min_{q \in Y^*} L^*(p - A^*q, q).$$

15.21 Problem 21 – Conjugate Functions of $x \to L(x, Ax)$

Let X be a Hilbert space, Y a finite-dimensional space, $A \in L(X, Y)$ and $L : X \times Y \to \mathbb{R} \cup \{+\infty\}$ a nontrivial, convex, lower semi-continuous function.

(a) Show that if

$$0 \in \mathrm{Int}((A \oplus -1)\, \mathrm{Dom}\, L) \qquad (1)$$

then the functions $q \to L^*(p - A^*q, q)$ are lower semi-compact when $q \in (1 \oplus A^*)\, \mathrm{Dom}\, L^*$.

(b) Deduce that the function

$$p \to \inf_{q \in Y^*} L^*(p - A^*q, q)$$

is convex, nontrivial and lower semi-continuous.

(c) Calculate the conjugate of this function and deduce the expression for the conjugate function of $x \to L(x, Ax)$.

15.22 Problem 22 – Hamiltonians and Partial Conjugates

We consider two Hilbert spaces X and Y and a nontrivial, convex, lower semi-continuous function $L : X \times Y \to \mathbb{R} \cup \{+\infty\}$. We set

$$H(x, q) := \sup_{y \in Y}(\langle q, y \rangle - L(x, y)). \tag{1}$$

Note: by convention, we set $H(x, q) := -\infty$ on

$$K := \{x \in X | \forall y \in Y, L(x, y) = +\infty\}.$$

(a) Verify that

(i) $\forall x \in X, \ q \to H(x, q)$ is convex and lower semi-continuous
(ii) $\forall q \in Y^*, \ x \to H(x, q)$ is concave. $\tag{2}$

Deduce that

(i) $$L(x, y) = \sup_{q \in Y^*}(\langle q, y \rangle - H(x, q))$$

(ii) $$L^*(p, q) = \sup_{x \in X}(\langle p, x \rangle + H(x, q)). \tag{3}$$

(b) Show that if we suppose that

$$\forall q \in Y^*, \ x \to L(x, q) \text{ is upper semi-continuous} \tag{4}$$

then

$$H(x, q) = \inf_{p \in X^*}(L^*(p.q) - \langle p, x \rangle). \tag{5}$$

(c) Show that the following conditions are equivalent

(i) $(p, q) \in \partial L(x, y)$
(ii) $p \in \partial_x(-H)(x, q)$ and $y \in \partial_q H(x, q)$ $\tag{6}$

where $\partial_x(-H)$ and $\partial_q H$ denote the subdifferentials of the functions $x \to -H(x, q)$ and $q \to H(x, q)$.

(d) Show that the following conditions are equivalent

(i) $0 \in \partial_x(-H)(\bar{x}, \bar{q})$ and $0 \in \partial_q H(\bar{x}, \bar{q})$

(ii) $(\bar{x}, \bar{q})$ is a saddle point of the function $(x, q) \to H(x, q)$.

15.23 Problem 23 – Lack of Convexity and Fenchel's Theorem for Pareto Optima

We consider two finite-dimensional vector spaces X and Y together with

(i) a *linear* operator A from X to Y

(ii) two *nontrivial* functions $f : X \to \mathbb{R} \cup \{+\infty\}$ and $g : Y \to \mathbb{R} \cup \{+\infty\}$.

We set

$$L := \mathrm{Dom}\, f, \qquad M := \mathrm{Dom}\, g$$

and we suppose that

(i) L is convex (but f is not necessarily convex)

(ii) g is convex.

We set

$$v := \inf_{x \in L} (f(x) + g(Ax)) \tag{1}$$

and

$$w := \inf \left(\sum_{\text{finite}} \alpha_i f(x_i) + g \left(A \left(\sum_{\text{finite}} \alpha_i x_i \right) \right) \right) \tag{2}$$

where in w, the infimum is taken over all (finite) *convex combinations* of points x_i of L (with positive coefficients α_i such that $\sum_{\text{finite}} \alpha_i = 1$). We associate f with the number

$$\rho(f) := \sup \left(f \left(\sum_{\text{finite}} \alpha_i x_i \right) - \sum_{\text{finite}} \alpha_i f(x_i) \right) \tag{3}$$

where the supremum is taken over the (finite) convex combinations of elements of L.

(a) (i) Show that $\rho(f) \geq 0$.

(ii) What are the functions f such that $\rho(f) = 0$?

(iii) Show that

$$v \leq w + \rho(f). \tag{4}$$

(b) We set

(i) $\phi(x, y) := \{Ax - y, f(x) + g(y)\} \in Y \times \mathbb{R}$
(ii) $Q := \{0\} \times]0, \infty[\subset Y \times \mathbb{R}. \tag{5}$

Show that

$$(0, w) \notin \mathrm{co}(\phi(L \times M)) + Q \tag{6}$$

where 'co' denotes the convex hull.

(c) Deduce (carefully, but concisely) from (6) that there exist $c \geq 0$ and $q \in Y^*$ such that

$$cw \leq \inf_{x \in L, y \in M} (cf(x) + cg(y) + \langle q, Ax - y \rangle). \tag{7}$$

(d) Assuming that

$$0 \in \mathrm{Int}(A(L) - M) \tag{8}$$

show that $c > 0$.

(e) Deduce from (a), (c) and (d), that there exists $\bar{q} \in Y^*$ such that

$$v \leq -f^*(-A^*\bar{q}) - g^*(\bar{q}) + \rho(f). \tag{9}$$

(f) Deduce that if we set

$$v^* := \inf_{q \in Y^*} (f^*(-A^*q) + g^*(q)) \tag{10}$$

then

$$0 \leq v + v^* \leq \rho(f). \tag{11}$$

Do you recognize something when f is convex?

(g) Show that assumption (8) is satisfied if we suppose that

$$\exists x_0 \in L \text{ such that } g \text{ is continuous at } Ax_0. \tag{12}$$

(h) What does (e) become if we take g defined by

$$g(y) := \begin{cases} 0 & \text{if } y = u, \ u \text{ given in } \mathbb{R}^n \\ +\infty & \text{otherwise.} \end{cases} \tag{13}$$

15.24 Problem 24 – Duality in Linear Programming

We consider Hilbert spaces X and Y, a continuous linear operator $B \in L(X, Y)$, two convex closed cones $P \subset X$ and $Q \subset Y$ and two elements $u_0 \in Y$ and $p_0 \in X^*$. We introduce the linear program

$$v(u_0) := \inf_{\substack{x \in P \\ Bx \in Q + u_0}} \langle p_0, x \rangle \tag{1}$$

where we suppose that

$$\{x \in P | Bx \in Q + u_0\} \neq \emptyset. \tag{2}$$

(a) Show that the dual problem is

$$v_*(p_0) = \inf_{\substack{q \in Q^- \\ -B^*q \in P^- +p_0}} \langle q, u_0 \rangle. \tag{3}$$

(b) Show that $v(u_0)$ and $v_*(p_0)$ are finite if and only if (2) and

$$\{q \in Q^- \,|\, -B^*q \in P^- + p_0\} \neq \emptyset \tag{4}$$

are satisfied.

(c) Show that

$$u_o \in \text{Int}(BP + Q) \tag{5}$$

implies that there exists a solution $\bar{q}$ of the dual problem $v_*(p_0)$ and that the condition

$$-p_0 \in \text{Int}(P^- + B^*Q^-) \tag{6}$$

implies that there exists a solution $\bar{x}$ of the primal problem $v(u_0)$. Deduce that (5) and (6) imply that

(i) $\partial v(u_0)$ is the set of solutions of the dual problem $v_*(p_0)$
(ii) $\partial v_*(p_0)$ is the set of solutions of the primal problem $v(u_0)$.   (7)

15.25 Problem 25 – Lagrangian of a Convex Minimisation Problem

Suppose that X and Y are two Hilbert spaces and that g is a nontrivial, convex, lower semi-continuous function from $X \times Y$ to $\mathbb{R} \cup \{+\infty\}$. We introduce the marginal function

$$g(y) := \inf_{x \in X} f(x, y) \tag{1}$$

together with the partial conjugate

$$h(x, q) := \sup_{y \in Y}(\langle q, y \rangle - f(x, y)). \tag{2}$$

The function

$$\ell_y(x, q) := \langle q, y \rangle - h(x, q) \tag{3}$$

is called the *Lagrangian* of the family of minimisation problems $g(y)$.
    We fix a parameter y and suppose that there exists a solution $\bar{x}$:

$$g(y) = f(\bar{x}, y). \tag{4}$$

(a) Show that the following conditions are equivalent

(i) $\qquad\qquad\qquad\qquad \bar{q} \in \partial g(\bar{y})$
(ii) $\qquad\qquad\qquad\qquad (0, \bar{q}) \in \partial f(\bar{x}, \bar{y})$
(iii) $\qquad\qquad 0 \in \partial_x(-h)(\bar{x}, \bar{q})$ and $\bar{y} \in \partial_q h(\bar{x}, \bar{q}).$ $\qquad$ (5)

Hint: use Problem 22.

(b) Show that

$$g(y) = \inf_{x \in X} \sup_{q \in Y^*} \ell_y(x, q) \qquad (6)$$

and that

$$g^*(q) = \sup_{x \in X} h(x, q) = f^*(0, q). \qquad (7)$$

(c) Deduce that the marginal function g is lower semi-continuous if for all $y \in Y$,

$$\inf_{x \in X} \sup_{q \in Y^*} \ell_y(x, q) = \sup_{q \in Y^*} \inf_{x \in X} \ell_y(x, q). \qquad (8)$$

(d) We say that $\bar{q} \in Y^*$ is a 'Lagrange multiplier' for the minimisation problem $g(y)$ if and only if

$$g(y) = \inf_{x \in X} \ell_y(x, \bar{q}). \qquad (9)$$

Show that the set of Lagrange multipliers is the subdifferential $\partial g(y)$ of the marginal function g.

15.26 Problem 26 – Variational Principles for Convex Lagrangians

Suppose that we have

(i)   two Hilbert spaces X and Y,
(ii)   a continuous linear operator $A \in L(X, Y)$,
(iii)   a nontrivial, convex, lower semi-continuous function
$\qquad L : X \times Y \to \mathbb{R} \cup \{+\infty\}$ $\qquad\qquad\qquad\qquad\qquad\qquad\qquad$ (1)

We set

$$H(x, q) := \sup_{y \in Y}(\langle q, y \rangle - L(x, y)) \qquad (2)$$

and

$$A(x, q) = L(x, Ax) + L^*(-A^*q, q). \tag{3}$$

We say that $\bar{q}$ is a Lagrange multiplier for the problem

$$v = \inf_{x \in X} L(x, Ax) \tag{4}$$

if and only if

(i) $\qquad\qquad v^* := \inf_{q \in Y^*} L^*(-A^*q, q) = L^*(-A^*\bar{q}, \bar{q})$

(ii) $\qquad\qquad v + v^* = 0. \tag{5}$

(a) Show that the following conditions are equivalent

    (i) $\bar{x}$ minimises $x \to L(x, Ax)$ and $\bar{q}$ is a Lagrange multiplier for v
    (ii) $A(\bar{x}, \bar{q}) = 0 \;\; (= \min A(x, q))$
    (iii) $(-A^*\bar{q}, \bar{q}) \in \partial L(\bar{x}, A\bar{x})$
    (iv) $\bar{x}$ is a solution of the inclusion $0 \in (1 \oplus A^*)\partial L(\bar{x}, A\bar{x})$
    (v) $\bar{q}$ is a solution of the inclusion $0 \in (-A \oplus 1)\partial L^*(-A^*\bar{q}, \bar{q}). \tag{6}$

(b) Using the results of Problem 22, show that each of these conditions is equivalent to

$$A^*\bar{q} \in \partial_x H(\bar{x}, \bar{q}) \;\; \text{and} \;\; A\bar{x} \in \partial_q H(\bar{x}, \bar{q}). \tag{7}$$

15.27 Problem 27 – Variational Principles for Convex Hamiltonians

We consider

(i) two Hilbert spaces X and Y,

(ii) a nontrivial, convex, lower semi-continuous function
$H : X \times Y^* \to \mathbb{R} \cup \{+\infty\}. \tag{*}$

We associate these with the function L from $X \times Y$ to $\overline{\mathbb{R}}$ defined by

$$L(x, y) := \sup_{q \in Y^*} \langle q, y \rangle - H(x, q). \tag{**}$$

We define the function B on $X \times Y^*$ by

$$B(x, q) = H(x, q) - \langle q, Ax \rangle + H^*(A^*q, Ax) - \langle A^*q, x \rangle. \tag{***}$$

(a) Deduce from the results of Problem 22 that the following conditions are equivalent

    (i) $B(\bar{x}, \bar{q}) = 0 \;\; (= \min B(x, q))$
    (ii) $0 \in -\partial_x(-L)(\bar{x}, A\bar{x}) + A^*\partial_y L(\bar{x}, A\bar{x})$
    (iii) $(A^*\bar{q}, A\bar{x} \in \partial H(\bar{x}, \bar{q}). \tag{1}$

(b) Show that any solution $(\bar{x}, \bar{q})$ of the 'law of least action'

$$w = \inf_{(x,q) \in X \times Y^*} (H(x,q) - \langle q, Ax \rangle) \qquad (2)$$

is a solution of (1).

(c) We suppose that

$$0 \in \mathrm{Int}\,(\mathrm{Im}\,(A^* \times A) - \mathrm{Dom}\,H^*). \qquad (3)$$

Show that one solution $(\bar{x}, \bar{q})$ of the 'law of least action dual problem'

$$w^* = \inf_{(x,q) \in X \times Y^*} (H^*(A^*q, Ax) - \langle A^*q, x \rangle) \qquad (4)$$

is a solution of the dual problem (1).

15.28 Problem 28 – Approximation to Fermat's Rule

Suppose that X is a Hilbert space and that $f : X \to \mathbb{R}_+ \cup \{+\infty\}$ is a nontrivial, convex, positive, lower semi-continuous function.

(a) Fix $\varepsilon > 0$, $\lambda > 0$ and $x_0 \in \mathrm{Dom}\,f$ such that

$$f(x_0) \le \inf_{x \in X} f(x) + \varepsilon \lambda.$$

Show that there exist $x_\varepsilon \in \mathrm{Dom}\,f$ and $p_\varepsilon \in \partial f(x_\varepsilon)$ such that

$$f(x_\varepsilon) \le f(x_0), \quad \|x_0 - x_\varepsilon\| \le \lambda \text{ and } \|p_\varepsilon\| \le \varepsilon$$

(use Ekeland's Theorem, Theorem 1.2).

(b) Suppose now that $f : X \to \mathbb{R} \cup \{+\infty\}$ is nontrivial, convex and lower semi-continuous. Show that for any $x \in \mathrm{Dom}\,f$, there exists a sequence of elements $x_n \in X$ such that

$$x_n \to x, \quad f(x_n) \to f(x) \text{ and } \partial f(x_n) \neq \emptyset.$$

Deduce that the set of points at which f is subdifferentiable is dense in $\mathrm{Dom}\,f$. Compare with Theorem 4.3.

15.29 Problem 29 – Transposes of Convex Processes

We consider a set-valued map F from a Hilbert space X to a Hilbert space Y and suppose that

$$\text{the graph of } F \text{ is a convex closed cone} \qquad (1)$$

(F is said to be a convex process).

Consider the set-valued map F^* from Y^* to X^* defined by

$$p \in F^*(q) \Leftrightarrow \sup_{x \in X} \sup_{y \in F(x)} (\langle p, x \rangle - \langle q, y \rangle) = 0. \qquad (2)$$

(a) Show that the graph of F^* is a closed convex cone and that $F = F^{**}$.

(b) Suppose that $B \in L(X, Z)$ is a continuous linear operator. Show that $\mathrm{Graph}(BF) = (1 \times B)\mathrm{Graph}(F)$ and deduce that

$$(BF)^* = F^*B^*. \tag{3}$$

(c) Let U be a Hilbert space and $A \in L(U, X)$. Suppose that

$$\mathrm{Im}\, A - \mathrm{Dom}\, F = X. \tag{4}$$

Show that $\mathrm{Graph}(AF) = (A \times 1)^{-1}\mathrm{Graph}(F)$ and deduce (using formula (70) of Chapter 3) that

$$(FA)^* = A^*F^*. \tag{5}$$

(d) Suppose that X, Y and Z are three Hilbert spaces and that $F : X \to Z$ and $G : Y \to Z$ are set-valued maps the graphs of which are convex closed cones. Show that the assumption

$$A\, \mathrm{Dom}\, F - \mathrm{Dom}\, G = Y \tag{6}$$

implies that

$$(F + GA)^* = F^* + A^*G^*.$$

(e) Suppose that $F : X \to Y$ is a set-valued map satisfying (1) and that $K \subset X$ is a convex closed cone. Show that the assumption

$$K - \mathrm{Dom}\, F = X \tag{7}$$

implies that $(F|_K)^*(q) = F^*(q) + K^-$.

(f) Show that

$$\mathrm{Im}\,(F)^- = -F^{*-1}(0). \tag{8}$$

Deduce that the image of F is dense in Y if and only if $F^{*-1}(0) = \{0\}$ and that F is surjective if and only if $\mathrm{Im}(F)$ is closed and $F^{*-1}(0) = \{0\}$.

(g) Deduce from the previous two questions that if K is a convex closed cone of X and that if (7) is satisfied, then

$$F(K)^- = -F^{*-1}(-K^-).$$

(h) We consider a set-valued map F from X to X^* satisfying (1) and

$$\exists c > 0 \text{ such that } \forall (x_i, y_i) \in \mathrm{Graph}(G), i = 1, 2$$
$$\langle y_1 - y_2, x_1 - x_2 \rangle \geq c\|x_1 - x_2\|^2. \tag{9}$$

Show that $\mathrm{Im}(F)$ is closed and that its inverse F^{-1} is one-to-one and Lipschitz with constant c^{-1}. Deduce that if $\mathrm{Dom}(F) = X$ then F is surjective.

15.30 Problem 30 – Cones with a Compact Base

Let P be a convex closed cone of a Hilbert space X and P^+ its positive polar cone.

(a) Show that if $p_0 \in \text{Int } P^+$ then

$$\forall x \in P, \quad x \neq 0, \quad \langle p_0, x \rangle > 0 \tag{1}$$

and deduce that the set

$$S := \{x \in P | \langle p_0, x \rangle = 1\} \tag{2}$$

generates the cone S in the sense that

$$P = \bigcup_{\lambda \geq 0} \lambda S. \tag{3}$$

(b) Deduce that S is a convex, closed, bounded set (which does not contain zero).

(c) Conversely, if K is a convex, closed, bounded set which does not contain 0 and which generates P in the sense that $P = \cup_{\lambda \geq 0} \lambda K$, show that $\text{Int } P^+ \neq \emptyset$.

(d) Show that

$$T_S(x) := \{v \in T_P(x) | \langle p_0, v \rangle = 0\} \tag{4}$$

that

$$T_P(x) = \text{closure}(P + \mathbb{R}x) \quad \text{and} \quad N_P(x) = P^- \cap \{x\}^\perp \tag{5}$$

and that

$$\begin{aligned}
p &\in N_P(x) \Leftrightarrow x \in N_{P^-}(p) \Leftrightarrow x \in P, \\
p &\in P^- \quad \text{and} \quad \langle p, x \rangle = 0.
\end{aligned} \tag{6}$$

15.31 Problem 31 – Regularity of Tangent Cones

Suppose that $K \subset X$ is a convex subset of a Hilbert space X and that $x \in K$.

(a) Set

$$S_K(x) := \bigcup_{h>0} \frac{1}{h}(K - x). \tag{1}$$

Show that

$$\forall v \in S_K(x), \exists h > 0 \text{ such that } \forall t \in [0, h], \ x + tv \in K. \tag{2}$$

(b) Consider the set

$$
C_K(x) := \bigcap_{\varepsilon>0} \bigcup_{\alpha,\beta>0} \bigcap_{\substack{h\in]0,\alpha[\\y\in B_K(x,\beta)}} \left(\frac{1}{h}(K-y)+\varepsilon B\right)
\tag{3}
$$

where $B_K(x,\alpha) := K\cap(x+\alpha B)$. Show that

$$
C_K(x) = T_K(x) \qquad (:= \text{closure}(S_K(x))).
$$

(c) Let π_K be the projector of best approximation onto the set K, which we assume to be closed (Problem 17). Show that

(i) $\qquad\qquad\qquad N_K(x) = \pi_K^{-1}(x) - x$

(ii) $\qquad v \in T_K(x) \Leftrightarrow \forall y \in \pi_K^{-1}(x),\ \langle y-x,v\rangle \le 0.$ $\qquad$ (4)

(d) Suppose that K is closed and show that the graph of the set-valued map $x \to N_K(x)$ is closed. Deduce from Problem 5 that if X is finite dimensional then $x \to T_K(x)$ is a lower semi-continuous set-valued map.

(e) Suppose that $\text{Int } K \ne \emptyset$ and show that

$$
\text{Int } T_K(x) = \bigcup_{h>0} \frac{1}{h}(\text{Int } K - x).
\tag{5}
$$

(f) Deduce that $x \to \text{Int } T_K(x)$ has an open graph.

15.32 Problem 32 – Tangent Cones to an Intersection

(a) Take

$$
K_1 = (-1,0) + B, \qquad K_2 = \mathbb{R}_+^2.
$$

Verify that $(0,0) \notin \text{Int}(K_1 - K_2)$ and calculate

$$
T_{K_1}(0,0) \cap T_{K_2}(0,0) \text{ and } T_{K_1\cap K_2}(0,0).
$$

(b) Take $K_1 = [-1,0] \times [-1,+1]$, $K_2 = [0,1] \times [-1,+1]$, verify that $(0,0) \notin \text{Int}(K_1 - K_2)$ and calculate

$$
T_{K_1}(0,0) \cap T_{K_2}(0,0) \text{ and } T_{K_1\cap K_2}(0,0).
$$

(c) Consider n convex closed subsets K_i ($i = 1, 2, \ldots, n$) and take $x \in \cap_{i=1}^n K_i$. Suppose that there exists $\gamma > 0$ such that

$$
\forall v_i \in \gamma B\ (i = 1,\ldots,n),\ \cap_{i=1}^n(K_i - v_i) \ne \emptyset.
\tag{1}
$$

Show that

$$
\forall x \in \cap_{i=1}^n K_i,\ T_K(x) = \cap_{i=1}^n T_{K_i}(x).
\tag{2}
$$

Hint: consider $A := x \in X \to (x, x, \ldots, x) \in X^n$. Show that $K = A^{-1}(\prod_{i=1}^n K_i)$ and apply formula (50) of Chapter 4.

15.33 Problem 33 – Derivatives of Set-valued Maps with Convex Graphs

Suppose that $F : X \to Y$ is a set-valued map.

(a) Show that the graph of F is convex if and only if

$$\forall x, y \in \text{Dom}(F), \quad \forall \alpha \in [0, 1],$$
$$\alpha F(x) + (1 - \alpha) F(y) \subset F(\alpha x + (1 - \alpha) y). \tag{1}$$

(b) Let (x_0, y_0) be an element of the graph of F. We define the set-valued maps $DF(x_0, y_0)$ from X to Y and $DF(x_0, y_0)^*$ from Y^* to X^* by

$$v \in DF(x_0, y_0)(u) \Leftrightarrow (u, v) \in T_{\text{Graph}(F)}(x_0, y_0) \tag{2}$$

and

$$p \in DF(x_0, y_0)^*(q) \Leftrightarrow (p, -q) \in N_{\text{Graph}(F)}(x_0, y_0). \tag{3}$$

Show that the following conditions are equivalent

(i) $\qquad\qquad\qquad\qquad p \in DF(x_0, y_0)^*(q)$
(ii) $\qquad \forall x \in X, \quad \forall y \in F(x), \quad \langle q, y_0 - y \rangle \le \langle p, x_0 - x \rangle$
(iii) $\qquad \forall u \in X, \quad \forall v \in DF(x_0, y_0)(u), \quad \langle p, u \rangle \le \langle q, v \rangle. \tag{4}$

(c) Show that if $p_i \in DF(x_i, y_i)^*(q_i)$, $(i = 1, 2)$, then

$$\langle q_1 - q_2, y_1 - y_2 \rangle \le \langle p_1 - p_2, x_1 - x_2 \rangle. \tag{5}$$

(d) Suppose that K is a closed convex subset of X and that ϕ_K is the set-valued map from X to Y defined by

$$\phi_K(x) = \{0\} \text{ if } x \in K \text{ and } \phi_K(x) = \emptyset \text{ if } x \notin K. \tag{6}$$

Prove that

$$\forall x \in K, \quad D\phi_K(x, 0) = \phi_{T_K}(x). \tag{7}$$

(e) Show that $h \to d(v, \frac{F(x+hu)-y}{h})$ is increasing and deduce that $v_0 \in DF(x_0, y_0)(u_0)$ if and only if

$$\liminf_{\substack{u \to u_0 \\ h \to 0+}} \inf d\left(v_0, \frac{F(x_0 + hu) - y_0}{h}\right) = 0. \tag{8}$$

(f) Show that if $x_0, x \in \text{Dom}(F)$, $y_0 \in F(x_0)$, then

$$F(x) - y_0 \subset DF(x_0, y_0)(x - x_0). \tag{9}$$

(g) Suppose that P is a convex closed cone of X. Show that the following conditions on $(x_0, y_0) \in \text{Graph}(F)$ are equivalent:

(i) $\qquad\qquad \forall x \in K, \quad F(x) \subset y_0 + P$

(ii) $\qquad\qquad \forall u_0 \in X, \quad DF(x_0, y_0)(u_0) \subset P$

(iii) $\qquad\qquad \forall q \in P^+, \quad 0 \in DF(x_0, y_0)^*(q).$ $\qquad$ (10)

Hint: show that (ii) $\Rightarrow$ (i) $\Rightarrow$ (iii) $\Rightarrow$ (ii).

(h) Suppose that X and Y are two Hilbert spaces, that $L \subset X$ and $M \subset Y$ are two convex closed subsets and that $A \in L(X, Y)$ is a continuous linear operator. We associate A with the set-valued map F defined by

$$F(x) := \begin{cases} Ax - M & \text{if } x \in L \\ \emptyset & \text{if } x \notin L. \end{cases} \qquad (11)$$

Show that the graph of F is convex and closed, that

$$F^{-1}(y) = L \cap A^{-1}(M + y)$$

and that

$$DF(x, y)(u) = \begin{cases} Au - T_M(Ax - y) & \text{if } u \in T_L(x) \\ \emptyset & \text{if } u \notin T_L(x). \end{cases} \qquad (12)$$

Deduce that

$$DF(x, y)^*(q) = \begin{cases} A^*q + N_L(x) & \text{if } q \in N_M(Ax - y) \\ \emptyset & \text{if } q \notin N_M(Ax - y). \end{cases}$$

15.34 Problem 34 – Epiderivatives of Convex Functions

We consider a function $f : X \to \mathbb{R} \cup \{+\infty\}$ with which we associate a set-valued map F_+ from X to $\mathbb{R}$ defined by

$$F_+(x) = f(x) + \mathbb{R}_+ \text{ if } f(x) < +\infty, \qquad F_+(x) = \emptyset \text{ if } f(x) = +\infty. \qquad (1)$$

(a) Prove that $\text{Dom} f = \text{Dom} F_+$, that $\text{Graph}(F_+) = \text{Ep}(f)$ and that $F_+^{-1}(\lambda) = S(f, \lambda)$. Deduce that f is lower semi-continuous (respectively convex) if and only if the graph of F_+ is closed (respectively convex).

(b) We recall (see Problem 33) that the set-valued maps $DF_+(x, \lambda) : X \to \mathbb{R}$ and $DF_+(x, \lambda)^* : \mathbb{R} \to X$ are defined by

(i) $\qquad v \in DF_+(x, \lambda)(u) \Leftrightarrow (u, v) \in T_{\text{Graph}(F_+)}(x, \lambda)$

(ii) $\qquad p \in DF_+(x, \lambda)^*(q) \Leftrightarrow (p, -q) \in N_{\text{Graph}(F_+)}(x, \lambda).$ $\qquad$ (2)

We set

$$D_+f(x)(u_0) := \liminf_{u \to u_0} Df(x)(u) \tag{3}$$

(the *epiderivative* of f at x in the direction u_0). Show that

(i) $\qquad DF_+(x, f(x))(u_0) = D_+f(x)(u_0) + \mathbb{R}_+$

(ii) $\qquad DF_+(x, f(x))^*(q) = \begin{cases} q\partial f(x) & \text{if } q > 0 \\ \emptyset & \text{if } q < 0. \end{cases} \tag{4}$

Deduce that

(i) $\qquad \mathrm{Ep}\,(D_+f(x)) = T_{\mathrm{Ep}(f)}(x, f(x))$

(ii) $\qquad p \in \partial f(x) \Leftrightarrow (p, -1) \in N_{\mathrm{Ep}\,(f)}(x, f(x)). \tag{5}$

15.35 Problem 35 – Subdifferentials of Marginal Functions

We consider a nontrivial, convex, lower semi-continuous function $f : X \to \mathbb{R} \cup \{+\infty\}$ and a set-valued map F from X to Y with a *convex, closed graph*. The marginal function h is defined on Y by

$$h(y) := \inf_{x \in F^{-1}(y)} f(x). \tag{1}$$

We also suppose that

$$0 \in \mathrm{Int}(\mathrm{Dom}f - \mathrm{Dom}(F)). \tag{2}$$

(a) If $\bar{x} \in F^{-1}(y)$ is a solution of $h(y)$ $(h(y) = f(\bar{x}))$, show that $q \in \partial h(y)$ if and only if there exists $\bar{p} \in \partial f(\bar{x})$ such that

$$(-\bar{p}, q) \in N_{\mathrm{Graph}(F)}(\bar{x}, y). \tag{3}$$

(b) Deduce from the definition of $DF(\bar{x}, y)^*$ (Problem 33 (b)) that

$$\partial h(y) = -DF(\bar{x}, y)^{*-1}(-\partial f(\bar{x})). \tag{4}$$

15.36 Problem 36 – Values of a Game Associated with a Covering

Suppose $f : X \times Y \to \mathbb{R}$. Consider a covering of $\mathcal{A}$ by subsets K, L of F satisfying

$$\text{if } K \text{ and } L \in \mathcal{A}, \text{ then } K \cup L \in \mathcal{A}. \tag{1}$$

We associate this covering $\mathcal{A}$ with the 'value'

$$v^{\natural}(\mathcal{A}) := \sup_{K \in \mathcal{A}} \inf_{x \in E} \sup_{y \in K} f(x, y). \tag{2}$$

(a) Show that if $\mathcal{A} \subset \mathcal{B}$, then

$$v^{\natural} \leq v^{\natural}(\mathcal{A}) = v^{\natural}(\mathcal{B}) \tag{3}$$

(we recall that $v^{\natural} = v^{\natural}(\mathcal{S})$, see formula (18), Chapter 8).

(b) If $\mathcal{U}$ is the family of all the subsets of F, show that for all $\mathcal{A}$,

$$v^{\natural}(\mathcal{A}) \leq v^{\natural}(\mathcal{U}) = v^{\natural} \qquad (:= \inf_{x \in E} \sup_{y \in F} f(x, y)). \tag{4}$$

(c) We consider the order relation on $\mathcal{A} \times \mathbb{N}$ given by $(K_1, n_1) \leq (K_2, n_2) \Leftrightarrow K_1 \subset K_2$ and $n_1 \leq n_2$. Show that there exists a generalised sequence $(K, n) \to x_{K,n} \in E$ such that

$$\forall K_0 \in \mathcal{A}, \quad \limsup_{(K,n) \geq (K_0, 1)} \left(\sup_{y \in K_0} f(x_{K,n}, y) \leq v^{\natural}(\mathcal{A}) \right). \tag{5}$$

(d) If the covering $\mathcal{A}$ is countable, show that there exists a sequence of elements $x_p \in E$ such that

$$\forall K_0 \in \mathcal{A}, \exists p_0 \text{ such that } \forall p \geq P_0, \limsup_{p \geq P_0} \sup_{y \in K} f(x_p, y) \leq v^{\natural}(\mathcal{A}). \tag{6}$$

(e) We suppose that

   (i) $\forall y \in F, x \to f(x, y)$ is lower semi-continuous
   (ii) $\exists K_0 \in \mathcal{A}$ such that $\{x \mid \sup_{y \in K_0} f(x, y) \leq v^{\natural}(\mathcal{A})\}$ is compact. (7)

Show that there exists $\bar{x} \in E$ such that $\sup_{y \in K} f(\bar{x}, y) \leq v^{\natural}(\mathcal{A}) \leq v^{\natural}$.

15.37 Problem 37 – Minimax Theorems with Weak Compactness Assumptions

(a) Let f be a function from $X \times Y$ to $\overline{\mathbb{R}} := \mathbb{R} \cup \{-\infty\} \cup \{+\infty\}$. We suppose that the sets

$$E := \{x \in X \mid \sup_{y \in Y} f(x, y) < +\infty\} \tag{1}$$

and

$$F := \{y \in Y \mid \inf_{x \in X} f(x, y) > -\infty\} \tag{2}$$

are non-empty and that X is finite dimensional (if not, use the weak topology). We introduce the family $\mathcal{A}$ of subsets

$$K_n := \{y \in F \mid \|y\| \leq p \text{ and } \inf_{x \in X} f(x, y) \geq -p\} \tag{3}$$

and use the definition of $v^{\natural}(\mathcal{A})$ given in formula (2) of Problem 36. Show that if

(i) $\forall y \in E$, $x \to f(x,y)$ is convex
(ii) $\forall x \in E$, $y \to f(x,y)$ is concave and upper semi-continuous   (4)

then

$$v^{\natural}(\mathcal{A}) = v^{\flat} \qquad (:= \sup_{y \in F} \inf_{x \in E} f(x,y)). \tag{5}$$

Hint: use the minimax theorem, Theorem 8.1.

(b) We suppose further that

$$\forall y \in F, \ x \to f(x,y) \text{ is lower semi-continuous} \tag{6}$$

and that if we set

$$f_y^*(p); = \sup_{x \in X}(\langle p, x \rangle - f(x,y)) \tag{7}$$

then

$$0 \in \mathrm{Int}\left(\cup_{y \in F}\mathrm{Dom}\, f_y^*\right). \tag{8}$$

Prove that there exists $\bar{x} \in E$ such that

$$\sup_{y \in F} f(\bar{x}, y) = v^{\natural}(\mathcal{A}). \tag{9}$$

Hint: use Problem 36 (d).

Deduce that under the assumptions (3), (6) and (8), we have

$$\sup_{y \in F} f(\bar{x}, y) = v^{\flat} = v^{\sharp}. \tag{10}$$

Thus, Theorem 8.1 remains true if the compactness of E is replaced by the weaker assumption (8).

15.38 Problem 38 – Minimax Theorems for Finite Topologies

Suppose that F is a convex subset of the infinite-dimensional vector space Y. We associate any finite subset $K = \{y_1, \ldots, y_n\}$ of F with the mapping β_K from the simplex M^n to F defined by

$$\forall \lambda \in M^n, \ \ \beta_K(\lambda) = \sum_{i=1}^{n} \lambda_i y_i. \tag{1}$$

The 'finite topology' on F is the 'ultimate' topology, the strongest of all topologies, in which the mappings β_K from M^n to F as K ranges over the family $\mathcal{S}$ of finite subsets $K \subset F$ are continuous. We recall that a mapping f from F to a topological space G is continuous if and only if

$$\forall K \in \mathcal{S}, \ \ f\beta_K : M^n \to G \text{ is continuous.} \tag{2}$$

(a) Show that the finite topology on a convex subset F is stronger than any vector-space topology.

(b) Show that any affine mapping A from F to a vector space Z is continuous when F and Z have the finite topology.

(c) Show that in the proof of Theorem 8.4, we may give F the finite topology and thus reduce the space $\mathcal{C}(F, F)$ of continuous decision rules; whence, we may obtain a stronger version of equation (48) of Chapter 8.

(d) Do the same for Theorem 8.5.

15.39 Problem 39 – Ky Fan's Inequality

We consider a *convex compact* subset K of a finite-dimensional vector space and a function $\varphi : K \times K \to \mathbb{R}$.

(a) We set

$$T(y) := \{x \in K | \varphi(x, y) > 0\} \tag{1}$$

and we suppose that

$$\forall y \in K, \quad x \to \varphi(x, y) \text{ is lower semi-continuous.} \tag{2}$$

Show that the negation of the property

$$\exists \bar{x} \in K \text{ such that } \sup_{y \in K} \varphi(x, y) \leq 0 \tag{3}$$

implies the existence of $y_1, \ldots, y_n \in K$ such that

$$K = \bigcup_{i=1}^{n} T(y_i). \tag{4}$$

(b) We suppose further that

$$\forall x \in K, \ y \to \varphi(x, y) \text{ is concave.} \tag{5}$$

Deduce from Brouwer's Fixed-point Theorem that there exists $\bar{x} \in K$ such that

$$\varphi(\bar{x}, \bar{x}) > 0. \tag{6}$$

Hint: use a continuous partition of unity subordinate to the covering of K by the $T(y_i)$.

(c) Deduce that Brouwer's Theorem implies the Ky Fan Inequality, namely that assumptions (2) and (5) together with

$$\forall y \in K, \ \varphi(y, y) \leq 0 \tag{7}$$

imply assertion (3).

(d) Prove that the Ky Fan Inequality remains true if K is a convex compact subset of a Hilbert space. Hint: use Proposition 8.2.

(e) Prove the converse, namely that the Ky Fan Inequality implies that any continuous mapping from a convex compact subset of a Hilbert space to itself has a fixed point.

15.40 Problem 40 – Ky Fan's Inequality for Monotone Functions

Suppose that

$$K \text{ is a convex compact subset} \tag{1}$$

and that $\varphi : K \times K \to \mathbb{R}$ is a function satisfying

(i) $\forall y \in K, x \to \varphi(x, y)$ is lower semi-continuous for the finite topology (see Problem 38)

(ii) $\forall x \in K, y \to \varphi(x, y)$ is concave and upper semi-continuous

(iii) $\forall y \in K, \varphi(y, y) \leq 0$. \hfill (2)

We suppose further that φ is monotone in the sense that

$$\forall x, y \in K, \ \varphi(x, y) + \varphi(y, x) \geq 0. \tag{3}$$

(a) Verify that

$$v^\natural := \sup_{K \in \mathcal{S}} \inf_{x \in E} \sup_{y \in K} \varphi(x, y) \leq 0. \tag{4}$$

(b) Assuming that $\bar{x} \in K$ satisfies

$$0 < \sup_{y \in K} \varphi(\bar{x}, y) \tag{5}$$

show that there exist $\bar{y}$ and $\bar{t} \in]0, 1[$ such that

$$0 < \varphi(\bar{x} + \bar{t}(\bar{y} - \bar{x}), \bar{y}). \tag{6}$$

(c) Problem 36 (c) and the compactness of K imply that there exists a generalised sequence of elements x_μ of K converging to an element $\bar{x}$ and satisfying

$$\forall y \in K, \ \exists \mu(y) \text{ such that } \limsup_{\mu \geq \mu(y)} \varphi(x_\mu, y) \leq v^\natural. \tag{7}$$

Use the fact that φ is monotone to show that

$$0 < \varphi(\bar{x} + \bar{t}(\bar{y} - \bar{x}), \bar{x} + \bar{t}(\bar{y} - \bar{x})). \tag{8}$$

(d) Deduce from this inequality and the fact that φ is monotone that

$$\exists \bar{x} \in K \text{ such that } \sup_{y \in K} \varphi(\bar{x}, y) \leq 0. \tag{9}$$

15.41 Problem 41 – Generalisation of the Gale–Nikaïdo–Debreu Theorem

We consider

(i) a compact metric space K,

(ii) a set-valued map F from K to a Hilbert space Y (identified with its dual),

(iii) a convex closed cone P of Y and its negative polar cone P^-. $\hfill$ (1)

(a) We suppose that

$$F \text{ is upper hemi-continuous with convex compact values} \tag{2}$$

and that there exists a continuous mapping C from P^- to K such that

$$\sigma(F(C(y), y)) \geq 0 \ \forall y \in P^-. \tag{3}$$

Show that there exists $\bar{x} \in K$ such that $0 \in F(\bar{x}) + P$. Hint: adapt the proof of Theorem 9.2.

(b) Show that we may weaken the assumptions by supposing only that

(i) $\forall y \in P^-$, $x \to \sigma(F(x), y)$ is upper semi-continuous,

(ii) $\forall x \in K$, $F(x) + P$ is convex and closed,

(iii) $\forall \varepsilon > 0$, $\exists C_\varepsilon$, a continuous mapping from P^- to K such that $\sigma(F(C_\varepsilon(y), y)) \geq -\varepsilon, \forall y \in P^-$. $\hfill$ (4)

(c) We now suppose that

$$K \text{ is a convex compact subset of } Y \tag{5}$$

and that F satisfies (2) and

$$\forall y \in Y, \ \sigma(F(\pi_K(y), y)) \geq 0 \qquad (\pi_K \text{ is a projector onto } K). \tag{6}$$

Deduce that there exists $\bar{x} \in K$ such that $0 \in F(\bar{x})$.

(d) Consider the unit ball B of a finite-dimensional vector space Y and a set-valued map F from B to Y satisfying (2) and

$$\forall x \in B, \quad \sigma(F(x), x) \geq 0. \tag{7}$$

Deduce that there exists $\bar{x} \in B$ such that $0 \in F(\bar{x})$.

15.42 Problem 42 – Equilibrium of Coercive Set-valued Maps

We consider

(i) a convex closed subset K of a finite-dimensional space X,

(ii) an upper hemi-continuous function with convex closed values,   (1)

satisfying the tangential condition

$$\forall x \in K, \quad F(x) \cap T_K(x) \neq \emptyset. \tag{2}$$

(a) Show that the condition

$$\lim_{\|x\| \to \infty, x \in K} \sigma(F(x), x) < 0 \tag{3}$$

implies the existence of an equilibrium of F, $\bar{x} \in K$.

(b) Show that the condition

$$\lim_{\|x\| \to \infty, x \in K} \frac{\sigma(F(x), x)}{\|x\|} = -\infty \tag{4}$$

implies that

$$\forall y \in K, \quad \exists \hat{x} \in K, \text{ solution of } y \in \hat{x} - F(\hat{x}). \tag{5}$$

15.43 Problem 43 – Eigenvectors of Set-valued Maps

Suppose that P is a convex cone of a finite-dimensional space X such that $\operatorname{Int} P^+ \neq \emptyset$. Suppose that F is a set-valued map from P to X which is upper semi-continuous with convex compact values and satisfies

$$\forall x \in P, \quad F(x) \cap T_P(x) \neq \emptyset. \tag{1}$$

(a) We take $p_0 \in \operatorname{Int} P^+$ and $S := \{x \in P | \langle p_0, x \rangle = 1\}$ which is convex, closed and bounded (whence compact, see Problem 30). Show that the set-valued map G from S to X defined by

$$G(x) := \{v - \langle p_0, v \rangle x\}_{v \in F(x)} \tag{2}$$

satisfies the tangential condition

$$\forall x \in S, \quad G(x) \cap T_S(x) \neq \emptyset. \tag{3}$$

(b) Deduce that there exists an eigenvector $\bar{x} \in P$ of F, in other words a solution of

$$\bar{x} \in S, \quad \bar{\lambda}\bar{x} \in F(\bar{x}).$$

Hint: apply Theorem 9.4 to the set-valued map G.

15.44 Problem 44 – Positive Eigenvectors of Positive Set-valued Maps

We consider

- a mapping $F : M^n \to \mathbb{R}^m$ with convex lower semi-continuous components f_i (1)

and

- an upper semi-continuous set-valued map G from M^n to $\mathbb{R}^m_+$ with non-empty compact values and a closed graph (2)

We suppose that

(i) $\exists \tilde{p} \in M^n$ such that $\forall x \in M^n$, $\langle \tilde{p}, F(x) \rangle > 0$
(ii) $\exists \tilde{x} \in M^n$ such that $\forall p \in M^m$, $\sigma(G(\tilde{x}), p) > 0$. (3)

We set

$$\frac{1}{\delta} := \sup_{p \in M^m} \inf_{x \in M^n} \frac{\langle p, F(x) \rangle}{\sigma(G(x), p)}. \tag{4}$$

(a) Show that the set $(G - \delta F)(M^n) - \mathbb{R}^m_+$ is convex and closed.

(b) Deduce that there exists $\bar{x}$ such that

$$\delta F(\bar{x}) \in G(\bar{x}) - \mathbb{R}^m_+. \tag{5}$$

(c) Deduce that there exists $\bar{p} \in M^m$ such that

$$\frac{1}{\delta} = \inf_{x \in M^n} \frac{\langle \bar{p}, F(x) \rangle}{\sigma(G(x), \bar{p})} = \sup_{p \in M^m} \frac{\langle p, F(\bar{x}) \rangle}{\sigma(G(\bar{x}), p)}. \tag{6}$$

(d) Show that if there exist $\tilde{x} \in M^n$ and $\mu > 0$ such that

$$\mu F(x) \in G(x) - \mathbb{R}^n_+, \text{ then } \mu \geq \delta. \tag{7}$$

(e) Take $\mu > \delta$ and $y \in \text{Int}(\mathbb{R}^m_+)$ and set

$$\beta := \sup_{p \in M^m} \inf_{x \in M^n} \frac{\mu \langle p, F(x) \rangle + \sigma(-G(x), p)}{\langle p, y \rangle} \tag{8}$$

Show that $\beta > 0$ and that there exists $\hat{x}$ such that

$$\beta y \in \mu F(\hat{x}) - G(\hat{x}) + \mathbb{R}^m_+. \tag{9}$$

15.45 Problem 45 – Some Variational Principles

We consider a nontrivial, convex, lower semi-continuous function f from X to $\mathbb{R} \cup \{+\infty\}$ and a mapping A from $\mathrm{Dom}\, f$ to X. We set

$$\phi(y) := f(y) + f^*(-Ay) + \langle A(y), y \rangle. \tag{1}$$

(a) Show that $\forall y \in \mathrm{Dom}\, f,\ \phi(y) \geq 0$.

(b) Show that the following problems are equivalent

(i) $\exists \bar{x} \in \mathrm{Dom}\, f$ such that $0 \in A(\bar{x}) + \partial f(\bar{x})$
(ii) $\exists \bar{p} \in \mathrm{Dom}\, f^*$ such that $0 \in \bar{p} + A\partial f^*(\bar{p})$
(iii) $\exists \bar{x} \in \mathrm{Dom}\, f$ such that
       $\forall y \in \mathrm{Dom}\, f,\ \langle A(\bar{x}), \bar{x} - y \rangle + f(\bar{x}) - f(y) \leq 0$
(iv) $\exists \bar{x} \in \mathrm{Dom}\, f$ such that $\phi(\bar{x}) = 0\ (= \min_{y \in \mathrm{Dom}\, f} \phi(y))$. $\tag{2}$

(c) Identify these problems when $f = \psi_K$ is the indicator function of a nonempty, convex, closed set K.

(d) Suppose now that A is a set-valued map from $\mathrm{Dom}\, f$ to X with convex compact values. We set

$$\phi(y) := f(y) + \inf_{u \in A(y)} (f^*(-u) + \langle u, y \rangle). \tag{3}$$

Show that $\forall y \in \mathrm{Dom}\, f,\ \phi(y) \geq 0$ and (b) remains true if we replace (2)(iii) by

$$\forall y \in \mathrm{Dom}\, f,\ -\sigma(A(\bar{x}), y - \bar{x}) + f(\bar{x}) - f(y) \leq 0. \tag{4}$$

(e) Show that a necessary condition for the existence of a solution to one of the equivalent problems (2) is that

$$0 \in \mathrm{Dom}\, f^* + A\, \mathrm{Dom}\, f. \tag{5}$$

15.46 Problem 46 – Generalised Variational Inequalities

We consider a finite-dimensional vector space X together with

(i) a nontrivial, convex, lower semi-continuous function $f : X \to \mathbb{R} \cup \{+\infty\}$,

(ii) a nontrivial, convex, lower semi-continuous function $\beta : X \to \mathbb{R}_+ \cup \{+\infty\}$,

(iii) an upper semi-continuous set-valued map $A : \mathrm{Dom}\, f \to X$ with compact values, which is β-monotone in the sense that

$$\forall (x, p), (y, q) \in \mathrm{Graph}(A), \quad \langle p - q, x - y \rangle \geq \beta(x - y). \tag{1}$$

We recall the results of Problem 45, namely that the problems

(i) $\qquad\qquad\qquad\qquad 0 \in A(\bar{x}) + \partial f(\bar{x})$

(ii) $\qquad \forall y \in \mathrm{Dom}\, f, \quad -\sigma(A(\bar{x}), y - \bar{x}) + f(\bar{x}) - f(y) \leq 0 \tag{2}$

are equivalent and that a necessary condition for the existence of a solution is that

$$0 \in \mathrm{Dom}\, f^* + A\mathrm{Dom}\, f. \tag{3}$$

(a) Show that $0 \in \mathrm{Dom}\, \beta^*$. Calculate $\mathrm{Dom}\, \beta^*$ when β is equal to one of the following three functions

$$\beta_0(z) \equiv 0, \quad \beta_1(z) := c\|z\|, \quad \text{and if } \alpha > 1, \ \beta_2(z) := \frac{c}{2}\|z\|^2 \tag{4}$$

and characterise the following sets

$$\mathrm{Int}\,(\mathrm{Dom}\, f^* + A\mathrm{Dom}\, f + \mathrm{Dom}\, \beta_i^*) \ \ (i = 0, 1, 2). \tag{5}$$

Describe these sets explicitly when f is the indicator function of a convex closed set K.

(b) Set

$$K_n := \{x \in \mathrm{Dom}\, f \,|\, f(x) \leq n \text{ and } \|x\| \leq n\}. \tag{6}$$

Show that there exists $x_n \in K_n$ such that

$$\forall y \in K_n, \quad -\sigma(A(x_n), y - x_n) + f(x_n) - f(y) \leq 0. \tag{7}$$

(c) Suppose that

$$0 \in \mathrm{Int}(\mathrm{Dom} f^* + A\mathrm{Dom} f + \mathrm{Dom} \beta^*). \tag{8}$$

Show that a subsequence of the sequence x_n is bounded.

(d) Deduce that assumption (8) implies the existence of a solution of problem (2)(i). Deduce sufficient conditions for the existence of a solution of the variational inequality

$$\bar{x} \in K \ \text{ and } \ \forall y \in K, \ \langle A(\bar{x}), \bar{x} - y \rangle \leq 0. \tag{9}$$

15.47 Problem 47 – Monotone Set-valued Maps

Let X be a Hilbert space (identified with its dual). A set-valued map A from X to X is said to be *monotone* if

$$\forall (x,p) \in \text{Graph}(A), \ \forall (y,q) \in \text{Graph}(A), \ \langle p - q, x - y \rangle \geq 0. \tag{1}$$

More generally, if $\beta : X \to \mathbb{R}_+ \cup \{+\infty\}$ is a nontrivial, convex, lower semi-continuous function, A is set to be *β-monotone* if

$$\forall (x,p) \in \text{Graph}(A), \forall (y,q) \in \text{Graph}(A), \langle p - q, x - y \rangle \geq \beta(x - y). \tag{2}$$

(a) Show that if F is non-expansive in the sense that

$$\forall (x,p), (y,q) \in \text{Graph}(F), \ \|p - q\| \leq \|x - y\| \tag{3}$$

then $A := 1 - F$ is monotone.

(b) Show that A is monotone if and only if

$$\forall \lambda > 0, \ \forall (x,p), (y,q) \in \text{Graph}(A),$$
$$\|x - y\| \leq \|x - y + \lambda(p - q)\|. \tag{4}$$

(c) Suppose that $f : X \to \mathbb{R} \cup \{+\infty\}$ is a nontrivial convex function. Show that $A := \partial f$ is monotone.

(d) Set

$$J_\lambda := (1 + \lambda A)^{-1}, \ \ A_\lambda := \frac{1}{\lambda}(1 - J_\lambda). \tag{5}$$

Show that if A is monotone then J_λ and A_λ are (one-to-one) mappings from $\text{Im}(1 + \lambda A)$ to X satisfying

$$A_\lambda(x) \in A(J_\lambda(x)) \tag{6}$$

and

$$\|J_\lambda(y_1) - J_\lambda(y_2)\| \leq \|y_1 - y_2\|$$
$$\|A_\lambda(y_1) - A_\lambda(y_2)\| \leq \frac{1}{\lambda}\|y_1 - y_2\|. \tag{7}$$

(e) Show that A_λ is monotone.

(f) Suppose from now on that

$$\forall \lambda > 0, \ (1 + \lambda A) \text{ is surjective.} \tag{8}$$

Show that A satisfies the property

$$\forall (y,v) \in \text{Graph}(A), \ \langle u - v, x - y \rangle \geq 0 \Rightarrow u \in A(x). \tag{9}$$

(g) Deduce that

$$\text{the images of } A \text{ are convex and closed} \tag{10}$$

and that

$$\text{the graph of } A \text{ is closed.} \tag{11}$$

(h) We denote the element of $A(x)$ with minimum norm by $m(A(x))$ (the projection of 0 onto the convex closed subset $A(x)$). Show that for all $x \in \text{Dom}(A)$,

$$\|A_\lambda(x) - m(A(x))\|^2 \leq \|m(A(x))\|^2 - \|A_\lambda(x)\|^2 \tag{12}$$

and deduce that

$$\|x - J_\lambda(x)\| \leq \lambda \|m(A(x))\|. \tag{13}$$

(i) Show that

$$A_{\mu+\lambda}(x) = (A_\mu)_\lambda(x) \tag{14}$$

and deduce that

$$(J_\lambda(x), A_\lambda(x)) \text{ converges to } (x, m(A(x))) \text{ as } \lambda \to 0. \tag{15}$$

(j) If $f : X \to \mathbb{R} \cup \{+\infty\}$ is nontrivial, convex and lower semi-continuous, show that $A = \partial f$ satisfies (8) and that $(\partial f)_\lambda(x) = \nabla f_\lambda(x)$ and complete Theorem 5.2 by showing that $\nabla f_\lambda(x)$ converges to the element $\partial f(x)$ of minimum norm.

15.48 Problem 48 – Walrasian Equilibrium for Set-valued Demand Maps

We consider a convex closed set L (consumption set) and a continuous function r from the price simplex M^ℓ to $\mathbb{R}$ (yield function) and we suppose that

$$\forall p \in M^\ell, \ \exists \tilde{x} \in L \text{ such that } \langle p, \tilde{x} \rangle < r(p). \tag{1}$$

(a) Show that the set-valued budgetary map $p \to B(p, r(p))$ defined by

$$B(p, r(p)) := \{y \in L | \langle p, y \rangle \leq r(p)\} \tag{2}$$

has a closed graph and is lower semi-continuous.

(b) Suppose that f is a (loss) function from $L \times M^\ell$ to $\mathbb{R}$. Suppose further that

$$L \text{ is compact and } f \text{ is continuous.} \tag{3}$$

Using Problem 6, show that the set-valued demand map defined by

$$D(p, r(p)) := \left\{ x \in B(p, r(p)) | f(x) = \min_{y \in B(p, r(p))} f(y) \right\} \tag{4}$$

is upper semi-continuous.

(c) We suppose that there are n consumers such that

(i)   the n consumption sets are convex and compact
(ii)   the n loss functions $f_i : L_i \times M^\ell \to \mathbb{R}$
     are continuous and convex in x_i (5)

and we consider

$$n \text{ compact sets } M_i^0 \text{ such that } M_i = M_i^0 - \mathbb{R}_+^\ell \text{ is convex} \tag{6}$$

and such that

$$\forall i = 1, \ldots, n, \quad 0 \in \text{Int}(L_i - M_i). \tag{7}$$

16. Solutions to Problems

16.1 Problem 1 – Solution. Set-valued Maps with a Closed Graph

(a) Consider a sequence of elements (x_n, y_n) of Graph(F) converging to (x, y). Since F is upper semi-continuous, given any $\varepsilon > 0$ there is an integer $N(\varepsilon)$ such that

$$\forall n \geq N(\varepsilon), \ y_n \in F(x_n) \subset F(x) + \varepsilon B. \tag{1}$$

It follows that $y_n \in \overline{F(x)} = F(x)$.

(b) Let V be an open neighbourhood of $F(x_0)$ and K its complement which is compact and disjoint from $F(x_0)$. Since for all $y \in K$, the pair (x_0, y) does not belong to the graph of F, which is closed, there exist neighbourhoods $U_y(x_0)$ of x_0 and $W(y)$ of y such that

$$\text{Graph}(F) \cap (U_y(x_0) \times W(y)) = \emptyset. \tag{2}$$

Since the set K is compact, it is covered by n neighbourhoods $W(y_i)$. Then the neighbourhood $U(x_0) := \cap_{i=1}^{n} U_{y_i}(x_0)$ is such that

$$\forall x \in N(x_0), \ F(x) \cap (\cup_{i=1}^{n} W(y_i)) = \emptyset. \tag{3}$$

It follows that

$$\forall x \in N(x_0), \ F(x) \subset V \tag{4}$$

whence that F is upper semi-continuous at x_0.

16.2 Problem 2 – Solution. Upper Semi-continuous set-valued Maps

(a) The proof for part (a) is self-evident.

(b) If $y \in K$, then (x_0, y) does not belong to Graph(G). Since the latter is closed, there exists neighbourhoods $U_y(x_0)$ and $W(y)$ such that

$$\text{Graph}(G) \cap (U_y(x_0) \times W(y)) = \emptyset.$$

Thus (5) holds.

(c) We cover K by n open sets $W(y_i)$ and set $M := \cup_{i=1}^n W(y_i)$ and $U_0(x_0) := \cap_{i=1}^n U_{y_i}(x_0)$, so that

$$\forall x \in U_0(x), \quad G(x) \cap M = \emptyset. \tag{$*$}$$

Since F is upper semi-continuous at x_0, there exists a neighbourhood $U_1(x_0)$ such that

$$\forall x \in U_1(x_0), \quad F(x) \subset M \cap N, \text{ an open neighbourhood of } F(x_0). \tag{$**$}$$

Consequently, $(*)$ and $(**)$ imply that

$$\forall x \in U(x_0) := U_1(x_0) \cap U_2(x_0), \quad F(x) \cap G(x) \subset N. \tag{$***$}$$

16.3 Problem 3 – Solution. Image of a Set-valued Map

(a) We consider a sequence x_n converging to x_0 and take $y_0 = f(x_0, u_0) \in F(x_0)$ where $u_0 \in G(x_0)$, Since G is lower semi-continuous, there exists a sequence $u_n \in G(x_n)$ converging to u_0. The sequence $y_n := f(x_n, u_n)$ converges to y_0 since f is continuous.

(b) We take $x_0 \in \text{Dom}(G)$ and $\varepsilon > 0$ consider the neighbourhood V of $F(x_0)$. It is a neighbourhood of each of the points $f(x_0, u)$ where $u \in G(x_0)$. The continuity of f implies that there exist $\eta_u > 0$ and $\delta_u > 0$ such that

$$f(x, v) \in V \text{ when } x \in B(x_0, \eta_u) \text{ and } v \in B(u, \delta_u).$$

Since $G(x_0)$ is compact, it can be covered by p balls $B(u_i, \delta_i)$ $(i = 1, \ldots, p)$. Since G is upper semi-continuous, there exists $\eta_0 > 0$ such that

$$\forall x \in B(x_0, \eta_0), \quad F(x) \subset \bigcup_{i=1}^p B(u_i, \delta u_i).$$

We take $\eta := \min(\eta_0, \min_{i=1,\ldots,p} \eta_i) > 0$. Then

$$\forall x \in B(x_0, \eta), \quad F(x) \subset V.$$

16.4 Problem 4 – Solution. Inverse Image of a Set-valued Map

(a) The proof for part (a) is easy.

(b)

(i) We take $x_n \to x_0$ and $u \in S(x_0)$. From (2), there exists $u_n \in F(x_n)$ such that $u_n \to u$. Since $(x_n, f(x_n, u_n))$ converges to $(x_0, f(x_0, u))$ and, since the graph of $x \to \text{Int } T(x)$ is open, it follows that $f(x_n, u_n) \in \text{Int } T(x_n)$ for n sufficiently large. Thus u_n belongs to $S(x_n)$ and converges to u.

(ii) Since $S(x) \subset R(x)$, we have $\overline{S(x)} \subset R(x)$. Conversely, we take $u \in R(x)$. Since there exists $u_0 \in S(x)$, it follows that $\theta u_0 + (1-\theta)u$ belongs to $S(x)$ for all $\theta > 0$ since

$$f(x, \theta u_0 + (1-\theta)u) = \theta f(x, u_0) + (1-\theta)f(x, u)$$

belongs to the interior of $T(x)$ because $f(x, u_0)$ does. Letting θ tend to 0, it follows that u belongs to $\overline{S(x)}$.

(iii) Since $x \to S(x)$ is lower semi-continuous, the same is true of $x \to \overline{S(x)} = R(x)$.

(c) We consider a sequence $x_n \to x_0$ and $u \in R(x_0)$. Since F and T are lower semi-continuous, there exist sequences $u_n \in F(x_n)$ and $y_n \in T(x_n)$ which converge to u and $f(x, u)$, respectively. Since f is continuous

$$\varepsilon_n := \|f(x_n, u_n) - y_n\|$$

converges to 0 and

$$\theta_n := \frac{\gamma}{\gamma + \varepsilon_n} \in]0, 1[$$

converges to 1.

Since $\theta_n \varepsilon_n = (1-\theta_n)\gamma$, it follows that

$$
\begin{aligned}
\theta_n(f(x_n, u_n) - y_n) &\in \varepsilon \theta_n B \\
&= (1-\theta_n)\gamma B \\
&\subset (1-\theta_n)(f(x_n, F(x_n)) - T(x_n))
\end{aligned}
$$

by virtue of assumption (7). Thus, there exist $\hat{u}_n \in F(x_n)$ and $\hat{y}_n \in T(x_n)$ such that

$$\theta_n(f(x_n, u_n) - y_n) = (1-\theta_n)(f(x_n, \hat{u}_n) - \hat{y}_n).$$

This implies that $v_n := \theta_n u_n + (1-\theta_n)\hat{u}_n$ belongs to $R(x_n)$ since

$$f(x_n, v_n) = \theta_n y_n + (1-\theta_n)\hat{y}_n \in T(x_n).$$

Moreover, $u_n - v_n = (1-\theta_n)(u_n - \hat{u}_n)$. Since $u_n - \hat{u}_n \in F(x_n) - F(\hat{x}_n)$ belongs to a bounded set by virtue of (8), it follows that $u_n - v_n$ tends to 0. Thus, u is the limit of $u_n \in R(x_n)$.

16.5 Problem 5 – Solution. Polars of a Set-valued Map

(a) We consider a sequence (x_n, p_n) converging to (x, p) such that $p_n \in T(x_n)^-$ for all n. We take an arbitrary $y \in T(x)$. Since T is lower semi-continuous, there exists a sequence $y_n \in T(x_n)$ which converges to y. Since $\langle p_n, y_n \rangle \leq 0$, it follows that $\langle p, y \rangle \leq 0$. This is true for all $y \in T(x)$ and consequently $p \in T(x)^-$.

(b) We consider a sequence x_n converging to x and take $y \in T(x)$. Suppose that $\pi_{T(x_n)}$ is the orthogonal projector onto $T(x_n)$. It is sufficient to show that $y_n := \pi_{T(x_n)}(y)$ converges to y. Then $p_n := y - y_n$ is the projector of y onto $T(x_n)^-$. Thus, $\langle p_n, y_n \rangle = 0$ and $\|p_n\| \leq \|y\|$. Consequently, a subsequence (again denoted by) p_n converges to $\bar{p}$, since Y is finite dimensional. This element $\bar{p}$ belongs to $T(x)^-$, since the graph of $T(\cdot)^-$ is closed, and satisfies $\langle \bar{p}, y - \bar{p} \rangle = \lim_{n \to \infty} \langle p_n, y_n \rangle = 0$. Thus, $\|\bar{p}\|^2 = \langle \bar{p}, y \rangle \leq 0$ since $\bar{p} \in T(x)^-$ and $y \in T(x)$. Consequently $\bar{p} = 0$ and y_n converges to $y = y - \bar{p}$.

16.6 Problem 6 – Solution. Marginal Functions

(a) See Proposition 9.3.

(b) We take $\varepsilon > 0$ and $x_0 \in X$. Since g is upper semi-continuous, for any $y \in Y$, there exist neighbourhoods $V(y)$ and $U_y(x_0)$ such that

$$\forall z \in V(y), \ \forall x \in U_y(x_0), \ g(x, y) \leq g(x_0, y) + \varepsilon. \tag{$*$}$$

Since $F(x_0)$ is compact, it is covered by n neighbourhoods $V(y_i)$. Let $N := \cup_{i=1}^n V(y_i)$ denote a neighbourhood of $F(x_0)$ and $U_0(x_0) := \cap_{i=1}^n U_{y_i}(x_0)$ a neighbourhood of x_0. Since F is upper semi-continuous at x_0, there exists a neighbourhood $U_1(x_0)$ of x_0 such that

$$\forall x \in U_1(x_0), \ F(x) \subset N. \tag{$**$}$$

Then $(*)$ and $(**)$ imply that $\forall x \in U(x_0) := U_0(x) \cap U_1(x_0)$ and for all $y \in F(x)$, there exists an element y_i such that

$$g(x, y) \leq g(x_0, y_i) + \varepsilon \leq f(x_0) + 2\varepsilon. \tag{$***$}$$

Consequently, $f(x) \leq f(x_0) + 2\varepsilon$ for all $x \in U(x_0)$.

(c) The set-valued map M is the intersection of the set-valued map F and the set-valued map G defined by

$$G(x) := \{y \in Y | f(x) = g(x, y)\}.$$

Since f and g are continuous, the graph of G is closed. We now apply Problem 2.

(d) Fix $\varepsilon > 0$, $x_1, x_2 \in X$ and $y_1 \in F(x_1)$ such that $f(x_1) \leq g(x_1, y_1) + \varepsilon$. Since F is Lipschitz, there exists $y_2 \in F(x_2)$ such that $\|y_1 - y_2\| \leq d(y_1, F(x_2)) + \varepsilon \leq c\|x_1 - x_2\| + \varepsilon$. Whence, since $g(x_2, y_2) \leq f(x_2)$, we have

$$
\begin{aligned}
f(x_1) - f(x_2) &\leq g(x_1, y_1) - g(x_2, y_2) + \varepsilon \\
&\leq \ell(\|x_1 - x_2\| + \|y_1 - y_2\|) + \varepsilon \\
&\leq \ell(c + 1)\|x_1 - x_2\| + (\ell + 1)\varepsilon.
\end{aligned}
$$

It now suffices to let ε tend to 0.

16.7 Problem 7 – Solution. Generic Continuity of a Set-valued Map with a Closed Graph

(a) The proof for part (a) is easy.

(b) Since F is upper semi-continuous (see Problem 1), the points of discontinuity of F are those where F is not lower semi-continuous, in other words those such that

$$
\text{if } K_n \ni x, \text{ then } x \in \partial K_n := K_n \cap \text{comp Int} (K_n)
$$

(c) Since K_n is closed, the interior of ∂K_n is empty. Then, Baire's theorem implies that the interior of $\cup_{n=1}^{\infty} \partial K_n$ is empty. Thus, the interior of the set of points of discontinuity is empty.

16.8 Problem 8 – Solution. Approximate Selection of an Upper Semi-continuous Set-valued Map

(b) We may associate any x with neighbourhoods $U(x)$ of x such that

$$
\forall y \in U(X), \quad F(y) \subset F(x) + \varepsilon B. \tag{$*$}
$$

Since X is compact, it is covered by n neighbourhoods $U(x_i)$. We introduce a continuous partition of unity $\{a_i\}_{i=1,\ldots,n}$ associated with this covering, we choose points $y_{x_i} \in F(x_i)$ and define f_ε by

$$
f_\varepsilon(x) := \sum_{i \in I(x)} a_i(x) y_{x_i}, \quad I(x) := \{i \,|\, a_i(x) > 0\}. \tag{$**$}
$$

If $i \in I(x)$, then $x \in U(x_i)$ and consequently, $F(x_i) \subset F(x) + \varepsilon B$. Thus, $f_\varepsilon(x) \subset \text{co}(F(x) + \varepsilon B) = F(x) + \varepsilon B$.

16.9 Problem 9 − Solution. Continuous Selection of a Lower Semi-continuous Set-valued Map

(a) Given any $x \in X$ and $y_x \in F(x)$ there exists an open neighbourhood $V(x)$ of x such that

$$\forall x_1 \in V(x), \ (y_x + \varepsilon B) \cap F(x_1) \neq \emptyset. \qquad (*)$$

Since X is compact, it is covered by n neighbourhoods $V(x_i)$. We associate a continuous partition of unity $\{a_i(\cdot)\}_{i=1,\dots,n}$ with this covering and define f_ε by

$$\forall x \in K, \ f_\varepsilon(x) := \sum_{i \in I(x)} a_i(x) y_{x_i} \qquad (**)$$

where $I(x) = \{i = 1, \dots, n | a_i(x) > 0\} \neq \emptyset$. If $i \in I(x)$, then $x \in V(x_i)$ and, by virtue of $(*)$, $y_{x_i} \in F(x) + \varepsilon B$. Multiplying these inclusions by $a_i(x)$ and taking the sum, we obtain

$$f_\varepsilon(x) \in \mathrm{co}(F(x) + \varepsilon B) = F(x) + \varepsilon B.$$

(b) We begin the induction with $n = 1$, by applying (a) with $\varepsilon = 1/2$. We suppose assumption (2) holds for n and prove it holds for $n + 1$. We set

$$F_{n+1}(x) := \left(f_n(x) + \tfrac{1}{2^n} \overset{\circ}{B} \right) \cap F(x). \qquad (***)$$

By virtue of (2)(i) , $F_{n+1}(x)$ is non-empty. It is easy to show that F_{n+1} is lower semi-continuous. Applying (a) to F_{n+1} and taking $\varepsilon = \frac{1}{2^{n+1}}$, we deduce that there exists a continuous mapping f_{n+1} from X to Y with

$$d(f_{n+1}(x), F_{n+1}(x)) \leq \frac{1}{2^{n+1}}.$$

This implies (2) for $n + 1$.

(c) The inequalities (2)(ii) imply that the sequence of the f_n is a Cauchy sequence in the space of continuous mappings from X to Y. This space is complete, since Y is complete. Thus, the sequence converges uniformly to a continuous mapping f from X to Y and property (3) follows from (2)(ii), since the images $F(x)$ are closed.

16.10 Problem 10 − Solution. Interior of the Image of a Convex Closed Cone

(a) We set

$$K_n := K \cap nB \quad \text{and} \quad T := A(K_1) = A(K \cap B).$$

Since K is convex and $0 \in K$, it follows that

$$\frac{1}{n}K_n \subset K_1$$

whence that

$$A(K_n) \subset nA(K_1) := nT.$$

Since

$$0 \in \operatorname{Int} A(K) = \operatorname{Int}\left(\cup_{n=1}^\infty A(K_n)\right) \subset \operatorname{Int}\left(\cup_{n=1}^\infty n\overline{T}\right).$$

Baire's theorem implies that the interior of one of these closed sets $n\overline{T}$ is non-empty, and thus that the interior of $\overline{T}$ is non-empty. Consequently, there exists $y_0 \in \overline{T}$ and $\delta > 0$ such that

$$y_0 + \delta B \subset \overline{T}. \tag{$*$}$$

Moreover, since there exists $\gamma > 0$ such that $\gamma B \subset A(K) = \cup_{n=1}^\infty A(K_n)$, there exists $n > 0$ such that $-\gamma \frac{y_0}{\|y_0\|} \in A(K_n)$. Whence,

$$-\gamma \frac{y_0}{n\|y_0\|} \in A(K_1) = T \subset \overline{T}. \tag{$**$}$$

We set $\lambda := \gamma/(\gamma + n\|y_0\|) \in]0,1[$ and multiply $(*)$ by λ and $(**)$ by $(1-\lambda)$. Then the convexity of T implies that

$$\lambda\delta B = \lambda y_0 - (1-\lambda)\frac{\gamma y_0}{\|y_0\|} + \lambda\delta B \subset \lambda\overline{T} + (1-\lambda)\overline{T} \subset \overline{T}.$$

Thus, $0 \in \operatorname{Int}(\overline{T})$.

(b) We take $y := \frac{1}{2}\sum_{k=0}^\infty 2^{-k}A(x_k) \in \frac{1}{2}\sum_{k=0}^\infty 2^{-k}T$, where $x_k \in K_1$. We set $\alpha_n := 1/\sum_{k=0}^n 2^{-k}$. Thus, $\alpha_n \to \frac{1}{2}$ and $y_n := \alpha_n \sum_{k=0}^n 2^{-k}A(x_k) = A(\alpha_n \sum_{k=0}^n 2^{-k}x_k)$ converges to y. But, since K_1 is contained in B, the sequence of the $u_n := \alpha_n \sum_{k=0}^n 2^{-k}x_k$ is Cauchy and so converges to an element u. Since K_1 is convex and closed, the u_n (and thus also u) belong to K_1. Since A is continuous, $y = Au$ belongs to $A(K_1) = T$. Thus,

$$\frac{1}{2}\sum_{k=0}^\infty 2^{-k}T \subset T.$$

(c) We suppose that T satisfies property (3) and that $0 \in \operatorname{Int}(\overline{T})$. Thus, there exists $\gamma > 0$ such that $2\gamma B \subset \overline{T}$. Then for all $k > 0$ we have $2 \cdot 2^{-k}\gamma B \subset 2^{-k}\overline{T}$. We take $y \in \gamma B$. Then there exists $v_0 \in T$ such that

$$2y - v_0 \in 2 \cdot 2^{-1}\gamma B \subset 2^{-1}\overline{T}$$

since $2y \in \overline{T}$. We suppose that we have constructed a sequence of $v_k \in T$ such that

$$2y - \sum_{k=0}^{n-1} 2^{-k} v_k \in 2 \cdot 2^{-n} \gamma B.$$

Since $2 \cdot 2^{-n} \gamma B \subset 2^{-n} \overline{T}$, we can find $v_n \in T$ such that

$$2y - \sum_{k=0}^{n-1} 2^{-k} v_k - 2^{-n} v_n \in 2 \cdot 2^{-(n+1)} \gamma B \subset 2^{-(n+1)} \overline{T}.$$

Thus, we have constructed a sequence of $v_k \in T$ such that

$$y \in \frac{1}{2} \sum_{k=0}^{\infty} 2^{-k} v_k \in \frac{1}{2} \sum_{k=0}^{\infty} 2^{-k} T.$$

Consequently, $\gamma B \subset T$ and $0 \in \text{Int}(T)$. We have proved that assumption (1) implies that there exists $\gamma > 0$ such that

$$\gamma B \subset A(K \cap B) \qquad (*)$$

If $K = X$ this is just Banach's Theorem for the open image.

(d) If K is a convex closed cone, the statement $A(K) = Y$ implies that $0 \in \text{Int}(A(K))$. Thus, there exists $\gamma > 0$ such that $\gamma B \subset A(K \cap B)$, in other words, for $c := \frac{1}{\gamma}$, we obtain

$$\forall x \in Y, \exists x \in K \text{ such that } y = Ax \text{ and } \|x\| \le c\|y\|. \qquad (*)$$

This shows that A^{-1} is Lipschitz, in the sense that: $\forall x_1 \in K$, and all $y_2 \in Y$, there exists $x_2 \in X$, a solution of $y_2 = Ax_2$, such that $\|x_1 - x_2\| \le c\|Ax_1 - y_2\|$.
    In fact, we associate $y := y_2 - Ax_1$ with the solution $x \in K$ given by $(*)$. Thus,

$$
\begin{aligned}
x_2 &:= x_1 + x \in K + K \subset K \text{ is such that} \\
Ax_2 &= Ax_1 + Ax = y_2 \text{ and} \\
\|x_2 - x_1\| &= \|x\| \le c\|y\| = c\|Ax_1 - y_2\|.
\end{aligned}
$$

(e) Suppose that $A \in L(X \times Y, Y)$ is the operator defined by

$$A(x, y) := Lx - y$$

and that $K = P \times Q$. Since $A(K) = L(P) - Q$ is the whole space, it follows from (d) that for fixed x_0 in P, y in Y and $y_0 := \pi_Q(Lx_0 - y)$, there exists a solution $(x_1, y_1) \in P \times Q$ of the equation $A(x_1, y_1) = y$ such that

$$
\begin{aligned}
\|x_1 - x_0\| &\le \max(\|x_1 - x_0\|, \|y_1 - y_0\|) \le c\|y - A(x_0, y_0)\| \\
&= cd_Q(Lx_0 - y).
\end{aligned}
$$

To say that $(x_1, y_1) \in P \times Q$ is a solution of $A(x_1, y_1) = Lx_1 - y_1 = y$ implies that $x_1 \in M(y)$.

(f) We take $K := \text{Graph}(F)$ a convex closed cone in $X \times Y$ and $A := \pi_Y$ the canonical projection from $X \times Y$ onto Y. Since $A(K) := \pi_Y(\text{Graph}F) = \text{Im}(F)$ is the whole space, there exists $\gamma > 0$ such that, for all $(x_1, y_1) \in \text{Graph}(F)$ and all $y_2 \in Y$, there exists a solution $(x_2, \tilde{y}_2) \in K$ of the equation $\tilde{y}_2 := \pi_Y(x_2, \tilde{y}_2) = y_2$ satisfying the inequality

$$\begin{aligned} \|x_2 - x_1\| &\leq \max(\|x_2 - x_1\|, \|y_2 - y_1\|) \\ &\leq c\|\pi_Y(x_1, y_1) - y_2\| = c\|y_1 - y_2\|. \end{aligned}$$

Thus, F^{-1} is Lipschitz.

(g) In the case where K is not a cone, but only a convex closed set (containing 0), the condition $0 \in \text{Im}\, A(K)$ implies that there exists $\gamma > 0$ such that

$$2\gamma B \subset A(K \cap B).$$

Suppose that $x_0 \in K$ and $y \in \gamma B$ are given, where $y \neq Ax_0$. Then

$$y + \frac{\gamma}{\|y - Ax_0\|}(y - Ax_0) \in 2\gamma B \subset A(K \cap B)$$

and thus there exists $u \in K \cap B$ such that

$$y + \frac{\gamma}{\|y - Ax_0\|}(y - Ax_0) = Au. \tag{$*$}$$

We take $\lambda := \frac{\|y - Ax_0\|}{\gamma + \|y - Ax_0\|} \in]0, 1[$ so that $\lambda \frac{\gamma}{\|y - Ax_0\|} = (1 - \lambda)$. Multiplying $(*)$ by λ, we obtain $\lambda y + (1 - \lambda)(y - Ax_0) = A(\lambda u)$. Thus, $y = Ax$, where $x = (1 - \lambda)x_0 + \lambda u$ belongs to K (since x_0 and u belong to K). Furthermore

$$\begin{aligned} \|x - x_0\| &= \lambda(\|u - u_0\|) \\ &= \frac{\|y - Ax_0\|}{\gamma + \|y - Ax_0\|}(\|u - x_0\|) \\ &\leq \frac{\|x_0\| + 1}{\gamma}\|y - Ax_0\|. \end{aligned}$$

(h) Suppose that $A \in L(X \times Y, Y)$ is the operator defined by

$$A(x, y) := Lx - y$$

and that $K := P \times Q - (\bar{x}, L\bar{x})$ is a convex closed set containing $(0, 0)$ (since $\bar{x} \in P$ and $L\bar{x} \in Q$). Since $A(\bar{x}, L\bar{x}) = 0 \in \text{Int}(L(P) - Q) = \text{Int}\, A(K)$, we may apply the results of (g). Suppose that $\gamma > 0$ is the constant such that for all $(u_0, v_0) \in K$, $z \in \gamma B$, there exists a solution $(u, v) \in K$ of the equation $A(u, v) = z$ satisfying

$$\max(\|u - u_0\|, \|v - v_0\|) \leq \frac{1}{\gamma}(1 + \max(\|u_0\|, \|v_0\|))\|z - A(u_0, v_0)\|. \tag{$*$}$$

Fix x_0 in K and z in γB. We take $u_0 := x_0 - \bar{x}$ and $v_0 := \pi_Q(Lx_0 - z) - L\bar{x}$ so that $(u_0, v_0) \in K$. We set $x := \bar{x} + u \in P$, $y = L\bar{x} + v \in Q$ and prove that $Lx = L\bar{x} + Lu = L\bar{x} + v + A(u, v) = L\bar{x} + v + z = y + z \in Q + z$. Thus, $x \in M(z)$. Since $\|z - A(u_0, v_0)\| = \|Lx_0 - z - \pi_Q(Lx_0 - z)\| = d_Q(Lx_0 - z)$, since $\|u_0\| = \|x_0 - \bar{x}\|$, $\|v_0\| \leq \|Lx_0 - z - L\bar{x}\| \leq \|L\|\|x_0 - \bar{x}\| + \|z\| \leq \|L\|\|x_0 - \bar{x}\| + \gamma$, inequality (22) follows from $(*)$.

(i) We take the canonical projection π_Y from $X \times Y$ onto Y as the operator A and consider the convex closed set $K := \mathrm{Graph}(F) - (\bar{x}, \bar{y})$ which contains $(0, 0)$. Since $0 \in \mathrm{Int}(\mathrm{Im}F - \bar{y}) = \mathrm{Int}(\pi_Y(K))$, it follows from (g) that there exists a constant $\gamma > 0$ such that for all $(u_0, v_0) \in K$ and $v \in \gamma B$, there exists a solution $(u, \tilde{v}) \in K$ of the equation $\tilde{v} =: \pi_Y(u, \tilde{v}) = v$, satisfying

$$\max(\|u - v_0\|, \|v - v_0\|) \leq \frac{1}{\gamma}(1 + \max\|u_0\|, \|v_0\|)\|v - v_0\|. \qquad (*)$$

If $(x_0, y_0) \in \mathrm{Graph}(F)$, we take $u_0 := x_0 - \bar{x}$, $v_0 := y_0 - \bar{y}$, $v = y - \bar{y}$ and we set $x := \bar{x} + u$. Since $(u, v) \in \mathrm{Graph}(F) - (\bar{x}, \bar{y})$, then $(x, y) \in \mathrm{Graph}(F)$, in other words, $x \in F^{-1}(y)$ and

$$\begin{aligned}
\max(\|x - x_0\|, \|y - y_0\|) &= \max(\|u - u_0\|, \|v - v_0\|) \\
&\leq \frac{1}{\gamma}(1 + \max(\|u_0\|, \|v_0\|)(\|v - v_0\|) \\
&= \frac{1}{\gamma}(1 + \max(\|x_0 - \bar{x}\|, \|y_0 - \bar{y}\|)\|y - y_0\|.
\end{aligned}$$

Whence

$$d(x_0, F^{-1}(y)) \leq \frac{1}{\gamma}(1 + \max(\|x_0 - \bar{x}\|, \|y_0 - \bar{y}\|)\|y - y_0\|.$$

16.11 Problem 11 – Solution. Discrete Dynamical Systems

(a) See Theorem 1.4.

(b) If $f(x)$ is finite, the assumption implies that

$$d(g^{(t+1)}(x), g^t(x)) \leq f(g^t(x)) - f(g^{(t+1)}(x)).$$

Taking the sum from $t = 0$ to T, we obtain

$$\sum_{t=0}^{T} d(g^{(t+1)}(x), g^t(x)) \leq f(x) - f(g^{(T+1)}(x)) \leq f(x)$$

(since $f(y) \geq 0$). Letting T tend to infinity, we deduce that $f_g(x) \leq f(x)$.

Moreover,

$$
\begin{aligned}
f_g(x) &= d(g(x), x) + \sum_{t=1}^{\infty} d(g^{(t+1)}(x), g^t(x)) \\
&= d(g(x), x) + f_g(g(x)).
\end{aligned}
$$

If there exists x_0 such that $f_g(x_0) < +\infty$, then there exists an equilibrium to which the solution based on x_0 converges.

(c) See Theorem 1.5.

(d) See Theorem 1.4.

(e) It suffices to show that $x \in H(x)$ (reflexivity) and that if $y \in H(x)$, then $H(y) \subset H(x)$ (transitivity). This is self-evident.

(f)

(i) The set of limit points of x^t is the set

$$
L(x^t) := \bigcap_{t \geq 0} \left(\bigcup_{s \geq T} x^s \right).
$$

Since $\cup_{s \geq t} x^s \subset H(x^t)$, it follows that $L(x^t) \subset H(x)$.

(ii) Since $G(H(x)) \subset H(x)$, it follows that when G is lower semi-continuous, $G(\overline{H(x)}) \subset \overline{H(x)}$. Suppose that $u = \lim_{n \to \infty} u_n$, where $u_n \in H(x)$ and that $v \in G(u)$. Since G is lower semi-continuous, there exist $v_n \in G(u_n) \subset H(x)$ such that v_n converges to v. Consequently, v belongs to $H(\bar{x})$. It then follows that $G(L(x^t)) \subset L(x^t)$.

(g)

(i) Assumption (8) implies that

$$
\forall y \in H(x), \quad d(x, y) \leq f(x) - f(y).
$$

Actually, $y = x^t$, where $x^t \in G(x^{t-1}), \dots x^1 \in G(x)$. Thus,

$$
\begin{aligned}
d(x, y) &\leq \sum_{s=1}^{t} d(x^s, x^{s-1}) \\
&\leq \sum_{s=1}^{t} (f(x^s) - f(x^{s-1})) \\
&\leq f(x^t) - f(x^0) \\
&= f(y) - f(x).
\end{aligned}
$$

From the definition (10) of the function f, it follows that

$$\forall y \in H(x), \quad d(x,y) \leq f(x) - v(x),$$

whence that

$$\text{diameter}(H(x)) \leq 2(f(x) - v(x)).$$

We now choose the sequence x_n satisfying (9) and show that

$$v(x_{n+1}) \leq f(x_{n+1}) \leq v(x_n) + 2^{-n} \leq v(x_{n+1}) + 2^{-n}.$$

Thus, the decreasing sequence of sets $\overline{H(x_n)}$, the diameter of which tends to 0, reduces to a single point $\bar{x}$. Since $\bar{x} \in H(x_n)$, whence $H(\bar{x}) \subset H(x_n)$, it follows that

$$H(\bar{x}) \subset \cap_{n \geq 0} \overline{H(x_n)} = \{\bar{x}\} \subset H(\bar{x}).$$

(ii) Since G is lower semi-continuous, whence, by virtue of (f)(ii) $G(\overline{H(x_n)}) \subset \overline{H(x_n)}$, it follows that

$$G(\bar{x}) \subset \cap_{n \geq 0} \overline{H(x_n)} = \{\bar{x}\}.$$

16.12 Problem 12 – Solution. Fixed Points of Contractive Set-valued Maps

(a) We suppose that f satisfies (2) and construct a solution (x_n) based on x, satisfying

$$x_{n+1} \in G(x_n) \quad \text{and} \quad d(x_n, x_{n+1}) \leq f(x_n) - f(x_{n+1}).$$

It follows that, for all k,

$$\sum_{n=0}^{k} d(x_n, x_{n+1}) \leq f(x) - f(x_{k+1}) \leq f(x),$$

whence that $f_G(x) \leq f(x)$.

(b) Inequality (5) is trivial. Moreover, for all $\varepsilon > 0$, there exists a solution (x_n) based on x of the set-valued map $x \to B(G(x), \varepsilon)$ such that

$$\sum_{n=0}^{\infty} d(x_n, x_{n+1}) \leq f_\varepsilon(x) + \varepsilon.$$

Now,

$$d(x, x_1) + f_\varepsilon(x_1) \leq d(x, x_1) + \sum_{n=1}^{\infty} d(x_n, x_{n+1}) \leq f_\varepsilon(x) + \varepsilon.$$

Thus, we take $x_\varepsilon := x_1$.

(c) We fix x. Since $G(x)$ is compact, there is a subsequence $x_{\varepsilon_k} \in B(G(x), \varepsilon_k)$ which converges to $\bar{x} \in G(x)$. Since G is upper semi-continuous, for any $\delta > 0$, there exists an integer $k_\delta > 0$ such that

$$B(G(x_{t_k}), \varepsilon_k) \subset B(G(\bar{x}), \delta).$$

Consequently, any solution (x_n) of $B(G(x), \varepsilon_k)$ based on x_{ε_k} may be associated with the trajectory $\bar{x}, x_{\varepsilon_k}, x_1, \ldots$ which is a trajectory of $B(G(x), \delta)$ based on $\bar{x}$. Thus, taking into account (6),

$$
\begin{aligned}
f_\delta(\bar{x}) &\le d(\bar{x}, x_{\varepsilon_k}) + f_{\varepsilon_k}(x_{\varepsilon_k}) \\
&\le d(\bar{x}, x_{\varepsilon_k}) - d(x, x_{\varepsilon_k}) + f_{\varepsilon_k}(x) + \varepsilon_k.
\end{aligned}
$$

(d) Letting k tend to infinity, we obtain

$$f_\delta(\bar{x}) \le -d(x, \bar{x}) + f_0(x)$$

and letting δ tend to 0, we obtain $f_0(\bar{x}) + d(x, \bar{x}) \le f_0(x)$ with $\bar{x} \in G(x)$. Thus, f_0 satisfies (2) and following (a) and (5), we have $f_G(x) \le f_0(x) \le f_G(x)$.

Since G is upper semi-continuous with compact values, it follows that the function f_G defined by (1) satisfies property (2). We show that f_G is a finite function. Since $G(x)$ has compact values, we construct a trajectory $x_n \in T(x)$ satisfying

$$d(x_{n+1}, x_n) = d(x_n, G(x_n)).$$

Since G is a contraction we observe that

$$f_G(x) \le \sum_{n=0}^{\infty} d(x_n, x_{n+1}) \le \sum_{n=0}^{\infty} \lambda^n d(x, x_1) = \frac{1}{1-\lambda} d(x, x_1) < +\infty.$$

We then apply Theorem 1.4 to deduce that G has a fixed point.

16.13 Problem 13 – Solution. Approximate Variational Principle

(a) By virtue of Ekeland's Theorem (Corollary 1.3), there exist $\varepsilon > 0$, $\lambda > 0$, $x_0 \in X$ and $\bar{x} \in X$, such that

$$f(x_0) \le \inf_{x \in X} f(x) + \varepsilon\lambda, \quad f(\bar{x}) \le f(x_0), \quad d(x_0, \bar{x}) \le \lambda$$

and

$$f(\bar{x}) \le f(x) + \varepsilon\|x - \bar{x}\| \qquad \forall x \in X.$$

Taking $x = \bar{x} + hv$, we obtain

$$0 \le \frac{f(\bar{x} + hv) - f(x)}{h} + \varepsilon \|v\|.$$

Since f is Gâteaux differentiable, it follows that, letting h tend to 0,

$$0 \le \langle Df(\bar{x}), v \rangle + \varepsilon \|v\|.$$

Taking the infimum with respect to v on the unit circle

$$\|Df(\bar{x})\|_* \le \varepsilon.$$

(b) We take $\varepsilon = \lambda = \frac{1}{n}$. The previous question provides x_n such that $f(x_n) \le \inf_{x \in X} f(x) + \frac{1}{n^2}$ and $\bar{x}_n$ such that $f(\bar{x}_n) \le f(x_n)$, $d(x_n, \bar{x}_n) \le \frac{1}{n}$ and $\|Df(\bar{x}_n)\|_* \le \frac{1}{n}$.

(c) Since $\lim_{\|x\| \to \infty} \frac{f(x)}{\|x\|} = +\infty$, we may associate any $p \in X^*$ with a constant $c > 0$ such that $\forall \|x\| \ge c$, $f(x) \ge \|p\|\|x\|$. Let $m := \inf_{\|x\| \le c}(f(x) - \langle p, x \rangle)$, which is a finite number since f is lower semi-continuous. Then the function g defined by

$$g(x) := f(x) - \langle p, x \rangle \ge \inf(m, 0)$$

is bounded below. From the previous question, there exists a minimising sequence $\bar{x}_n$ for $\bar{g}$ such that $Dg(\bar{x}_n) = Df(\bar{x}_n) - p$ converges to 0. Thus p may be approximated by the $Df(\bar{x}_n)$.

16.14 Problem 14 – Solution. Open Image Theorem

(a) The proof follows by reduction to the absurd, assuming that $A(x_0)$ does not belong to the interior of $A(K)$, in other words, assuming that

$$\forall n \ge 0, y_n \notin A(K) \text{ such that } \|A(x_0) - y_n\| < \frac{1}{n^2}.$$

We apply Ekeland's Theorem (Theorem 1.2) with $\varepsilon = \frac{1}{n}$ to the function $x \to \|A(x) - y\|$ on the closed subset K of X. Thus, there exists $x_n \in K$ such that

$$\|A(x_n) - y_n\| + \frac{1}{n}\|x_n - x_0\| \le \|A(x_0) - y_n\| \le \frac{1}{n^2}$$

and

$$\forall x \in K, \quad \|A(x_n) - y_n\| \le \|A(x) - y_n\| + \frac{1}{n}\|x_n - x\|. \tag{$*$}$$

(b) Since $A(x_n) - y_n \neq 0$ and since the function $z \to \|z\|$ is Fréchet differentiable at any point $z \neq 0$, given any n, there exists a number $\eta_n > 0$
such that $\forall z \in \eta_n B$,

$$\|A(x_n + z) - y_n\| - \|A(x_n) - y_n\| - \langle p_n, Az \rangle \leq \frac{1}{n}\|z\| \qquad (**)$$

where

$$p_n := \frac{A(x_n) - y_n}{\|x_n - y_n\|} \in S, \text{ the unit sphere of } Y.$$

(c) We take $u \in T_K(x_0)$. Following problem 31, we may associate the sequence x_n which converges to x_0 with a sequence of elements $h_n \in]0, \eta_n/(\|u\| + 1)[$ and a sequence of elements u_n converging to u such that

$$\forall n \geq 0, \quad x_n + h_n v_n \in K.$$

Since $\|h_n u_n\| \leq \eta_n \|u_n\|/(\|u\| + 1) \leq \eta_n$, the inequalities $(*)$ with $x = x_n + h_n v_n$ and $(**)$ with $z = u_n$ imply that

$$0 \leq h_n \left(\langle p_n, Au_n \rangle + \frac{2}{n} \right).$$

Since Y is a finite-dimensional space, its unit sphere is compact and there exists a subsequence (again denoted by p_n) which converges to $p \in S$. Dividing by $h_n > 0$ and letting n tend to infinity, we obtain

$$0 \leq \langle p, Au \rangle \ \forall u \in T_K(x_0).$$

(d) Since $A(T_K(x_0)) = Y$, it follows that

$$0 \leq \langle p, v \rangle \ \forall v \in Y.$$

This implies that $p = 0$, although p is an element of norm 1. Thus, we have a contradiction to the assumption that $A(x_0) \notin \text{Int } A(K)$.

(e) If A is no longer linear, but continuously differentiable, the function $z \to \|A(x_n + z) - y_n\|$ is differentiable and we have

$$\forall z \in \eta_n B, \ \|\|A(x_n + z) - y_n\| - \|A(x_n - y_n)\|| - \langle p_n, A'(x_n)z \rangle| \leq \frac{1}{n}\|z\|.$$

We obtain the inequality

$$0 \leq h_n \langle p_n, A'(x_n)u_n \rangle + \frac{2}{n}$$

which implies that, after passing to the limit,

$$0 \leq \langle p, A'(x_0)u \rangle \qquad \forall u \in T_K(x_0).$$

(f) We apply the result (d) to the mapping A from $X \times Y$ to Y defined by $A(x,y) = Bx - y$ and to the set $K = L \times M$. It is easy to check that $T_K(x,y) = T_L(x) \times T_M(y)$. We take $x_0 \in L \cap B^{-1}(M)$ such that $A(x_0, Bx_0) = 0$.

(g) We consider the mapping A from $X \times \mathbb{R}^n$ defined by $A(x,y) = F(x) - y$, the set $K \times \mathbb{R}_+^n$ and the point $(x_0, 0)$. Suppose that condition (12) is false. Thus, we may write

$$\text{if } \lambda \in \mathbb{R}_+^n \text{ is such that } F'(x_0)^*\lambda \in N_K(x_0), \text{ then } \lambda = 0. \qquad (***)$$

Since $\mathbb{R}_+^n = N_{\mathbb{R}_+^n}(0)$, since $A'(x_0, 0)(u,v) = F'(x_0)u - v$ and thus, since $A'(x_0, 0)^*\lambda = (F'(x_0)^*\lambda, -\lambda)$, this condition may be written as

$$A'(x_0, 0)^{*-1} N_{K \times \mathbb{R}_+^n}(x_0, 0) = \{0\}$$

which is equivalent to

$$A'(x_0, 0) T_{K \times \mathbb{R}_+^n}(x_0, 0) \text{ is dense in } \mathbb{R}^n$$

and thus identical (since the dimensions are finite). This would then imply that

$$A(x_0, 0) := F(x_0) \in \text{Int}(F(K) + \mathbb{R}_+^n) = F(K) + \overset{\circ}{\mathbb{R}}_+^n,$$

which contradicts the fact that x_0 is a Pareto optimum.

16.15 Problem 15 – Solution. Asymptotic Centres

(a) The proof for part (a) is easy.

(b) Since the sequence is bounded in a finite-dimensional space E, it is relatively compact and thus has limit points. Thus, the set C is non-empty, convex and closed. By definition of the function v, there exists a subsequence x_s such that

$$v(\bar{y}) = \lim_{s \to \infty} \|x_s - \bar{y}\|^2.$$

Whence,

$$\limsup_{s \to \infty} \|x_s - \bar{x}\|^2 \le \limsup_{t \to \infty} \|x_t - \bar{x}\|^2 = v(\bar{x}).$$

Furthermore, there exists a subsequence x_n of x_s which converges to a limit point $z \in C$. Thus.

$$\limsup_{n \to \infty} \langle \bar{y} - \bar{x}, \bar{y} - x_n \rangle = \langle \bar{y} - \bar{x}, \bar{y} - z \rangle \le 0$$

since $\bar{y}$ is the projection of $\bar{x}$ onto C.

Since

$$\|x_n - \bar{x}\|^2 = \|\bar{x}_n - \bar{y}\|^2 + \|\bar{y} - \bar{x}\|^2 + 2\langle x_n - \bar{y}, \bar{y} - \bar{x} \rangle$$

it follows that

$$
\begin{aligned}
v(\bar{x}) &\geq v(\bar{y}) + \|\bar{y} - \bar{x}\|^2 + 2\langle z - \bar{y}, \bar{y} - \bar{x} \rangle \\
&\geq v(\bar{y}) + \|\bar{y} - \bar{x}\|^2 \\
&\geq v(\bar{y}).
\end{aligned}
$$

Thus, $\bar{y} = \bar{x}$ (since there is a unique minimum).

(c) The limit point w of x_t is the limit of a subsequence x_t. Passing to the limit as s tends to infinity in

$$\|x_s - z\|^2 = \|x_s - y\|^2 + \|y - z\|^2 + 2\langle x_s - y, y - z \rangle$$

we obtain

$$v(z) \geq v(y) + \|y - z\|^2 + 2\langle w - y, y - z \rangle$$

since as y belongs to the attractor N, $v(y) = \lim_{s \to \infty} \|x_s - y\|^2$. Since this inequality holds for all the limit points w of the sequence x_t, it holds for all $w \in C$ and in particular for $w = y$ which is assumed to belong there. Thus,

$$v(z) \geq v(y) + \|y - z\|^2 \geq v(y).$$

This implies that the unique minimum of v is attained at y, which coincides with the asymptotic centre of the sequence.

16.16 Problem 16 – Solution. Fixed Points of Non-expansive Mappings

(a) We have

$$\|f_t(x) - f_t(y)\| \leq \left(1 - \frac{1}{t}\right) \|f(x) - f(y)\| \leq \left(1 - \frac{1}{t}\right) \|x - y\|.$$

By virtue of the Banach–Picard Fixed-point Theorem for contractions (Theorem 1.5), there exists $x_t \in K$, such that $x_t = f_t(x_t) = (1 - \frac{1}{t})f(x_t) + \frac{1}{t}x_0$. Whence,

$$\|x_t - f(x_t)\| = \frac{1}{t}\|x_0 - f(x_0)\| \leq \frac{1}{t}\mathrm{diam}(K)$$

and, since K is bounded,

$$\lim_{t \to \infty} \|x_t - f(x_t)\| = 0.$$

(b) Since $x_t \in K$, which is bounded and thus (weakly) relatively compact, a subsequence x_s converges weakly to $w \in K$. Since f is not expansive, it follows that

$$
\begin{aligned}
\langle x_\lambda - f(x_\lambda) - (x_s - f(x_s)), x_\lambda - x_s \rangle \\
= \|x_\lambda - x_s\|^2 - \langle f(x_\lambda) - f(x_s), x_\lambda - x_s \rangle \\
\geq \|x_\lambda - x_s\|^2 - \|f(x_\lambda) - f(x_s)\|\|x_\lambda - x_s\| \\
\geq \|x_\lambda - x_s\|^2 - \|x_\lambda - x_s\|^2 = 0.
\end{aligned}
$$

Letting s tend to infinity and since $x_s - f(x_s)$ is known to tend (strongly) to 0, we obtain

$$
\langle x_\lambda - f(x_\lambda), x_\lambda - w \rangle \geq 0.
$$

Replacing x_λ by its value and dividing by $\lambda > 0$, we obtain

$$
\langle (1 - \lambda)w + \lambda f(w) - f(w + \lambda(f(w) - w)), f(w) - w \rangle \geq 0.
$$

Letting λ tend to 0, we obtain

$$
\|f(w) - w\|^2 \leq 0
$$

whence w is a fixed point of f.

(c) Since f is continuous, the set of its fixed points is closed. Suppose that x_0 and x_1 are two fixed points of f and show that $x_\lambda := (1 - \lambda)x_0 + \lambda x_1$ is also a fixed point. We have

$$
\|f(x_\lambda) - x_0\| = \|f(x_\lambda) - f(x_0)\| \leq \|x_\lambda - x_0\| = \lambda\|x_0 - x_1\|
$$

and similarly,

$$
\|f(x_\lambda) - x_1\| \leq (1 - \lambda)\|x_0 - x_1\|.
$$

Thus,

$$
\begin{aligned}
\|f(x_\lambda) - x_0\| + \|f(x_\lambda) - x_1\| &\leq \|x_0 - x_1\| \\
&\leq \|x_0 - f(x_\lambda)\| + \|f(x_\lambda) - x_1\|
\end{aligned}
$$

in other words,

$$
\|f(x_\lambda) - x_0\| = \lambda\|x_0 - x_1\| \quad \text{and} \quad \|f(x_\lambda) - x_1\| = (1 - \lambda)\|x_0 - x_1\|.
$$

Since E is a Hilbert space, it follows that

$$
f(x_\lambda) = (1 - \lambda)x_0 + \lambda x_1 = x_\lambda.
$$

(d)

(i) If $\bar{x}$ is a fixed point of f then it belongs to the attractor N of the sequence $f^t(x_0)$, since

$$\|f^t(x_0) - \bar{x}\| = \|f^t(x_0) - f(\bar{x})\| \leq \|f^{t-1}(x_0) - \bar{x}\|.$$

Thus, the sequence of the $\|f^t(x_0) - \bar{x}\|^2$ is decreasing and bounded below, whence convergent.

(ii) Consequently, if a limit point w of the sequence $f^t(x_0)$ is a fixed point of f, it belongs to $N \cap C$ and is thus equal to the asymptotic centre of this sequence.

(iii) The proof of (b) shows that condition (5) implies that all the limit points of the sequence $f^t(x_0)$ are fixed points of f and that, consequently, they are all equal to the asymptotic centre of $f^t(x_0)$, which is thus the limit of the sequence $f^t(x_0)$.

16.17 Problem 17 – Solution. Orthogonal Projectors onto Convex Closed Cones

(a) It is clear that (1) implies the variational inequality (19) of Chapter 2. Conversely, taking $y = 0$ and $y = 2x$ (which belong to K, since K is a cone) in the inequality (19) of Chapter 2, we obtain (1)(i) from which we deduce (1)(ii).

(b) We write

$$\langle \lambda x - \pi_K(\lambda_x), \pi_K(\lambda x) \rangle = \lambda^2 \langle x - \frac{1}{\lambda}\pi_K(\lambda x), \frac{1}{\lambda}\pi_K(\lambda x) \rangle = 0$$

$$\langle \lambda x - \pi_K(\lambda x), z \rangle \leq 0 \implies \langle x - \frac{1}{\lambda}\pi_K(\lambda x), z \rangle \leq 0$$

for all $z \in K$. Then the characterisation of (1) shows that $\frac{1}{\lambda}\pi_K(\lambda x) = \pi_K(x)$.

(c) (2) follows from (1)(i).

(d) Set $\bar{y} = x - \pi_K x$. Then $\bar{y} \in K^-$ (following (1)(ii)) and $\langle x - \bar{y}, \bar{y} \rangle = \langle \pi_K(x), x - \pi_K(x) \rangle = 0$ and $\langle x - \bar{y}, z \rangle = \langle \pi_K(x), z \rangle \leq 0$, $\forall z \in K^-$, since $\pi_K(x) \in K$. Thus (1) applied to K^- shows that $\bar{y} = \pi_{K^-}(x)$.

(e) If $\pi_K(x) = 0$, then $x = x - \pi_K(x) \in K^-$. Conversely, if $x \in K^-$, then $\|\pi_K x\| = \langle \pi_K x, \pi_K x \rangle = \langle \pi_K x, x \rangle \leq 0$ (following (1)(i)), whence $\pi_K(x) = 0$.

(f) Clearly, $y = \pi_K(x)$ and $z = \pi_{K^-}(x)$ are solutions of (6). Conversely, (6) implies that $y = \pi_K(x)$, following the characterisation of (1).

(g) If K is a closed vector subspace, then $K^- = K^\perp$ is the orthogonal subspace and π_K is characterised by

$$\langle x - \pi_k(x), z \rangle = 0 \ \forall z \in K \ (\text{or } x - \pi_K(x) \in K^\perp). \tag{$*$}$$

It follows from $(*)$ that π_K is linear. Moreover,

$$
\begin{aligned}
\langle \pi_K(x), y \rangle &= \langle \pi_K x, \pi_K y + \pi_{K^\perp} y \rangle \\
&= \langle \pi_K x, \pi_K y \rangle \\
&= \langle x - \pi_{K^\perp} x, \pi_K y \rangle \\
&= \langle x, \pi_K y \rangle.
\end{aligned}
$$

16.18 Problem 18 – Solution. Gamma-convex Functions

(a)

  (i) They are the convex sets.

  (ii) They are the convex functions.

(b) We have

$$
\begin{aligned}
f(\alpha(m)) &= f(\sum_{i=1}^{n} \alpha_i K_i) \\
&:= \inf_{x_i \in K_i} g(\sum_{i=1}^{n} \alpha_i x_i) \\
&\leq \inf_{x_i \in K_i} \sum_{i=1}^{n} \alpha_i g(x_i) \\
&= \sum_{i=1}^{n} \alpha_i \inf_{x_i \in K_i} g(x_i) \\
&= \sum_{i=1}^{n} \alpha_i f(K_i) \\
&= f^s(m).
\end{aligned}
$$

(c) The proof is analogous to that of Proposition 2.6.

(d) We take $m := \{(\alpha_i, x_i)\}_{i=1,\dots,n} \in \mathcal{M}(E)$. Then

$$
\begin{aligned}
\sum_{i=1}^{n} \alpha_i F(x_i) &= \sum_{i=1}^{n} \alpha_i (F(x_i) + 0) \in \operatorname{co}(F(E) + \mathbb{R}^n_+) \\
&= F(E) + \mathbb{R}^n_+ \ \text{(from assumption (9))}.
\end{aligned}
$$

Thus, there exist $x \in E$ and $p \in \mathbb{R}^n_+$ such that

$$\sum_{i=1}^{n} F(x_i) = F(x) + p. \tag{$*$}$$

We denote the set of such x by $\gamma(m)$ and check that the n functions f_i are γ-convex.

(e) The proof is analogous to that of Lemma 8.2.

(f) The proof is analogous to that of Proposition 12.3.

16.19 Problem 19 – Solution. Proper Mappings

(a) (1) clearly implies (2). Conversely, we assume (2) and consider a sequence x_n such that $f(x_n)$ converges to y. The set L of limit points of x_n is:

$$L := \bigcap_{N \geq 0} \overline{\{x_n\}_{n \geq N}}.$$

Since f is continuous and transforms closed sets into closed sets, we obtain

$$f(\overline{\{x_n\}_{n \geq N}}) = \overline{\{f(x_n)\}_{n \geq N}}.$$

Since $\{y\} = \cap_{N \geq 0} \overline{\{f(x_n)\}_{n \geq N}}$, it follows that

$$\forall N \geq 0, \quad L_n := \overline{\{x_n\}_{n \geq N}} \cap f^{-1}(y) \neq \emptyset.$$

Since $f^{-1}(y)$ is compact and since the L_n form a sequence of closed decreasing sets, their intersection L is non-empty.

(b) The proof for part (b) is easy.

(c) Suppose we have a sequence $x_n \in K$ such that Ax_n converges to y in Y. Since any $p \in X^*$ may be written as $p = A^*q + r$, where $q \in Y^*$ and $r \in b(K)$ by virtue of (4), it follows that

$$\langle p, x_n \rangle = \langle A^*q, x_n \rangle + \langle r, x_n \rangle = \langle p, Ax_n \rangle + \langle r, x_n \rangle < +\infty$$

(since the convergent sequence Ax_n is bounded and since $r \in b(K)$). If X is finite dimensional, it follows that the sequence x_n is bounded and that it has limit points in K, which is assumed to be closed. If X is infinite dimensional, it follows that the sequence x_n is weakly bounded and thus that it has weak limit points belonging to K which, being convex and closed, is weakly closed.

(d) We apply the previous question with $A \in L(X \times Y, Y)$ defined by $A(x, y) = Lx - y$ and $K = P \times Q$. Since $A^*q = (L^*q, -q)$ and $b(K) = P^- \times Q^-$, we need to check that for all $(p, q) \in X^* \times Y^*$, there exist $q_0 \in Y^*$ and $(p_1, q_1) \in P^- \times Q^-$ such that $(p, q) = (L^*q_0, -q_0) + (p_1, q_1)$. However, by virtue of (5), $p + L^*q \in L^*Q^- + P^-$ may be written in the form

$$p - L^*q = L^*q_1 + p_1 \text{ where } q_1 \in Q^-, p_1 \in P^-.$$

Thus, it is sufficient to check that $q_0 = q_1 - q$, where q_1 and p_1 satisfy the question.

Since $L(P) - Q = A(P \times Q)$, it follows that this convex cone is closed. Moreover, $A^{-1}(y) = \{(x, z) \in P \times Q | Lx - z = y\}$ is compact and its projection onto X, which is the set $\{x \in P | Lx \in Q + y\}$, is also compact.

(e) Suppose that $y^p = \sum_{i=1}^n x_i^p$ is a sequence converging to y, which is therefore bounded. Since L_i is bounded below,

$$u_i \leq x_i^p \leq y^p - \sum_{j \neq i} x_j^p \leq y^p - \sum_{j \neq i} u_j \leq v - \sum_{j \neq i} u_j$$

where $y^p \leq v$. Thus, the sequences x_i^p lie in compact sets and subsequences $x_i^{p'}$ converge to elements x_i of L_i. In particular, it follows that $\sum_{i=1}^n L_i$ is closed.

(f) Suppose that $z_n = x_n - y_n$ is a sequence converging to z. We shall show that y_n is bounded. If this is not the case, then there exists a subsequence (again denoted by) y_n such that $\|y_n\| \to \infty$. Thus, $v_n := y_n / \|y_n\|$ belongs to the unit sphere, which is *compact*. Thus, there exists a subsequence (again denoted by) v_n which converges to an element v. Since z_n converges, there exists an element w such that $z_n \leq a$. Thus, $y_n \geq x_n - z_n \geq u - a$. Whence, the inequalities $v_n \geq \frac{u-w}{\|y_n\|}$ imply that $v \geq 0$, in other words, that $v \in \mathbb{R}_+^\ell$. Moreover, since M is convex,

$$\frac{1}{\|y_n\|} y_n + \left(1 - \frac{1}{\|y_n\|}\right) w \in M.$$

Letting n tend to infinity, we obtain $v + w \in M$. Thus, $v \in (M - w) \cap \mathbb{R}_+^\ell = \{0\}$ (by virtue of (9)(ii)). Thus, we have obtained a contradiction.

16.20 Problem 20 – Solution. Fenchel's Theorem for the Functions $L(x, Ax)$

The proofs for parts (a) to (d) follow the proof of Theorem 3.2 (Fenchel's Theorem) for this particular case.

Subsidiary Question. Generalise the proof of Theorem 3.2 in the case of Hilbert spaces to the minimisation problem (1) stated in Problem 20.

16.21 Problem 21 – Solution. Conjugate Functions of $x \to L(x, Ax)$

(a) We set $f(q) := L^*(p - A^*q, q)$ and $K_\lambda = \{q | f(q) \le \lambda\}$. It suffices to show that K_λ is bounded, whence, that there exists $\gamma > 0$ such that for all $z \in \gamma B$,

$$\sup_{q \in K_\lambda} \langle q, z \rangle < +\infty.$$

Fenchel's inequality implies that $\langle q, z \rangle \le f(q) + f^*(z)$ and since $\sup_{q \in K_\lambda} f(q) \le \lambda$, it is sufficient to show that

$$\forall z \in \gamma B, \quad f^*(z) < +\infty.$$

Take $\gamma > 0$ such that $\gamma B \subset (A \oplus -1)\mathrm{Dom}\, L$. Then z may be written as $z = y_0 - Ax_0$ where $(x_0, y_0) \in \mathrm{Dom}\, L$, whence

$$
\begin{aligned}
f^*(z) &= \sup_{q \in Y^*} (\langle q, z \rangle - L^*(p - A^*q, q)) \\
&= \sup_{q \in Y^*} \inf_{x \in X} \inf_{y \in Y} (\langle q, z \rangle - \langle p - A^*q, x \rangle - \langle q, y \rangle + L(x, y)) \\
&\le \sup_{q \in Y^*} (\langle q, z \rangle - \langle p - A^*q, x_0 \rangle - \langle q, y_0 \rangle + L(x_0, y_0)).
\end{aligned}
$$

Since $\langle q, z \rangle - \langle -A^*q, x_0 \rangle - \langle q, y_0 \rangle = \langle q, z - (y_0 - Ax_0) \rangle = 0$, it follows that

$$f^*(z) \le -\langle p, x_0 \rangle + L(x_0, y_0) < +\infty.$$

(b) Set $g(p) := \inf_{q \in Y^*} L^*(p - Aq, q)$. This is convex following Proposition 2.5. When $p \in (1 \oplus A^*)\mathrm{Dom}\, L^*$, there exists $(r, q) \in \mathrm{Dom}\, L^*$ such that $p = r + A^*q$. Thus, $L^*(p - A^*q, q) < +\infty$ and $g(p) < +\infty$. Take $\lambda > g(p)$. Since K_λ is compact and since $g(p) = \inf_{q \in K_\lambda} L^*(p - A^*q, q)$, Proposition 1.7 implies that g is also lower semi-continuous and finite on $(1 \oplus A^*)\mathrm{Dom}\, L^*$, which, by assumption, is non-empty.

(c) We may apply Proposition 3.3 or use direct calculations, as follows:

$$
\begin{aligned}
g^*(x) &= \sup_{p \in X^*} (\langle p, x \rangle - \inf_{q \in Y^*} L^*(p - A^*q, q)) \\
&= \sup_{p, q} (\langle p - A^*q, x \rangle + \langle q, Ax \rangle - L^*(p - A^*q, q)) \\
&= L(x, Ax).
\end{aligned}
$$

Thus, $g(p) = g^{**}(p) = L(\cdot, A\cdot)^*(p)$.

16.22 Problem 22 – Solution. Hamiltonians and Partial Conjugates

(a) We note that $H(x, \cdot)$ is the conjugate function of $L(x, \cdot)$, whence that (2)(i) and (3)(i) follow from Theorem 3.1. Since the function $(x, y) \to \langle q, y \rangle - L(x, y)$ is concave, Proposition 2.5 implies (2)(ii).

Lastly,

$$
\begin{aligned}
L^*(p,q) &= \sup_{x,y}(\langle p,x\rangle + \langle q,y\rangle - L(x,y)) \\
&= \sup_{x}(\langle p,x\rangle + \sup_{y}(\langle q,y\rangle - L(x,y))) \\
&= \sup_{x}(\langle p,x\rangle + H(x,q)).
\end{aligned}
$$

(b) Since $p \to L^*(p,q)$ is the conjugate function of $x \to -H(x,q)$ (by (3)(ii)) and since this function is convex and lower semi-continuous (by (4)), Theorem 3.1 implies that

$$
-H(x,q) = \sup_{x \in X^*}(\langle p,x\rangle - L^*(p,q)).
$$

(c) We begin by noting that (6)(i) is equivalent to

$$
\langle p,x\rangle\langle q,y\rangle = L^*(p,q) + L(x,y) \tag{$*$}
$$

which may be written as

$$
(\langle p,x\rangle - (-H(x,q)) - L^*(p,q)) + (\langle q,y\rangle - L(x,y) - H(x,q)) = 0.
$$

Since $p \to L^*(p,q)$ is the conjugate function of $x \to -H(x,q)$ (by (3)(ii)) and since $g \to H(x,q)$ is the conjugate function of $y \to L(x,y)$ (by (1)), each term of this sum is negative, whence $(*)$ is equivalent to

(i) $\qquad\qquad \langle p,x\rangle - (-H(x,q)) - L^*(p,q) = 0$

(ii) $\qquad\qquad \langle q,y\rangle - L(x,y) - H(x,q) = 0.$ $\qquad\qquad$ $(**)$

We end by noting that $(**)$ is equivalent to (6)(ii).

(d) By virtue of the characterisation of the subdifferential, we have

$$
\begin{aligned}
0 \in \partial_x(-H(\bar{x},\bar{q})) &\Leftrightarrow \forall x \in X, \quad -H(\bar{x},\bar{q}) + H(\bar{x},\bar{q}) \le \langle 0, \bar{x} - x\rangle \\
&\Leftrightarrow \forall x \in X, \quad H(x,\bar{q}) \le H(\bar{x},\bar{q})
\end{aligned}
$$

and

$$
\begin{aligned}
0 \in \partial_q H(\bar{x},\bar{q}) &\Leftrightarrow \forall q \in Y^*, \quad H(\bar{x},\bar{q}) - H(\bar{x},q) \le \langle 0, \bar{q} - q\rangle \\
&\Leftrightarrow \forall q \in Y^*, \quad H(\bar{x},\bar{q}) \le H(\bar{x},q).
\end{aligned}
$$

16.23 Problem 23 – Solution. Lack of Convexity and Fenchel's Theorem for Pareto Optima

(a)

(i) Clearly $\rho(f) \ge f(x) - f(x) = 0$, where $x \in \mathrm{Dom}(f)$.

(ii) f is clearly convex if and only if $\rho(f) = 0$.

(iii) For any convex combination, we have

$$
\begin{aligned}
v &\leq f\left(\sum \alpha_i f(x_i)\right) + g\left(A\left(\sum \alpha_i x_i\right)\right) \\
&\leq \sum \alpha_i f(x_0) + g\left(A\left(\sum \alpha_i x_i\right)\right) + \rho(f).
\end{aligned}
$$

We obtain the inequality (4) by taking the infimum over the convex combinations.

(b) If (6) is false, there would exist finite sequences of $x_i \in L$, $y_i \in M$ and $\alpha_i \geq 0$ with $\sum \alpha_i = 1$ satisfying

$$
\begin{aligned}
(0, w) &= \left(\sum_{i=1}^{n} \alpha_i A x_i - \sum \alpha_i y_i, \sum \alpha_i f(x_i) + \sum \alpha_i g(y_i) + \rho\right) \\
&\in \operatorname{co}(\phi(L \times M)) + Q \quad (\text{where } \rho > 0).
\end{aligned}
$$

Thus, $\sum \alpha_i y_i = A\left(\sum \alpha_i x_i\right)$ and

$$
\begin{aligned}
w &> \sum \alpha_i f(x_i) + \sum \alpha_i g(y_i) \\
&\geq \sum \alpha_i f(x_i) + g\left(\sum \alpha_i y_i\right) \\
&\geq w
\end{aligned}
$$

which is impossible.

(c) Since Y is finite dimensional, we may use the Large Separation Theorem (Theorem 2.5). There exists a non-zero linear form $(c, q) \in \mathbb{R} \times Y^*$, such that

$$
cw \leq \inf_{x \in L, y \in M} (cf(x) + g(y) + \langle q, Ax - y \rangle) + \inf_{\rho > 0} \rho c.
$$

It follows that $c \geq 0$, whence inequality (7) holds.

(d) Suppose that $c = 0$, Then (7) implies that

$$
\begin{aligned}
0 &\leq \inf_{x \in L, y \in M} \langle q, Ax - y \rangle \\
&= \inf_{z \in A(L) - M} \langle q, z \rangle \\
&\leq \inf_{z \in \gamma B} \langle q, z \rangle \\
&= -\gamma \|q\|.
\end{aligned}
$$

This would give $q = 0$, which contradicts the fact that $(c, q) \neq (0, 0)$.

(e) Thus, $c > 0$ and, dividing by c and setting $\bar{q} = q/c$, we deduce from (7) that $w \leq -f^*(-A^*\bar{q}) - g^*(\bar{q})$. This inequality, together with (4) implies the inequality (9).

(f) Since $v_* \leq f^*(-A^*\bar{q}) + g^*(\bar{q}) \leq -w \leq -v + \rho(f)$ (from (4)), we obtain a majoration of $v+v_*$ by $\rho(f)$. If f is convex, $\rho(f) = 0$ and we rediscover Fenchel's Theorem (Theorem 3.2).

(g) The proof for part (g) is easy.

(h) In this case,

$$v = \inf_{Ax=u} f(x).$$

Assumption (8) may be written as

$$u \in \mathrm{Int}\, A(L)$$

and the conclusion (9) implies that there exists an approximate Lagrange multiplier $\bar{q}$ such that

$$v \leq -f^*(-A^*\bar{q}) + \rho(f).$$

16.24 Problem 24 – Solution. Duality in Linear Programming

(a) The linear programme may be written in the form

$$v = \inf(\langle p_0, x \rangle + \psi_{P \times Q - (0, u_0)}((1 \times B)x)).$$

We apply Fenchel's Theorem with

$$
\begin{aligned}
f(x) &:= \langle p_0, x \rangle, \\
g(x, y) &:= \psi_{P \times Q - (0, u_0)}(x, y), \\
Ax &:= (1 \times B)x.
\end{aligned}
$$

It follows that

$$
\begin{aligned}
f^*(p) &= \psi_{\{p_0\}}(p) \\
g^*(p, q) &= \psi_{P^-}(p) + \psi_{Q^-}(q) + \langle q, u_0 \rangle \\
A^*(p, q) &= p + B^* q.
\end{aligned}
$$

Thus,

$$
\begin{aligned}
v_* &= \inf(\psi_{\{p_0\}}(-p - B^* q) + \psi_{P^-}(p) + \psi_{Q^-}(q) + \langle q, u_0 \rangle) \\
&= \inf_{\substack{q \in Q^- \\ -B^* q \in P^- + q_0}} \langle q, u_0 \rangle \\
&= - \sup_{\substack{q \in Q^+ \\ B^* q \in P^- + q_0}} \langle q, u_0 \rangle.
\end{aligned}
$$

(b) The proof for part (b) is easy.

(c) The condition

$$(0,0) \in \mathrm{Int}(A \,\mathrm{Dom}\, f - \mathrm{Dom}\, g) = \mathrm{Int}\,(\mathrm{Im}\, A - P \times Q - (0, u_0))$$

is equivalent to the statement that

$$u_0 \in \mathrm{Int}(B(P) + Q)$$

and the condition $0 \in \mathrm{Int}(A^*\mathrm{Dom}\, g^* + \mathrm{Dom}\, f^*)$ may be written as

$$-p_0 \in \mathrm{Int}(B^*Q^- + P^-).$$

The first of the above statements implies the existence of a solution of the dual problem, whilst the second implies the existence of a solution of the primal problem.

(d) This is a consequence of Theorem 5.1.

16.25 Problem 25 – Solution. Lagrangian of a Convex Minimisation Problem

(a) The equivalence of (5)(i) and (5)(ii) is simply Proposition 4.3. The equivalence of (5)(ii) and (5)(iii) follows from Problem 22.

(b) Formula (2) implies that

$$f(x, y) = \sup_{q \in Y^*} (\langle q, y \rangle - h(x, q))$$

(since f is lower semi-continuous), whence

$$g(y) = \inf_{x \in X} \sup_{q \in Y^*} (\langle q, y \rangle - h(x, q)).$$

Moreover,

$$
\begin{aligned}
g^*(q) &= \sup_y (\langle q, y \rangle - \inf_{x \in X} f(x, y)) \\
&= \sup_{x \in X} \sup_{y \in Y} (\langle q, y \rangle - f(x, y)) \\
&= \sup_{x \in X} h(x, q).
\end{aligned}
$$

We also know that this is equal to $f^*(0, q)$ (Proposition 3.2).

(c) As a consequence of the above,

$$
\begin{aligned}
g^{**}(q) &= \sup_{q \in Y^*} (\langle q, y \rangle - g^*(q)) \\
&= \sup_{q \in Y^*} \inf_{x \in X} (\langle q, y \rangle - h(x, q)).
\end{aligned}
$$

Thus, the marginal function g is lower semi-continuous if and only if $g = g^{**}$; in other words, if and only if the minimax equality (8) holds.

(d) We know from (7) that $\bar{q} \in \partial g(y)$ if and only if

$$
\begin{aligned}
\langle \bar{q}, y \rangle &= g^*(\bar{q}) + g(y) \\
&= \sup_{x \in X} h(x, \bar{q}).
\end{aligned}
$$

This inequality may be rewritten in the form

$$
\begin{aligned}
g(y) &= \inf_{x \in X} (\langle \bar{q}, y \rangle - h(x, \bar{q})) \\
&= \inf_{x \in X} \ell_y(x, \bar{q})
\end{aligned}
$$

Thus, $\bar{q} \in \partial g(y)$ if and only if $\bar{q}$ is a Lagrange multiplier.

16.26 Problem 26 – Solution. Variational Principles for Convex Lagrangians

(a) We note firstly that

$$
\begin{aligned}
\mathcal{A}(x, q) &= L(x, Ax) + L^*(-A^*q, q) \\
&\geq \langle -A^*q, x \rangle + \langle q, Ax \rangle = 0
\end{aligned}
$$

and that the condition $\mathcal{A}(\bar{x}, \bar{q}) = 0$ implies that

$$
\mathcal{A}(\bar{x}, \bar{y}) = \min_{(x,q)} \mathcal{A}(x, q).
$$

The implications (i) $\Rightarrow$ (ii) $\Rightarrow$ (iii) $\Rightarrow$ (i) are self-evident. (6)(iii) is equivalent to (6)(iv) eliminating $\bar{q}$ and (6)(iii), which is equivalent to $(\bar{x}, A\bar{x}) \in \partial L^*(-A^*\bar{q}, \bar{q})$ is equivalent to (6)(v) eliminating $\bar{x}$.

(b) The equivalence of (6)(iii) and (7) follows from Problem 22.

16.27 Problem 27 – Solution. Variational Principles for Convex Hamiltonians

(a) We note that Fenchel's inequality implies that $\mathcal{B}(x, q) \geq 0$ for all x, q. The statement that $\mathcal{B}(\bar{x}, \bar{p}) = 0$ is equivalent to (1)(iii). The equivalence of (1)(ii) and (1)(iii) follows from Problem 22.

(b) It is easy to check that if $(\bar{x}, \bar{q})$ is a solution of (2), then

$$\forall (x, q) \in X \times Y^*, \quad \langle A^*\bar{q}, x \rangle + \langle q, A\bar{x} \rangle \leq DH(\bar{x}, \bar{q})(x, q).$$

This means that $(A^*\bar{q}, A\bar{x})$ belongs to $\partial H(\bar{x}, \bar{q})$.

(c) Suppose that $(\tilde{x}, \tilde{q})$ is a solution of (4). It follows that

$$\forall (x, q) \in X \times Y^*, \quad \langle A^*\tilde{q}, x \rangle + \langle q, A\tilde{x} \rangle \leq DH^*(A^*\tilde{q}, A^*\tilde{x})(A^*q, Ax).$$

Theorem 4.4 and assumption (3) then imply that

$$(A\tilde{x}, A^*\tilde{q}) \in (A \times A^*)\partial H^*(A^*\tilde{q}, A\tilde{x}).$$

Thus, there exists $(\bar{x}, \bar{y}) \in \partial H^*(A^*\tilde{q}, A\tilde{x})$ such that

$$(A\tilde{x}, A^*\tilde{q}) = (A\bar{x}, A^*\bar{q}).$$

Consequently, $(\bar{x}, \bar{q}) \in \partial H^*(A^*\bar{q}, Ax)$, which is equivalent to (1)(iii).

16.28 Problem 28 – Solution. Approximation to Fermat's Rule

(a) By Ekeland's Theorem (Corollary 1.3), there exist $\varepsilon > 0$, $\lambda > 0$ $x_0 \in \mathrm{Dom}\, f$ and x_ε such that

$$f(x_0) \leq \inf_{x \in X} f(x) + \varepsilon\lambda, \quad f(x_\varepsilon) \leq f(x_0), \quad \|x_0 - x_\varepsilon\| \leq \lambda$$

and

$$f(x_\varepsilon) \leq f(x) + \varepsilon\|x - x_\varepsilon\|$$

Fermat's rule then implies that

$$0 \in \partial(f + \varepsilon\|\cdot -x_\varepsilon\|)(x_\varepsilon).$$

Since $\mathrm{Dom}\|\cdot -x_\varepsilon\| = X$, Corollary 4.3 implies that

$$0 \in \partial f(x_\varepsilon) + \varepsilon\partial(\|\cdot -x_\varepsilon\|)(x_\varepsilon) = \partial f(x_\varepsilon) + \varepsilon B.$$

(b) By Theorem 3.1, there exists $p_0 \in \mathrm{Dom} f$. We set

$$g(x) := f(x) + f^*(p_0) - \langle p_0, x \rangle \geq 0.$$

Suppose that $x_0 \in \mathrm{Dom} f = \mathrm{Dom}\, g$. Since g is convex, lower semi-continuous and positive, we may apply the previous question to it with $\varepsilon_n = n(g(x_0) - \inf_{x \in X} g(x))$. Taking $\lambda = 1/n$, we find points $x_n \in \mathrm{Dom}\, g = \mathrm{Dom} f$ such that $g(x_n) \leq g(x_0)$, $\|x_0 - x_n\| \leq \frac{1}{n}$ and points $q_n \in \partial g(x_n)$ with $\|q_n\| \leq \varepsilon_n$. This implies that f is subdifferentiable at x_n, since $\partial f(x_n) = \partial g(x_n) + p_0$. Moreover, the inequality $g(x_n) \le g(x_0)$ may be written as

$$f(x_n) \le f(x_0) - \langle p_0, x_0 - x_n \rangle \ \forall n.$$

Since x_n converges to x_0, we obtain

$$\limsup_{n \to \infty} f(x_n) \le f(x_0).$$

Since f is lower semi-continuous at x_0, we have $f(x_0) \le \liminf_{n \to \infty} f(x_n)$. Thus, x_0 is the limit of a sequence of points x_n at which f is subdifferentiable. This restates the first part of Theorem 4.3, with a proof that is valid for Banach spaces.

16.29 Problem 29 – Solution. Transposes of Convex Processes

(a) We have $p \in F^*(q) \Leftrightarrow (p, -q) \in \mathrm{Graph}(F)^-$, which is a convex closed cone. Since $\mathrm{Graph}(F) = \mathrm{Graph}(F)^{--}$, we have $y \in F(x) \Leftrightarrow \langle p, x \rangle - \langle -q, y \rangle \le 0$ for all $(p, q) \in \mathrm{Graph}(F)^-$, in other words

$$\sup_{q \in Y^*} \sup_{p \in F^*(q)} (\langle p, x \rangle - \langle q, y \rangle) \le 0.$$

(b) This follows from formula (65) of Chapter 3.

(c) It suffices to show that the assumption (4) implies that

$$\exists \gamma > 0 \text{ such that } \gamma(B \times B) \subset \mathrm{Im}(A \times 1) + \mathrm{Graph}(F)$$

so that we may apply formula (70) of Chapter 3. Since $\mathrm{Graph}(F)$ is a cone, it suffices to show that $\mathrm{Im}(A \times 1) + \mathrm{Graph}(F) = X \times Y$. Then, $x = Ax_0 - x_1$, where $x_0 \in U$ and $x_1 \in \mathrm{Dom}F$ and we may write $(x, y) = (Ax_0, y_0) - (x_1, y_1) \in \mathrm{Im}(A \times 1) - \mathrm{Graph}(F)$ where $y_1 \in F(x_1)$ and $y_0 = y + y_1$.

(d) We may write

$$F + GA = B(F \times G)(A \times A)$$

where $(1 \times A)(x) = (x, Ax)$, $(F \times G)(x, y) = F(x) \times G(y)$ and $B(y, z) = y + z$. Assumption (6) implies assumption (4) where A is replaced by $(1 \times A)$. If $(u, v) \in X \times Y$, there exist $x \in \mathrm{Dom}\,F$ and $y \in \mathrm{Dom}\,G$ such that $Au - v = -Ax + y$, so that we may write

$$(u, v) = (z, Az) - (x, y) \in \mathrm{Im}(1 \times A) - \mathrm{Dom}(F \times G)$$

taking $z = u + x$. Questions (b) and (c) imply that

$$(F + GA)^* = (1 \times A)^*(F \times G)^* B^* = F^* + A^* G^*.$$

(e) We apply the previous question with $X = Y$, $A = 1$ and $G(x) = 0$ if $x \in K$ and $G(x) = \emptyset$ if $x \notin K$.

(f) Now $q \in \text{Im}(F)^-$ is equivalent to the statement that $(0, q) \in \text{Graph}(F)^-$ or that $(0, -q) \in \text{Graph}(F^*)$. Consequently, if $\text{Im}(F)$ is dense, then $\text{Im}(F)^- = X^- = \{0\}$. Conversely, suppose that $\text{Im}(F)^- = \{0\}$ and that $\text{Im}(F)$ is not dense. Then there would exist $y_0 \notin \overline{\text{Im}(F)}$ and the Separation Theorem would then imply the existence of $q \in Y^*$ such that $q \in \text{Im}(F)^-$ and $\langle q, y_0 \rangle > 0$. But then, we would have $q = 0$, which is a contradiction. Consequently, $\text{Im}(F) = Y$ if and only if $\text{Im}(F)$ is dense and closed.

(g) We apply question (f) to the set-valued map $F|_K$, which has image $F(K)$. Then

$$F(K)^- = \text{Im}(F|_K)^- = -(F|_K)^{*-1}(0).$$

But, from (e), $-q \in (F|_K)^{*-1}(0)$ if and only if $0 \in (F|_K)^*(q) = F^*(-q) + K^-$; in other words, if and only if $-q \in F^{*-1}(-K^-)$.

(h) The fact that F^{-1} is one-to-one follows from the inequality (9), taking $y = y_1 = y_2$ and x_1, x_2 in $F^{-1}(y)$. This inequality also implies that

$$c\|F^{-1}(y_1) - F^{-1}(y_2)\|^2 \leq c\|y_1 - y_2\| \|F^{-1}(y_1) - F^{-1}(y_2)\|$$

for all $y_1, y_2 \in \text{Im}(F)$.

To show that $\text{Im}(F)$ is closed, take a sequence of $y_n \in \text{Im}(F)$ converging to y. Now, the inequality (9) implies that $x_n := F^{-1}(y_n)$ is a Cauchy sequence in X, which converges to an element x since X is complete. Since $(x_n, y_n) \in \text{Graph}(F)$, which is closed, it follows that $y \in F(x) \subset \text{Im}(F)$.

To show that F is surjective, it is sufficient to show that if $x_0 \in F^{*-1}(0)$, then $x_0 = 0$. We take $y_0 \in F(x_0)$, Since $(0, -x_0) \in \text{Graph}(F)^-$, it follows that $\langle 0, x_0 \rangle - \langle x_0, y_0 \rangle \leq 0$. The inequality (9) then implies that

$$c\|x_0\|^2 = c\| - x_0 - 0\|^2 \leq \langle -x_0 - 0, y_0 - 0 \rangle = -\langle x_0, y_0 \rangle \leq 0.$$

16.30 Problem 30 – Solution. Cones with a Compact Base

(a) Suppose that $p_0 \in \text{Int} P^+$ is fixed. There exists $\varepsilon > 0$ such that $p_0 + \varepsilon B \subset P^+$, whence

$$\forall x \in P, \ x \neq 0, \ \langle p_0, x \rangle > 0.$$

If not, there would be an $x_0 \in P$, $x_0 \neq 0$, such that $\langle p_0, x_0 \rangle = 0$, whence $\forall p \in X^*, \langle p_0 + \frac{\varepsilon}{\|p\|} p, x_0 \rangle = \frac{\varepsilon}{\|p\|} \langle p, x_0 \rangle \geq 0$ and we would then have $x_0 = 0$. The set S generates P, since if $x \in P$ and $x \neq 0$, then $\langle p_0, x \rangle > 0$ and $x = \langle p_0, x \rangle y$ where $y := p_0/\langle p_0, x \rangle \in S$.

(b) Since $p_0 + \varepsilon B \subset P^+$, then for all $x \in S$

$$1 - \varepsilon\|x\| = \langle p_0, x \rangle + \varepsilon \inf_{p \in B} \langle p, x \rangle \geq \inf_{p \in P^+} \langle p, x \rangle \geq 0$$

whence $\|x\| \leq \frac{1}{\varepsilon}$. Thus, $S \subset \frac{1}{\varepsilon}B$ is a closed bounded set.

(c) Since $0 \notin K$, the Separation Theorem implies that there exists p_0 such that

$$0 \leq \inf_{x \in K} \langle p_0, x \rangle - \eta.$$

Suppose that $\|K\| := \sup_{z \in K} \|z\|$ and set $\gamma = \eta/\|K\|$. Then, for all $p \in \gamma B$, we have

$$\langle p_0 + p, x \rangle \geq \eta - \|p\|\|x\| \geq \eta - \gamma\|K\| = 0$$

which implies that $p_0 + p \in P^+$ for all $p \in \gamma B$. Thus, p_0 is in the interior of P^+.

(d) We consider the convex compact set S which may be written as $S = P \cap p_0^{-1}\{1\}$ where $p_0 \in L(X, \mathbb{R})$. Since $p_0(P) - \{1\} = \mathbb{R}_+ - \{1\}$ has a non-empty interior, formula (49) of Chapter 4 implies that

$$T_S(x) = T_P(x) \cap p_0^{-1} T_{\{1\}}(\{1\}) = T_P(x) \cap \operatorname{Ker} p_0.$$

Moreover, if P is a convex closed cone, then $p \in N_P(x)$ if and only if $\langle p, x \rangle = \sigma_P(p)$, in other words, if $p \in P^-$ and $\langle p, x \rangle = 0$. The formulae of (6) now follow. Consequently

$$T_P(x) = N_P(x)^- = (P^- \cap \{x\})^- = \operatorname{closure}(P + x\mathbb{R}).$$

16.31 Problem 31 – Solution. Regularity of Tangent Cones

indextangent cone—(

(a) We write $x + tv = (1 - \frac{t}{h})x + \frac{t}{h}(x + hv)$ where h is such that $x + hv \in K$ and use the convexity of K.

(b) It is clear that for an arbitrary $x_0 \in K$,

$$C_K(x_0) \subset T_K(x_0).$$

Suppose that $u_0 \in \operatorname{closure}(\cup \frac{1}{h}(K - x_0))$ and $\varepsilon > 0$ are fixed. There exist $y \in K$ and $\alpha > 0$ such that $u_0 - \frac{1}{\alpha}(y - x_0) \in \frac{\varepsilon}{2}B$. Take $\beta = \varepsilon\alpha/2$, $x \in B_K(x_0, \beta)$ and $h \in]0, \alpha]$ and set $u := \frac{y-x}{\beta}$. Then $x + hu = (1 - \frac{h}{\alpha})x + \frac{h}{\alpha}y$ belongs to the set K, since x and $y = x + \beta u$ belong to K and $\frac{h}{\alpha} \leq 1$. Consequently,

$$\|u - u_0\| \leq \frac{\|x - x_0\|}{\alpha} + \left\|u_0 - \frac{y - x_0}{\alpha}\right\| \leq \beta/\alpha + \varepsilon/2 = \varepsilon.$$

Thus, u_0 belongs to $C_K(x_0)$.

(c) We have

$$y \in \pi_K^{-1}(x) \quad \Leftrightarrow \quad \langle y - x, z - x \rangle \leq 0 \quad \forall z \in K$$
$$\Leftrightarrow \quad \langle y - x, v \rangle \leq 0 \quad \forall v \in S_K(x)$$
$$\Leftrightarrow \quad \langle y - x, v \rangle \leq 0 \quad \forall v \in T_K(x)$$
$$\Leftrightarrow \quad y - x \in N_K(x).$$

(d) Suppose that $(x_n, p_n) \in \mathrm{Graph} N_K(\cdot)$ is a sequence converging to (x, p). Then, for all $y \in K$, we have $\langle p_n, y \rangle \leq \langle p_n, x_n \rangle$ and, passing to the limit, $\langle p, y \rangle \leq \langle p, x \rangle$. Thus, $p \in N_K(x)$.

(e) It is clear that $\cup_{h>0} \frac{1}{h} (\mathrm{Int}\, K - x) \subset \mathrm{Int}\, T_K(x)$. Since $S_K(x)$ is convex, then $\mathrm{Int}\, T_K(x) = \mathrm{Int} S_K(x)$, whence $v \in \mathrm{Int} S_K(x)$ and there exists $h > 0$ such that $v \in \frac{1}{h}(\mathrm{Int} K - x)$. Suppose that $\eta > 0$ is such that $v + \eta B \subset S_K(x)$. If $x + v \in \mathrm{Int}(K)$, the question is proved. Otherwise, take $x_0 \in \mathrm{Int}(K)$ and $v_0 := x_0 - x$. Then $v - \frac{\eta}{\|v_0\|} v_0$ belongs to $S_K(x)$, whence, there exists $\eta_1 > 0$ such that $x + h(v - \frac{\eta_1}{\|v_0\|} v_0)$ belongs to K. Take $\alpha := \frac{h\eta_1}{h\eta_1 + \|v_0\|}$ and note that

$$x + (1 - \alpha)hv = \alpha x_0 + (1 - \alpha)\left(x + h\left(v - \frac{\eta_1}{\|v_0\|} v_0 \right) \right).$$

Since $x_0 \in \mathrm{Int} K$ and $x + h(v - \frac{\eta_1}{\|v_0\|} v_0) \in K$, it follows that $x + (1-\alpha)hv \in \mathrm{Int}(K)$ and that $v \in \cup_{k>0} \frac{1}{k}(\mathrm{Int}(K) - x)$.

(f) Suppose that $v_0 \in \mathrm{Int}\, T_K(x_0)$. Then there exists $h_0 > 0$ such that $v_0 \in \frac{1}{h_0}(\mathrm{Int}(K - x_0))$, whence, there exists $\varepsilon > 0$ such that

$$x_0 + h_0 v_0 + \varepsilon B = x_0 + h_0(v_0 + \frac{\varepsilon}{h_0} B) \subset \mathrm{Int}(K).$$

Thus,

$$(x_0 + \varepsilon/2 B) \times (v_0 + \frac{\varepsilon}{2h_0} B) \subset \mathrm{Graph}(\mathrm{Int} T_K(\cdot)).$$

16.32 Problem 32 – Solution. Tangent Cones to an Intersection

(a)

$$T_{K_1}(0,0) = \{x | x_1 \leq 0\}, \qquad T_{K_2}(0,0) = \mathbb{R}_+^2$$
$$K_1 \cap K_2 = \{0, 0\}, \qquad T_{K_1 \cap K_2}(0,0) = \{0, 0\}$$
$$T_{K_1}(0,0) \cap T_{K_2}(0,0) = \{x | x_1 = 0, x_2 = 0\} \neq T_{K_1 \cap K_2}(0,0).$$

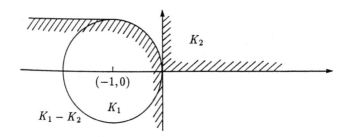

Fig. 16.1.

(b)

$$K_1 \cap K_2 = \{0\} \times [-1, +1] \text{ and } K_1 - K_2 = [-2, 0] \times [-2, +2].$$

Thus,

$$
\begin{aligned}
(0,0) &\in K_1 - K_2, & (0,0) &\notin \mathrm{Int}(K_1 - K_2) \\
T_{K_1 \cap K_2}(0,0) &= \{0\} \times \mathbb{R}, & T_{K_1}(0,0) &= {]-\infty, 0]} \times \mathbb{R} \\
T_{K_2}(0,0) &= [0, \infty[\times \mathbb{R} \text{ and } & T_{K_1 \cap K_2}(0,0) &= T_{K_1}(0,0) \times T_{K_2}(0,0).
\end{aligned}
$$

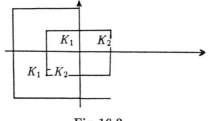

Fig. 16.2.

(c) The statement that $x \in K$ means that $Ax \in \prod_{i=1}^{n} JK_i$. Assumption (50) of Chapter 4 implies that in X^n,

$$0 \in \mathrm{Int}(\mathrm{Im}A - \prod_{i=1}^{n} K_i).$$

Thus, formula (50) of Chapter 4 implies that

$$T_K(x) = A^{-1}(T_{\prod_{i=1}^{n} K_i}(x_1, \ldots, x_n))$$

and formula (46) of Chapter 4 implies that

$$T_K(x) = A^{-1}(\prod_{i=1}^{n} T_{K_i}(x_i)).$$

indextangent cone—)

16.33 Problem 33 – Solution. Derivatives of Set-valued Maps with Convex Graphs

(a) The proof for part (a) is self-evident.

(b) The equivalence of (i) and (ii) follows from the definition of the normal cone and that of (i) and (iii) from the fact that the normal cone is the polar cone of the tangent cone.

(c) From the characterisation (ii), we have

$$\langle q_1, y_1 - y_2 \rangle \leq \langle p_1, x_1 - x_2 \rangle$$
$$\langle q_2, y_2 - y_1 \rangle \leq \langle p_2, x_2 - x_1 \rangle$$

and we add these inequalities.

(d) It is sufficient to observe that $\text{Graph}(\phi_K) = K \times \{0\}$ and that $T_{\text{Graph}\phi_K} = T_K \times \{0\} = \text{Graph}\phi_{T_K}$.

(e) Suppose that $h_1 \leq h_2$. The convexity of the graph of F implies that if $y \in F(x)$, then

$$\frac{h_1}{h_2} F(x + h_2 u) + \left(1 - \frac{h_1}{h_2}\right) y \subset F\left(\frac{h_1}{h_2}(x + h_2 u) + \left(1 - \frac{h_1}{h_2}\right) x\right)$$
$$= F(x + hu).$$

Thus,

$$\frac{F(x + h_2 u) - y}{h_2} \subset \frac{F(x + h_1 u) - y}{h_1}$$

whence

$$d\left(v, \frac{F(x + h_1 u) - y}{h_1}\right) \leq d\left(v, \frac{F(x + h_2 u) - y}{h_2}\right).$$

Thus,

$$\lim_{h \to 0+} d\left(v, \frac{F(x + hu) - y}{h}\right) = \inf_{h > 0} d\left(v, \frac{F(x + hu) - y}{h}\right)$$

exists in $\overline{\mathbb{R}}$.

The definition of the tangent cone to the graph of F at $(x_0, y_0) \in \text{Graph}(F)$ implies that $v_0 \in DF(x_0, y_0)(u_0)$ if and only if for all $(\varepsilon_1, \varepsilon_2) > 0$ we have

$$\inf_{\|u-u_0\| \leq \varepsilon_1} \inf_{h>0} d\left(v_0, \frac{F(x_0 + hu) - y_0}{h}\right) \leq \varepsilon_2$$

which is the desired formula.

(f) Take $y \in F(x)$. The convexity of the graph of F implies that

$$(1 - h)y_0 + hy \subset F(x_0 + h(x - x_0))$$

whence that

$$y - y_0 \in \frac{F(x_0 + h(x - x_0)) - y_0}{h}.$$

It suffices to let h tend to $0+$.

(g) We show that (ii) $\Rightarrow$ (i). From the previous question, we have $F(x) - y_0 \subset DF(x_0, y_0)(x - x_0) \subset P$ for all $x \in \text{Dom}(F)$. We now show that (i) $\Rightarrow$ (iii). From (b) we have, for all $q \in P^+$, $\forall x \in X$ and $\forall y \in F(x)$, $y \in y_0 + P$, whence $\langle q, y_0 - y \rangle \leq 0 = \langle 0, x_0 - x \rangle$, in other words, $0 \in DF(x_0, y_0)^*(q)$. Finally, we show that if $v_0 \in DF(x_0, y_0)(u_0)$, then $v_0 \in P = P^{++}$, since, by virtue of (iii), because $(0, -q) \in N_{\text{Graph}(F)}(x_0, y_0) = T_{\text{Graph}(F)}(x_0, y_0)^-$, we have

$$\langle (0, -q), (u_0, v_0) \rangle = -\langle q, v_0 \rangle \leq 0 \quad \forall q \in P^+$$

in other words, $v_0 \in P$.

(h) Take $v \in DF(x, y)(u)$. There exist sequences $u_n \to u$, $v_n \to v$ and $h_n > 0$ such that $y + h_n v_n \in F(x + h_n u_n)$ for all n. This implies that $x + h_n u_n \in L$ for all n (and thus that $u \in T_L(x)$) and that $Ax - y + h_n(Au_n - v_n) \in M$ for all n (and thus that $Au - v \in T_M(Ax - y)$). Conversely, take $u \in T_L(x)$ and $v \in Au - T_M(Ax - y)$. There exist sequences $u_n \to u$, $w_n \to Au - v$, $h_n^1 > 0$, $h_n^2 > 0$ such that $x + h_n^1 u_n \in L$ and $Ax - y + h_n^2 w_n \in M$ for all n. We set $h_n := \min(h_n^1, h_n^2)$ and $v_n := Au_n - w_n$, which converges to v. We then observe that $y + h_n v_n \in F(x + h_n u_n)$. The last formula is then clear.

16.34 Problem 34 – Solution. Epiderivatives of Convex Functions

(a) The proof for part (a) is self-evident.

(b) We note that $v_0 \in DF_+(x, f(x))(u_0)$ if and only if for all $\varepsilon_1, \varepsilon_2 > 0$, there exist $u \in u_0 + \varepsilon_1 B$, $v \in v_0 + \varepsilon_2 B$ and $h > 0$ such that $f(x) + hv \geq f(x + hu)$, in other words, that $v \geq \inf_{u \in u_0 + \varepsilon_1 B} DF(x)(u)$, or again that

$$v_0 \geq \sup_{\varepsilon_1 > 0} \inf_{u \in u_0 + \varepsilon_1 B} DF(x)(u).$$

Moreover, $DF_+(x, f(x))^*$ is a positively homogeneous set-valued map from $\mathbb{R}$ to X^*. It is sufficient to identify $DF_+(x, f(x))^*(-1)$ and $DF_+(x, f(x))^*(1)$. But, $p \in DF_+(x, f(x))^*(-1)$ if and only if

$$(p, 1) \in N_{\mathrm{Ep}(f)}(x, f(x))$$

which is impossible. However, $p \in DF_+(x, f(x))^*(1)$ if and only if $(p, -1) \in N_{\mathrm{Ep}(f)}(x, f(x))$:

$$\forall y \in X, \quad \langle p, y \rangle - f(y) \leq \langle p, x \rangle - f(x)$$

in other words, if and only if $p \in \partial f(x)$.

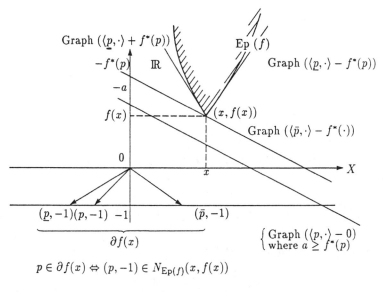

$$p \in \partial f(x) \Leftrightarrow (p, -1) \in N_{\mathrm{Ep}(f)}(x, f(x))$$

Fig. 16.3.

16.35 Problem 35 – Solution. Subdifferentials of Marginal Functions

(a) We identify f with the function $\hat{f}$ defined on $X \times Y$ by $\hat{f}(x, y) := f(x)$. Thus,

$$h(y) := \inf(\hat{f}(x, y) + \psi_{\text{Graph}(F)}(x, y)) = f(\bar{x})$$

where $\bar{x} \in F^{-1}(y)$.

We know from Proposition 4.3 that

$$q \in \partial h(y) \Leftrightarrow (0, q) \in \partial(\hat{f} + \psi_{\text{Graph}(F)})(\bar{x}, y).$$

Assumption (2) implies that

$$(0, 0) \;\in\; \text{Int}(\text{Dom}\hat{f} - \text{Dom}\psi_{\text{Graph}(F)})$$
$$= \text{Int}(\text{Dom}f \times Y - \text{Graph}(F)).$$

In fact, by virtue of (2), there exists $\gamma > 0$ such that any $u \in \gamma B_X$ may be written as $u = x_1 - x_2$ where $x_1 \in \text{Dom}f$ and $x_2 \in \text{Dom}F$. We take $v \in \gamma B_Y$ and $y_2 \in F(x_2)$ and set $y_1 = v + y_2$. Thus, $(u, v) = (x_1 - x_2, y_1 - y_2) \in \text{Dom}\hat{f} - \text{Dom}\psi_{\text{Graph}(F)}$. Following Corollary 4.3, we obtain

$$q \in \partial h(y) \Leftrightarrow (0, q) \in \partial f(\bar{x}) \times \{0\} + N_{\text{Graph}(F)}(\bar{x}, y).$$

Thus, there exists $\bar{p} \in \partial f(x)$ such that

$$(-\bar{p}, q) \in N_{\text{Graph}(F)}(\bar{x}, y).$$

(b) From the definition of $DF(\bar{x}, y)^*$, (3) may be written as

$$-\bar{p} \in DF(\bar{x}, y)^*(-q)$$

and formula (4) now follows.

16.36 Problem 36 – Solution. Values of a Game Associated with a Covering

The proofs for parts (a) and (b) are easy.

(c) From the definition of $v^\natural(\mathcal{A})$, the inequality

$$\inf_{x \in E} \sup_{y \in K} f(x, y) \leq v^\natural(\mathcal{A})$$

holds for all $K \in \mathcal{A}$. Thus, given any integer n, there exists an element $x_{K,n} \in E$ such that

$$\sup_{y \in K} f(x_{K,n}, y) \leq v^\natural(\mathcal{A}) + \frac{1}{n}$$

(definition of the infimum). Taking $(L, m) \geq (K, n)$, we obtain:

$$\sup_{y \in K} f(x_{L,m}, y) \leq \sup_{y \in L} f(x_{L,m}, y) \leq v^\natural(\mathcal{A}) + \frac{1}{m} \leq v^\natural(\mathcal{A}) + \frac{1}{n}.$$

Thus, if we fix $K_0 \in \mathcal{A}$ and $K \supset K_0$, it follows that

$$\sup_{(L,m) \geq (K,n)} \left(\sup_{y \in K_0} f(x_{L,m}, y) \right) \leq v^{\natural}(\mathcal{A}) + \frac{1}{n}$$

whence,

$$\limsup_{(K,n) \geq (K_0,1)} \left(\sup_{y \in K_0} f(x_{K,m}, y) \right) := \inf_{(K,n) \geq (K_0,1)} \sup_{(L,m) \geq (K,n)} \left(\sup_{y \in K_0} f(x_{K,m}, y) \right)$$
$$\leq v^{\natural}(\mathcal{A}).$$

(d) if $\mathcal{A}$ is countable, the same is true of $\mathcal{A} \times \mathbb{N}$ and the sequence $(K, n) \to x_{K,n}$ is a standard sequence.

(e) From (7)(ii), $\exists K_0 \in \mathcal{A}$ such that the (generalised) sequence of the $x_{K,n}$ lies in a compact set. Thus, a generalised subsequence converges to an element x_* and since $x \to f(x, y)$ is lower semi-continuous, it follows that

$$\forall y \in K, \quad f(x_*, y) \leq v^{\natural}(\mathcal{A}).$$

16.37 Problem 37 – Solution. Minimax Theorems with Weak Compactness Assumptions

(a) Since $y \to \inf_{x \in E} f(x, y)$ is concave and upper semi-continuous, the sets K_n are convex, closed and bounded (whence compact). Applying the minimax theorem (Theorem 8.1) to $-f$, we obtain, for all n

$$\inf_{x \in E} \sup_{y \in K_n} f(x, y) = \sup_{y \in K_n} \inf_{x \in E} f(x, y)$$

so that, since $F = \cup_n K_n$, we have

$$v^{\natural}(\mathcal{A}) = \sup_n \sup_{y \in K_n} \inf_{x \in E} f(x, y)$$
$$= \sup_{y \in F} \inf_{x \in E} f(x, y).$$

(b) Assumption (8) implies that there exists $\gamma > 0$ such that

$$\gamma B \subset \cup_{y \in F} \mathrm{Dom} f_y^*.$$

Thus, for all $p \in \gamma B$, there exists $y_p \in F$ such that $f_{y_p}^*(p) < +\infty$. From question (d) of Problem 36 with $K_0 = K_{n_p}$, there exists N_p such that

$$\forall n \geq N_p, \quad \langle p, x_n \rangle \leq f(x_n, y_p) + f_{y_p}^*(p)$$
$$\leq v^{\natural}(\mathcal{A}) + 1 + f_{y_p}^*(p)$$

(Fenchel's inequality). Since the sequence x_n is countable, it follows that

$$\sup_{n \geq 0} \langle p, x_n \rangle < +\infty$$

which implies that the sequence of the x_n is bounded. Thus, there exists a subsequence x_n which converges to $\bar{x}$ and, since $x \to f(x, y)$ is lower semi-continuous, it follows that

$$f(\bar{x}, y) \leq \liminf_{x_n \to \bar{x}} f(x_n, y) \leq v^\natural(\mathcal{A}).$$

Since $v^\natural(\mathcal{A}) = v^\flat$ (from the previous question), we have proved the minimax equality.

16.38 Problem 38 – Solution. Minimax Theorems for Finite Topologies

(a) We take $f = I$, the identity mapping from F to Y. For all $K \in \mathcal{S}$ the mapping $I\beta_K = \beta_K : M^n \to Y$ is continuous for all vector-space topologies on Y (under which addition and scalar multiplication are continuous). Thus, I is continuous from F (with the finite topology) to Y.

(b) Suppose that A is an affine mapping from F to Z. We have to show that for all $K \in \mathcal{S}$, $A\beta_K$ is continuous from F to Z. But, $A\beta_K = \beta_{A(K)}$ is associated with the finite subset $A(K)$ of Z and is thus continuous when Z has the finite topology.

Remark. . The *weak topology* on a Hilbert space Y is the *initial topology*, the weakest topology under which linear forms $f \in Y^*$ or linear operators from Y to finite-dimensional spaces are continuous (for the Hilbert space topology).

(c) In the proof of Theorem 8.4, the mapping D from E to F defined by $D(x) = \sum_{i=1}^{n} g_i(x) D_\varepsilon(x_i)$ is continuous when F has the weak topology, since D may be written in the form $\beta_K g$ where $K = \{D_\varepsilon(x_1), \ldots, D_\varepsilon(x_n)\}$ and where $g : x \to g(x) = (g_1(x), \ldots, g_n(x))$ is continuous. Thus, if F_0 denotes F with the finite topology and F_1 denotes the set F with a vector-space topology, then $\mathcal{C}(E, F_0) \subset \mathcal{C}(E, F_1)$, whence

$$\sup_{D \in \mathcal{C}(E, F_0)} \inf_{x \in E} f(x, D(x)) \leq \sup_{D \in \mathcal{C}(E, F_1)} \inf_{x \in E} f(x, D(x)) \leq v^\natural.$$

Thus, equation (48) of Chapter 8 is stronger with F_0 than with F_1.

(d) In the proof of Theorem 8.5, we may write $\varphi(u, \lambda) = \sum_{i=1}^n \lambda_i f(C\beta_K(\mu), y_i)$ in formula (55) of Chapter 8. The continuity of C from F to E when F has the finite topology implies the continuity of $\mu \to C\beta_K(\mu)$ and the lower semi-continuity of the functions $\mu \to \varphi(\mu, \lambda)$. Thus, in Theorem 8.5, we may replace the space F_1 (F with a vector-space topology) by the space F_0 (F with the finite topology). Since $\mathcal{C}(F_1, E) \subset \mathcal{C}(F_0, E)$, we obtain

$$\inf_{C \in \mathcal{C}(F_0, E)} \sup_{y \in F} f(C(y), y) \leq \inf_{C \in \mathcal{C}(F_1, E)} \sup_{y \in F} f(C(y), y) \leq v^\sharp.$$

Thus, equation (51) of Chapter 8 is stronger with F_0 than with F_1.

16.39 Problem 39 – Solution. Ky Fan's Inequality

(a) The negation of (3) may be written as

$$\forall x, \ \exists y \text{ such that } \varphi(x, y) > 0$$

or as

$$K \subset \cup_{y \in K} T(y).$$

Since K is compact, and since the sets $T(y)$ are open, (4) now follows using (2).

(b) We consider a continuous partition of unity $\{a_i(\cdot)\}$ which is subordinate to the covering of K by the open sets $T(y_i)$. We set

$$f(x) := \sum_{i=1}^n a_i(x) y_i.$$

Thus, f is a continuous set-valued map with values in K, since K is convex and the y_i are in K. Brouwer's Fixed-point Theorem (Theorem 7.1) implies the existence of a fixed point $\bar{x} = f(\bar{x}) = \sum_{i=1}^n a_i(\bar{x}) y_i$. Thus,

$$
\begin{aligned}
\varphi(\bar{x}, \bar{x}) &= \varphi\left(\bar{x}, \sum_{i=1}^n a_i(\bar{x}) y_i\right) \\
&\geq \sum_{i=1}^n a_i(\bar{x}) \varphi(\bar{x}, y_i) \qquad \text{(from (5))} \\
&= \sum_{i \in I(\bar{x})} a_i(\bar{x}) \varphi(\bar{x}, y_i)
\end{aligned}
$$

where

$$I(\bar{x}) := \{i = 1, \ldots, n | a_i(\bar{x}) > 0\} \neq \emptyset.$$

But, when $i \in I(x)$, $a_i(\bar{x}) \neq 0$ and $\bar{x}$ belongs to the support of a_i, which is contained in $T(y_i)$. Thus $i \in I(\bar{x})$ implies that $a_i(\bar{x}) > 0$ and $\varphi(\bar{x}, y_i) > 0$. Thus (6) is satisfied.

(c) Thus, Brouwer's Theorem shows that the negations of (3), (2) and (5) imply the negation of (7) and that (2), (5) and (7) imply (3).

(d) We use Proposition 8.2 to extend Ky Fan's Inequality to the case in which K is a convex compact subset of a Hilbert space (or a general topological vector space). There exists $\bar{x} \in K$ such that

$$\sup_{y \in K} \varphi(\bar{x}, y) \leq v^\natural := \sup_{\{y_1, \dots, y_n\}} \inf_{x \in K} \max_{i=1, \dots, n} \varphi(x, y_i).$$

But, for any fixed $x \in K$,

$$\max_{i=1, \dots, n} \varphi(x, y_i) = \sup_{\lambda \in M^n} \sum_{i=1}^{n} \lambda_i \varphi(x, y_i)$$

$$\leq \sup_{\lambda \in M^n} \varphi\left(x, \sum_{i=1}^{n} \lambda_i y_i\right)$$

$$= \sup_{y \in \mathrm{co}(y_1, \dots, y_n)} \varphi(x, y).$$

Since K is convex,

$$\mathrm{co}(y_1, \dots, y_n) \subset K$$

and (c) implies that

$$\inf_{x \in \mathrm{co}(y_1, \dots, y_n)} \sup_{y \in \mathrm{co}(y_1, \dots, y_n)} \varphi(x, y) \leq 0.$$

Thus,

$$\inf_{x \in K} \max_{i=1, \dots, n} \varphi(x, y_i) \leq \inf_{x \in \mathrm{co}(y_1, \dots, y_n)} \max_{i=1, \dots, n} \varphi(x, y_i)$$

$$\leq \inf_{x \in \mathrm{co}(y_1, \dots, y_n)} \sup_{y \in \mathrm{co}(y_1, \dots, y_n)} \varphi(x, y) \leq 0$$

whence $v^\natural \leq 0$.

(e) See the remarks following the proof of Theorem 8.6.

16.40 Problem 40 – Solution. Ky Fan's Inequality for Monotone Functions

(a) We may write

$$v^\natural \leq \sup_{K \in S} \inf_{x \in \mathrm{co}(K)} \sup_{y \in \mathrm{co}(K)} \varphi(x, y)$$

since $\mathrm{co}(K) \subset E$ and $K \subset \mathrm{co}(K)$. Applying Ky Fan's Theorem (Theorem 8.6) in finite-dimensional spaces (since $K = \{y_1, \dots, y_n\}$ is finite, $\mathrm{co}(K)$ is in a finite-dimensional space) we know from (2)(iii) that there exists $\bar{x} \in \mathrm{co}(K)$ such that $\sup_{y \in \mathrm{co}(K)} \varphi(\bar{x}, y) \leq \sup_{y \in \mathrm{co}(K)} \varphi(y, y) \leq 0$. Thus, $v^\natural \leq 0$.

(b) Thus, there exists $\bar{y} \in K$ such that

$$0 < \varphi(\bar{x}, \bar{y}) \qquad (*)$$

and since, from assumption (2)(i), the function $t \to \varphi(\bar{x} + t(\bar{y} - \bar{x}), \bar{y})$ is lower semi-continuous, there exists $\bar{t} \in]0, 1[$ satisfying the desired inequality.

(c) We set $y := \bar{x} + \bar{t}(\bar{y} - \bar{x})$. The fact that φ is monotone implies that

$$
\begin{aligned}
0 &\leq \limsup_{\mu \geq \mu(y)}(\varphi(x_\mu, y) + \varphi(y, x_\mu)) \\
&\leq \limsup_{\mu \geq \mu(y)} \varphi(x_\mu, y) + \limsup_{\mu} \varphi(y, x_\mu) \\
&\leq \varphi(y, \bar{x}) \qquad\qquad\qquad\qquad (**)
\end{aligned}
$$

because $\limsup_{\mu \geq \mu(y)} \varphi(x_\mu, y) \leq v^\natural \leq 0$ from Problem 36 and

$$\limsup_{\mu \geq \mu(y)} \varphi(y, x_\mu) \leq \varphi(y, \bar{x})$$

since $z \to \varphi(y, z)$ is upper semi-continuous. Since this function is also concave, the inequalities (6) and (**) imply that

$$
\begin{aligned}
0 &< \bar{t}\varphi(\bar{x} + \bar{t}(\bar{y} - \bar{x}), \bar{y}) + (1 - \bar{t})\varphi(y, \bar{x}) \\
&\leq \varphi(\bar{x} + \bar{t}(\bar{y}, \bar{x}), (1 - t)\bar{x} + \bar{t}\bar{y}). \qquad (***)
\end{aligned}
$$

The inequality (***) contradicts the assumption of monotonicity (3). It follows that assumption (5) is false, whence (9) is true.

Remark. In other words, the assumption of monotonicity (3) allowed us to replace the assumption of lower semi-continuity of $x \to \varphi(x, y)$ for the initial topology with that of lower semi-continuity for the finite topology, which is stronger. This is very useful since, despite appearances, there are many interesting examples in which it is not possible to assume that $x \to \varphi(x, y)$ is lower semi-continuous for the topology of an infinite-dimensional space.

16.41 Problem 41 – Solution. Generalisations of the Gale–Nikaïdo–Debreu Theorem

(a) We introduce the function φ, defined on $K \times P^-$ by

$$\varphi(x, y) := -\sigma(F(x), y)$$

which satisfies the assumptions of Theorem 8.5. Thus, there exists $\bar{x} \in K$ such that

$$\sup_{y \in P^-} \varphi(\bar{x}, y) \leq \sup_{y \in P^-} \varphi(C(y), y)$$

where, from (3), the second term is negative or zero. Thus,

$$\forall y \in P^-, \ 0 \le \sigma(F(\bar{x}), y) = \sigma(F(\bar{x}), y) + \sigma(P, y),$$

whence

$$\forall y \in Y, \ 0 \le \sigma(F(\bar{x}) + P, y).$$

Since $F(\bar{x}) + P$ is closed, it follows that $0 \in F(\bar{x}) + P$.

(b) The proof for part (b) is self-evident.

(c) We take $C = \pi_K$ and $P = \{0\}$ (whence $P^- = Y$).

(d) We take $P = \{0\}$, $K = B$ (which is compact, since Y is finite dimensional) and C to be the function defined by $C(y) = y$ if $y \in B$, $C(y) = \frac{y}{\|y\|}$ if $y \notin B$. Then $\sigma(F(C(y), y)) = \|y\|\sigma(F(C(y), y)) \ge 0$ if $y \notin B$ and $\sigma(F(C(y), y)) = \sigma(F(y), y) \ge 0$ if $y \in B$.

16.42 Problem 42 – Solution. Equilibrium of Coercive Set-valued Maps

(a) Condition (3) implies that $\forall \varepsilon > 0, \exists a > 0$ such that

$$\sup_{\|x\| \ge a, x \in K} \sigma(F(x), x) \le -\varepsilon < 0.$$

This implies that $F(x) \subset T_{aB}(x)$, where aB denotes the ball of radius a. Taking a sufficiently large that $aB \cap K \ne \emptyset$, and since

$$T_{K \cap aB}(x) = T_K(x) \cap T_{aB}(x)$$

we see that F satisfies the tangential condition on the convex compact subset $K \cap aB$. Theorem 9.4 then implies that F has a zero $\bar{x} \in K \cap aB$.

(b) We replace the set-valued map F by the set-valued map G defined by

$$G(x) := F(x) + y - x$$

and note that

$$\begin{aligned} \sigma(G(x), x) &= \sigma(F(x), x) - \|x\|^2 + \langle y, x \rangle \\ &\le \sigma(F(x), x) + \|y\|\|x\|. \end{aligned}$$

Condition (4) implies that for all $A > 0$ there exists a such that for all $\|x\| \ge a$, $\sigma(F(x), x) \le -A\|x\|$. Taking $A > \|y\|$, it follows that

$$\sigma(G(x), x) \le -(A - \|y\|)a < 0$$

whenever $\|x\| \ge a$. Thus G satisfies condition (3) and we may apply the previous question.

16.43 Problem 43 – Solution. Eigenvectors of Set-valued Maps

(a) From Problem 30 we have

$$T_S(x) = \{v \in T_P(x) | \langle p_0, x \rangle = 0\}. \tag{$*$}$$

We see that each element $u \in G(x)$ satisfies $\langle p_0, u \rangle = 0$. Let v be an element of $F(x) \cap T_P(x)$. Then $u := v - \langle p_0, v \rangle x$ belongs to $G(x)$. Since $v \in T_P(x) = $ closure$(P + \mathbb{R}x)$, the same is true of $v - \langle p_0, v \rangle x$. Thus, $u \in T_P(x) \cap G(x)$.

(b) G is clearly upper hemi-continuous, since because F is upper semi-continuous with compact values, for x in a neighbourhood of x_0 we have:

$$
\begin{aligned}
\sigma(G(x), p) &= \sigma(F(x), p - \langle p, x \rangle p_0) \\
&\leq \sigma(F(x_0), p - \langle p, x \rangle p_0) + \varepsilon(\|p\| + |\langle p, x \rangle| \|p_0\|) \\
&\leq \sigma(F(x_0), p - \langle p, x_0 \rangle p_0) + \|F(x_0)\| |\langle p, x_0 - x \rangle| \\
&\quad + \varepsilon(\|p\| + \langle p, x \rangle \|p_0\|).
\end{aligned}
$$

The images $G(x)$ are clearly convex and compact since they are images of the convex compact set $F(x)$ under $v \to v - \langle p_0, v \rangle x$. Thus, we may apply Theorem 9.4; there exists $\bar{x} \in S$ such that $0 \in G(\bar{x})$. Thus, there exists $\bar{v} \in F(\bar{x})$ such that

$$0 = \bar{v} - \langle p_0, \bar{v} \rangle \bar{x}.$$

16.44 Problem 44 – Solution. Positive Eigenvectors of Positive Set-valued Maps

(a) The graph of the set-valued map $x \to G(x) - \delta F(x) - \mathbb{R}_+^n$ is clearly convex; whence, the set $(G - \delta F)(M^n) - \mathbb{R}_+^m$ is convex. To show that it is closed, we take a sequence of elements $z_k \in G(x_k) - \delta F(x_k) - \mathbb{R}_+^m$ converging to z. Then, for all $p \in \mathbb{R}_+^m$, we have $\langle p, z_k \rangle \leq \sigma(G(x_k), p) - \delta \langle p, F(x_k) \rangle$ and, since M^n is compact, a subsequence $x_{k'}$ converges to $x \in M^n$. Since $x \to \sigma(G(x), p) - \delta \langle p, F(x) \rangle$ is upper semi-continuous, it follows that

$$\langle p, z \rangle \leq \sigma(G(x), p) - \delta \langle p, F(x) \rangle.$$

Since this inequality holds for all $p \in \mathbb{R}_+^m$, it follows that $z \in G(x) - \delta F(x) - \mathbb{R}_+^m$. The proofs for the remaining parts are similar to those of Theorem 11.2.

16.45 Problem 45 – Solution. Some Variational Principles

(a) This follows trivially from Fenchel's inequality.

(b)

(i) If $0 \in A\bar{x} + \partial f(\bar{x})$, then $\bar{p} = -A\bar{x}$ belongs to $\partial f(\bar{x})$ or again $\bar{x} \in \partial f^*(\bar{p})$. Thus,

$$0 = \bar{p} - \bar{p} \in \bar{p} - A\partial f^*(\bar{p}).$$

Conversely, if $\bar{p} \in A\partial f^*(\bar{p})$, there exists $\bar{x} \in \partial f^*(\bar{p})$ such that $\bar{p} = A\bar{x}$ and so $0 \in A\bar{x} + \partial f(\bar{x})$.

(ii) $-A\bar{x} \in \partial f(\bar{x})$ if and only if, following Proposition 4.2, we have

$$f(\bar{x}) - f(y) \leq \langle -A\bar{x}, \bar{x} - y \rangle \qquad \forall y \in X$$

or again if

$$f(\bar{x}) + f^*(-A\bar{x}) + \langle A\bar{x}, \bar{x} \rangle = \phi(\bar{x}) = 0.$$

(c) If $f(x) = \psi_K(x)$, any solution $\bar{x}$ of (2)(iii) is a solution of

$$\bar{x} \in K \text{ and } \forall y \in K, \ \langle A(\bar{x}), \bar{x} - y \rangle \leq 0$$

in other words a solution of a *variational inequality*.

(d) We consider the case in which A is a set-valued map with convex compact values. Question (b)(i) is the same. Suppose that $\bar{x}$ is a solution of $0 \in A(\bar{x}) + \partial f(\bar{x})$. Then, there exists $\bar{u} \in A(\bar{x})$ such that $-\bar{u} \in \partial f(\bar{x})$, whence, such that $\forall y \in \text{Dom} f$,

$$f(\bar{x}) - f(y) \leq \langle -\bar{u}, \bar{x} - y \rangle \leq \sigma(A(\bar{x}), y - \bar{x}).$$

Conversely, the inequality (4) may be written in the form

$$\sup_{y \in \text{Dom} f} \inf_{u \in A(\bar{x})} (f(\bar{x}) - f(y) + \langle u, \bar{x} - y \rangle) \leq 0.$$

Since $\text{Dom} f$ is convex, $A(\bar{x})$ is convex and compact and $(u, y) \to f(\bar{x}) - f(y) + \langle u, \bar{x} - y \rangle$ is lower semi-continuous and convex in u and concave in y. The minimax theorem (Theorem 8.1) then implies that there exists $\bar{u} \in A(\bar{x})$ such that

$$\forall y \in \text{Dom} f, \ f(\bar{x}) - f(y) + \langle \bar{u}, \bar{x} - y \rangle \leq 0$$

in other words $\bar{u}$ such that $-\bar{u} \in \partial f(\bar{x})$. Thus, $0 = \bar{u} - \bar{u} \in A(\bar{x}) + \partial f(\bar{x})$.

Finally, if $\bar{u} \in A(\bar{x})$ is such that $-\bar{u} \in \partial f(\bar{x})$, in other words, such that $f(\bar{x}) + f^*(-\bar{u}) + \langle \bar{u}, \bar{x} \rangle = 0$, then $\phi(\bar{x}) = 0$. Conversely, if $\phi(\bar{x}) = 0$, there exists an element $\bar{u} \in A(\bar{x})$ such that

$$0 = \phi(\bar{x}) = f(\bar{x}) + f^*(-\bar{u}) + \langle \bar{u}, \bar{x} \rangle$$

since $A(\bar{x})$ is compact and $u \to f^*(-u) + \langle u, \bar{x} \rangle$ is lower semi-continuous. Thus, $-\bar{u} \in \partial f(\bar{x})$ and $\bar{u} \in A(\bar{x})$.

(e) From (2)(iii), there exists $\bar{p} \in \mathrm{Dom} f^*$ such that

$$0 \in \bar{p} + A\partial f(\bar{p}) \subset \mathrm{Dom} f^* + A\bar{x} \subset \mathrm{Dom} f^* + A\mathrm{Dom} f$$

since $\bar{x} \in \partial f^*(\bar{p}) \subset \mathrm{Dom} f$.

16.46 Problem 46 – Solution. Generalised Variational Inequalities

(a) Since $\beta(z) \geq 0$, $\beta^*(0) = \sup_z(\langle 0, z \rangle - \beta(z)) = -\inf_z \beta(z)$ is finite. Thus $0 \in \mathrm{Dom}\, \beta^*$. We recall that

$$
\begin{array}{llll}
\beta_0^* &=& \psi_{\{0\}}, & \mathrm{Dom}\, \beta_0^* &=& \{0\} \\
\beta_1^* &=& \psi_{\frac{1}{c}B_*}, & \mathrm{Dom}\, \beta_1^* &=& \dfrac{1}{c}B_* \\
\beta_2^* &=& \dfrac{1}{2c}\|\cdot\|_*^2, & \mathrm{Dom}\, \beta_2^* &=& X^*.
\end{array}
$$

Consequently

$$
\begin{array}{rcl}
\mathrm{Int}(\mathrm{Dom}\, f^* + A\mathrm{Dom}\, f + \mathrm{Dom}\, \beta_0^*) &=& \mathrm{Int}(\mathrm{Dom}\, f^* + A\mathrm{Dom}\, f) \\
\mathrm{Int}(\mathrm{Dom}\, f^* + A\mathrm{Dom}\, f + \mathrm{Dom}\, \beta_1^*) &\supset& \mathrm{Dom}\, f^* + A\mathrm{Dom}\, f \\
\mathrm{Int}(\mathrm{Dom}\, f^* + A\mathrm{Dom}\, f + \mathrm{Dom}\, \beta_2^*) &=& X^*.
\end{array}
$$

If $f = \psi_K$, then $\mathrm{Dom}\, f = K$, $\mathrm{Dom}\, f^* = b(K)$ and

$$
\mathrm{Int}(b(K) + A(K) + \mathrm{Dom}\, \beta_i^*) \supset \begin{cases} \mathrm{Int}(b(K) + A(K)) & \text{if } i = 0 \\ b(K) + A(K) & \text{if } i = 1 \\ X^* & \text{if } i = 2 \end{cases}.
$$

(b) The subset $K_n := \{x \in \mathrm{Dom}\, f \,|\, f(x) \leq n \text{ and } \|x\| \leq n\}$ is clearly compact. The function φ defined by

$$\varphi(x, y) = -\sigma(A(x), y - x) + f(x) - f(y)$$

is concave in y and satisfies $\varphi(y, y) = 0$. The first part of the proof of Theorem 9.9 shows that $x \to \sigma(A(x), y - x)$ is upper semi-continuous. Thus, φ is lower semi-continuous in x and Ky Fan's Inequality (Theorem 8.6) implies that there exists a solution $x_n \in K_n$ such that

$$\forall y \in K_n, \quad \varphi(x_n, y) \leq 0.$$

(c) Following (8), there exists $\eta > 0$ such that

$$\eta B \subset \mathrm{Dom}\, f^* + A\mathrm{Dom}\, f + \mathrm{Dom}\, \beta^*.$$

Thus, for all $p \in X^*$, there exist $q \in \mathrm{Dom}\, f^*$, $y \in \mathrm{Dom}\, f$, $u \in A(y)$ and $r \in \mathrm{Dom}\, \beta^*$ such that

$$\frac{\eta p}{\|p\|} = r + q + u.$$

We choose $n(p)$ such that $y \in K_n$. Thus,

$$\frac{\eta}{\|p\|}\langle p, x_n \rangle = \langle r, x_n - y \rangle + \langle q, x_n \rangle + \langle u, x_n - y \rangle + \langle r + u, y \rangle$$
$$\leq \beta^*(r) + \beta(x_n - y) + f^*(q) + f(x_n) + \langle u, x_n - y \rangle + \langle r + u, y \rangle.$$

Since A is β-monotone, it follows that for all $v \in A(x_n)$,

$$\beta(x_n - y) \leq \langle x_n - y, v - u \rangle$$

whence that

$$\sigma(A(x_n), y - x_n) := \sup_{v \in A(x_n)} \langle v, y - x_n \rangle \leq -\langle u, x_n - y \rangle - \beta(x_n, y).$$

Adding these inequalities, we obtain:

$$\frac{\eta}{\|p\|}\langle p, x_n \rangle \leq -\sigma(A(x_n), y - x_n) + f(x_n) - f(y)$$
$$+ (\beta^*(r) + f^*(q) + f(y) + \langle r + u, y \rangle)$$

and following (7)

$$\langle p, x_n \rangle \leq \frac{\|p\|}{\eta}(\beta^*(r) + f^*(q) + f(y) + \langle r + u, y \rangle) < +\infty.$$

It follows that the sequence x_n is bounded.

(d) Since X is finite dimensional, there exists a subsequence (still denoted by) x_n which converges to $\bar{x}$. We fix $y \in \mathrm{Dom}\, f$ and choose n large enough so that y belongs to K_n. Then since A is upper semi-continuous with convex compact values, the first part of the proof of Theorem 9.9 shows that $x \to \sigma(A(x), y - x)$ is upper semi-continuous. Thus,

$$-\sigma(A(\bar{x}), y - \bar{x}) + f(\bar{x}) - f(y) \leq \liminf_{x_n \to \bar{x}} (-\sigma(A(x_n), y - x_n) + f(x_n) - f(y))$$
$$\leq 0.$$

This implies that $\bar{x} \in \mathrm{Dom}\, f$ and that $\bar{x}$ is a solution of (2)(iii) and thus also of (2)(i).

Taking $f = \psi_K$, the problem (2) is equivalent to the variational inequality (9). In this case, assumption (8) may be written as

$$0 \in \mathrm{Int}(b(K) + A(K) + \mathrm{Dom}\, \beta^*). \tag{$*$}$$

If K is bounded ($b(K) = X^*$), if A is surjective ($A(K) = X^*$) or if A is β_2-monotone ($\mathrm{Dom}\, \beta_2^* = X^*$), then there always exists a solution $\bar{x}$ of the variational inequality (9). If A is β_1-monotone, it is sufficient to assume that

$$0 \in b(K) + A(K),$$

while if A is only monotone, it is sufficient to assume that

$$0 \in \mathrm{Int}(b(K) + A(K)).$$

16.47 Problem 47 – Solution. Monotone Set-valued Maps

(a) We have

$$
\begin{aligned}
\langle x - p - (y - q), x - y \rangle &= \|x - y\|^2 - \langle p - q, x - y \rangle \\
&\leq \|x - y\|^2 - \|p - q\|\|x - y\| \geq 0.
\end{aligned}
$$

(b) We note that

$$
\|x - y + \lambda(p - q)\|^2 = \|x - y\|^2 + \lambda^2\|p - q\|^2 + 2\lambda\langle p - q, x - y \rangle.
$$

Thus, if A is monotone we obtain the inequality (4). Conversely, the inequality (4) together with the above equation imply that

$$
\lambda^2\|p - q\|^2 + 2\lambda\langle p - q, x - y \rangle \geq 0.
$$

We obtain the monotonicity by dividing by $\lambda > 0$ and letting λ tend to 0.

(c) Suppose that $p \in \partial f(x)$ and $q \in \partial f(y)$ are given. The inequalities

$$
\begin{aligned}
f(x) - f(y) &\leq \langle p, x - y \rangle \\
f(y) - f(x) &\leq \langle q, y - x \rangle
\end{aligned}
$$

imply that ∂f is monotone.

(d) Suppose that $x_i \in J_\lambda(y_i)$, $i = 1, 2$. Then there exist $v_i \in A(y_i)$ such that $y_i = x_i + \lambda v_i \in x_i + \lambda A(x_i)$. It follows that

$$
\begin{aligned}
\|y_1 - y_2\|^2 &= \|x_1 - x_2 + \lambda(v_1 - v_2)\|^2 \\
&= \|x_1 - x_2\|^2 + \lambda^2\|v_1 - v_2\|^2 + 2\lambda\langle v_1 - v_2, x_1 - x_2 \rangle \\
&\geq \|x_1 - x_2\|^2 + \lambda^2\|v_1 - v_2\|^2.
\end{aligned}
$$

It follows that

(i) $\qquad\qquad\qquad\qquad \|x_1 - x_2\| \leq \|y_1 - y_2\|$

(ii) $\qquad\qquad\qquad\qquad \|v_1 - v_2\| \leq \dfrac{1}{\lambda}\|y_1 - y_2\|.$

If $y_1 = y_2$, then $x_1 = x_2$, which shows that J_λ is one-to-one. Then $x_i = J_\lambda(y_i)$ and $v_i = \frac{1}{\lambda}(1 - J_\lambda)(y_i)$. The previous inequalities imply that J_λ and A_λ are Lipschitz with constants 1 and $1/\lambda$, respectively.

(e) We take y_1 and y_2.

$$\langle A_\lambda(y_1) - A_\lambda(y_2), y_1 - y_2 \rangle$$
$$= \langle A_\lambda(y_1) - A_\lambda(y_2), J_\lambda(y_1) - J_\lambda(y_2) + \lambda(A_\lambda(y_1) - A_\lambda(y_2)) \rangle$$
$$= \langle A_\lambda(y_1) - A_\lambda(y_2), J_\lambda(y_1) - J_\lambda(y_2) \rangle + \lambda \| A_\lambda(y_1) - A_\lambda(y_2) \|^2.$$

The first term is positive since A is monotone and $(J_\lambda(y_i), A_\lambda(y_i))$ belongs to the graph of A for $i = 1, 2$. Thus,

$$\langle A_\lambda(y_1) - A_\lambda(y_2), y_1 - y_2 \rangle \geq 0.$$

(f) We suppose that the pair (x, u) satisfies

$$\forall (y, v) \in \mathrm{Graph}(A), \quad \langle u - v, x - y \rangle \geq 0.$$

For y, we take the solution y_0 of the inclusion $u + x \in y_0 + A(y_0)$ (this choice is possible by virtue of (8)). Let $v_0 \in A(y_0)$ be such that $u + x = y_0 + v_0$. Then

$$\langle u - v_0, x - y_0 \rangle = -\| x - y_0 \|^2 \geq 0$$

since A is monotone. Then $x = y_0$ and $u = u_0 \in A(y_0) = A(x)$.

(g) If A satisfies (9), then $A(x)$ is the intersection of the closed half spaces $\{ u \in X | \langle u - v, x - y \rangle \geq 0 \}$ for (y, v) in the graph of A. Thus, it is a convex closed set.

We consider a sequence x_n converging to x and a sequence $u_n \in A(x_n)$ converging (weakly) to u. We take (y, v) in $\mathrm{Graph}(A)$. The inequalities

$$\langle u_n - v, x_n - y \rangle \geq 0$$

imply, after passing to the limit, that

$$\langle u - v, x - y \rangle \geq 0.$$

It follows from (9) that $u \in A(x)$.

(h) We calculate

$$\| A_\lambda(x) - m(A(x)) \|^2 = \| A_\lambda(x) \|^2 + \| m(A(x)) \|^2 - 2 \langle A_\lambda(x), m(A(x)) \rangle.$$

In addition, since A is monotone and since $(x, m(A(x)))$ and $(J_\lambda(x), A_\lambda(x))$ belong to the graph of A, we obtain the inequality

$$\langle m(A(x)) - A_\lambda(x), x - J_\lambda(x) \rangle = \lambda \langle m(A(x)) - A_\lambda(x), A_\lambda(x) \rangle \geq 0.$$

These two inequalities imply that

$$\| A_\lambda(x) - m(A(x)) \|^2 \leq \| m(A(x)) \|^2 - \| A_\lambda(x) \|$$

whence that

$$\| x - J_\lambda(x) \| \leq \lambda \| m(A(x)) \|.$$

(i) Let $y = A_{\mu+\lambda}(x)$ be a solution of the inclusion $y \in A(x - \lambda y - \mu y)$, which shows that y is a solution of the equation $y = A_\mu(x - \lambda y)$ and is thus equal to $(A_\mu)_\lambda(x)$. Since A_μ is monotone and $m(A_\mu(x)) = A_\mu(x)$, the inequality (12) implies that

$$\|A_{\mu+\lambda}(x) - A_\mu(x)\| \leq \|A_\mu(x)\|^2 - \|A_{\lambda+\mu}(x)\|^2.$$

This inequality shows that the sequence $\mu \to \|A_\mu(x)\|$ is an increasing sequence bounded by $\|m(A(x))\|$. Thus, it converges to a limit α. The same inequality shows that

$$\lim_{\lambda,\mu \to 0} \|A_{\mu+\lambda}(x) - A_\mu(x)\|^2 = 0$$

whence, the Cauchy criterion implies that $A_\lambda(x)$ converges strongly to an element v. Since $(J_\lambda(x), A_\lambda(x))$ belongs to the graph of A, which is closed, it follows that $v \in A(x)$. Moreover, since $\|A_\lambda(x)\| \leq \|m(A(x))\|$, it follows that $\|v\| \leq \|m(A(x))\|$. Since $m(A(x))$ is the unique projection of 0 onto the convex closed subset $A(x)$, it follows that $v = m(A(x))$.

(j) If $f : X \to \mathbb{R} \cup \{+\infty\}$ is a nontrivial, convex, lower semi-continuous function, Theorems 2.2 and 4.3 show that the set-valued map $A := \partial f$ satisfies the assumption (8). Thus, we complete the proof of Theorem 5.2 by showing that $\nabla f_\lambda(x) = A_\lambda(x)$ converges to the element $\partial f(x)$ of minimum norm.

Remark. Conversely, Minty's theorem shows that any monotone set-valued map satisfying (9) also satisfies (8). Condition (9) says that the graph of A is maximal over all monotone set-valued maps. This is why monotone set-valued maps A satisfying (8) and (9) are called *maximum monotone set-valued maps*.

16.48 Problem 48 – Solution. Walrasian Equilibrium for Set-valued Demand Maps

(a) The graph of B is clearly closed. To show that B is lower semi-continuous at $p_0 \in M^\ell$, we take fixed $x_0 \in B(p_0, r(p_0))$ and $\varepsilon > 0$. Suppose that $\tilde{x} \in L$ satisfies

$$-c := \langle p_0, \tilde{x} \rangle - r(p_0) < 0.$$

It follows that there exists θ such that $x_\theta := \theta \tilde{x} + (1 - \theta)x \in x_0 + \varepsilon B$. Since $\langle p_0, x_0 \rangle - r(p_0) \leq 0$, we have

$$\langle p_0, x_\theta \rangle - r(p_0) \leq -c\theta < 0.$$

Take $\eta = \frac{1}{4}c\theta$. Since r is lower semi-continuous, there exists a neighbourhood V of p_0 such that

$$\forall p \in V, \quad \langle p, x \rangle - r(p) \leq \langle p_0, x_\theta \rangle - r(p_0) + \varepsilon + \langle p - p_0, x_\theta \rangle$$
$$\leq -\frac{1}{2}c\theta < 0.$$

This implies that x_θ belongs to $B(p, r(p))$ for all $p \in V$, whence that $p \rightarrow B(p, r(p))$ is lower semi-continuous.

(b) If L is compact, the sets $B(p, r(p))$ lie in a fixed compact set. Since its graph is closed, $B(\cdot, r(\cdot))$ is upper semi-continuous whence continuous (see Problem 2). It follows that $D(p, r(p))$ is an upper semi-continuous set-valued map.

(c) We apply Theorem 9.2 (Gale–Nikaïdo–Debreu) to the set-valued map $C :$ $M^\ell \rightarrow \mathbb{R}^\ell$ defined by

$$C(p) := M^0 - \sum_{i=1}^{n} D_i(p, r_i(p)) \qquad (*)$$

where the functions $r_i(p)$ are continuous since they are the support functions of compact sets.

Appendix

17. Compendium of Results

17.1 Nontrivial, Convex, Lower Semi-continuous Functions

Definitions. A function $f : X \to \mathbb{R} \cup \{+\infty\}$ is said to be:

- *nontrivial* if there exists $x_0 \in X$ such that $f(x_0) < +\infty$;

- *lower semi-continuous* at x_0 if

$$\forall \lambda < f(x_0), \quad \exists \eta > 0 \text{ such that } \forall x \in B(x_0, \eta), \quad \lambda \leq f(x);$$

- *lower semi-compact* if $\forall \lambda \in \mathbb{R}$, the sets $S(f, \lambda) := \{x \in X | f(x) \leq \lambda\}$ are relatively compact;

- *convex* if for any convex combination $x := \sum_{i=1}^{n} \lambda_i x_i$

$$f\left(\sum_{i=1}^{n} \lambda_i x_i\right) \leq \sum_{i=1}^{n} \lambda_i f(x_i);$$

- *strictly convex* if for all x, y such that $f(x) < +\infty$, $f(y) < +\infty$ we have

$$f\left(\frac{x+y}{2}\right) < \frac{1}{2}(f(x) + f(y));$$

- *locally Lipschitz* on an open set Ω if for all $x_0 \in \Omega$, there exist $\eta > 0$ and $c > 0$ such that

$$\forall x, y \in B(x_0, \eta), \quad |f(x) - f(y)| \leq c\|x - y\|.$$

We make the following definitions:

- Dom $f := \{x \in X | f(x) < +\infty\}$ is the *domain* of f.

- Ep $f := \{(x, \lambda) \in X \times \mathbb{R} | f(x) \leq \lambda\}$ is the *epigraph* of f.

- $S(f, \lambda) := \{x \in X | f(x) \leq \lambda\}$ are the *sections* of f.

- ψ_K defined by

$$\psi_K(x) := \begin{cases} 0 & \text{if } x \in K \\ +\infty & \text{if } x \notin K \end{cases}$$

is the *indicator function* of K.

We note that

$$\text{Ep}\left(\sup_{i \in I} f_i\right) = \bigcap_{i \in I} \text{Ep}\,(f_i) \tag{1}$$

and

$$S\left(\sup_{i \in I} f_i, \lambda\right) = \bigcap_{i \in I} S(f_i, \lambda)$$

and that

$$\left\{\bar{x} \in X | f(\bar{x}) = \inf_{x \in X} f(x)\right\} = \bigcap_{\lambda > \inf_X f(x)} S(f, \lambda). \tag{2}$$

Suppose that $f : X \to \mathbb{R} \cup \{+\infty\}$ is a nontrivial function.

A function f is lower semi-continuous at x_0 if and only if

$$f(x_0) \leq \liminf_{x \to x_0} f(x) = \sup_{\eta > 0} \inf_{x \in B(x_0, \eta)} f(x) \tag{3}$$

The following properties are equivalent

(a) f is lower semi-continuous;
(b)   the epigraph of f is closed;
(c)   all the sections $S(f, \lambda)$ of f are closed.   (1.4)

If f, g, f_i $(i \in I)$ are lower semi-continuous functions then following are lower semi-continuous:

- $f + g$;
- αf, $\forall \alpha > 0$;
- $\inf(f, g)$;
- $\sup_{i \in I} f_i$;
- $f \circ A$, where A is a continuous mapping from Y to X.   (1.5)

If $K \subset X$ is closed and if $f : K \to \mathbb{R}$ is lower semi-continuous, then $f_K : X \to \mathbb{R} \cup \{+\infty\}$ defined by

$$f_K(x) = \begin{cases} f(x) & \text{if } x \in K \\ +\infty & \text{if } x \notin K \end{cases} \tag{6}$$

is lower semi-continuous.

If $K \subset Y$ is compact and if $g : X \times K \to \mathbb{R} \cup \{+\infty\}$ is lower semi-continuous, then $f : X \to \mathbb{R} \cup \{+\infty\}$ defined by

$$f(x) := \inf_{y \in K} g(x, y) \tag{7}$$

is lower semi-continuous.

If f is both lower semi-continuous and lower semi-compact, then the set M of elements at which f attains its minimum is non-empty and compact. In particular, this is the case if $K \subset X$ is compact and if $f : K \to \mathbb{R}$ is lower semi-continuous. (1.8)

Suppose that E is complete and that $f : E \to \mathbb{R}_+ \cup \{+\infty\}$ is nontrivial, positive and lower semi-continuous.

Consider $x_0 \in \mathrm{Dom}\, f$ and $\varepsilon > 0$. There exists $\bar{x} \in \mathrm{Dom}\, f$ such that

(i)
(ii)
$$f(\bar{x}) + \varepsilon d(x_0, \bar{x}) \leq f(x_0)$$
$$\forall x \neq \bar{x}, \quad f(\bar{x}) < f(x) + \varepsilon d(\bar{x}, x). \tag{1.9}$$

17.2 Convex Functions

A function f is convex if and only if the epigraph of f is convex. In this case, all the sections $S(f, \lambda)$ are convex. (2.1)

If $f, g, f_i (i \in I)$ are convex, then:

- $f + g$ is convex;
- $\forall \alpha > 0$, αf is convex;
- if $A : Y \to X$ is affine, then $f \circ A$ is convex;
- $\sup_{i \in I} f_i$ is convex;
- if $g : X \times Y \to \mathbb{R} \cup \{+\infty\}$ is convex then $f : X \to \mathbb{R} \cup \{+\infty\}$
  defined by $f(x) := \inf_{y \in Y} g(x, y)$ is convex. (2.2)

If K is convex, then $f : K \to \mathbb{R}$ is convex if and only if f_K is convex. (2.3)

If the functions $f_i : X \to \mathbb{R} \cup \{+\infty\}$ are convex, and if we set

$$F(x) := (f_1(x), \ldots, f_n(x)) \in \mathbb{R}^n, \quad K := \bigcap_{i=1}^n \mathrm{Dom}\, f_i \tag{4}$$

then the sets $F(K) + \mathbb{R}_+^n$ and $F(K) + \mathring{\mathbb{R}}_n^+$ are convex.

If f is convex, the set M of elements at which f attains its minimum is convex. If f is strictly convex, this set contains at most one point. (2.5)

If f is convex, the following conditions are equivalent:

(a) f is bounded above on an open subset

(b) f is locally Lipschitz (whence continuous) on the interior of its domain.

In particular:

- if X is finite dimensional, any convex function is continuous on the interior of its domain;
- if X is a Hilbert space, any convex lower semi-continuous function is continuous on the interior of its domain. (2.6)

17.3 Conjugate Functions

Definitions. The function $f^* : X^* \to \mathbb{R} \cup \{+\infty\}$ associated with a nontrivial function $f : X \to \mathbb{R} \cup \{+\infty\}$ by the formula

$$\forall p \in X^*, \quad f^*(p) := \sup_{x \in X}(\langle p, x \rangle - f(x)) \tag{1}$$

is called the *conjugate function* of f. The function $f^{**} : X \to \{-\infty\} \cup \mathbb{R} \cup \{+\infty\}$ defined by

$$\forall x \in X, \quad f^{**}(x) := \sup_{p \in X^*}(\langle p, x \rangle - f^*(p))$$

is called the *biconjugate function* of f.

We note that

$$\forall x \in X, \quad \forall p \in X^*, \quad \langle p, x \rangle \le f(x) + f^*(p) \qquad \text{(Fenchel's inequality)}$$

that

$$\forall x \in X, \quad f^{**}(x) \le f(x)$$

and that

$$- f^*(0) = \inf_{x \in X} f(x). \tag{2}$$

A nontrivial function $f : X \to \mathbb{R} \cup \{+\infty\}$ is convex and lower semi-continuous if and only if $f = f^*$. In this case, f^* is also nontrivial. (3.3)

- If $f \le g$ then $g^* \le f^*$.
- If $A \in L(X, X)$ is an isomorphism, then $(f \circ A)^* = f^* \circ A^{*-1}$.
- If $g(x) := f(x - x_0) + \langle p_0, x \rangle + a$, then

$$g^*(p) = f^*(p - p_0) + \langle p, x_0 \rangle - (a + \langle p_0, x_0 \rangle).$$

- If $g(x) := f(\lambda x)$, then $g^*(p) = f^*(\frac{p}{\lambda})$ and if $h(x) := \lambda f(x)$, then $h^*(p) = \lambda f^*(\frac{p}{\lambda})$.
- If $f : X \times Y \to \mathbb{R} \cup \{+\infty\}$ and if $g(y) := \inf_{x \in X} f(x, y)$, then $g^*(q) = f^*(0, q)$. (3.4)

If $f : X \to \mathbb{R} \cup \{+\infty\}$ and $g : Y \to \mathbb{R} \cup \{+\infty\}$ are nontrivial, convex, lower semi-continuous functions and if $A \in L(X, Y)$ satisfies $0 \in \mathrm{Int}\,(A\mathrm{Dom}\,f - \mathrm{Dom}\,g)$, then $\forall p \in A^*\mathrm{Dom}\,g^* + \mathrm{Dom}f^*$, $\exists \bar{q} \in Y^*$ such that

$$(f + g \circ A)^*(p) = f^*(p - A^*\bar{q}) + g^*(\bar{q}) = \inf_{q \in Y^*} (f^*(p - A^*q) + g^*(q)) \quad (5)$$

If, in particular, we suppose that $0 \in \mathrm{Int}\,(\mathrm{Im}\,A - \mathrm{Dom}\,g)$, then $\forall p \in A^*\mathrm{Dom}\,g^*$, $\exists \bar{q} \in \mathrm{Dom}\,g^*$ satisfying

$$A^*\bar{q} = p \quad \text{and} \quad (g \circ A)^*(p) = g^*(\bar{q}) = \min_{A^*q=p} g^*(q).$$

If $K \subset X$ is convex and closed and satisfies

$$0 \in \mathrm{Int}\,(\mathrm{Dom}\,f - K)$$

then $\forall p \in \mathrm{Dom}(f|_K)^*$, $\exists \bar{q} \in b(K)$ such that

$$(f|_K)^*(p) = f^*(p - \bar{q}) + \sigma_K(\bar{q}).$$

If

$$0 \in \mathrm{Int}\,(\mathrm{Dom}\,f - \mathrm{Dom}\,g)$$

then $\forall p \in \mathrm{Dom}\,f^* + \mathrm{Dom}\,g^*$, $\exists \bar{q} \in \mathrm{Dom}\,g^*$ such that

$$(f + g)^*(p) = f^*(p - \bar{q}) + g^*(\bar{q}) = \inf_{q \in X^*} (f^*(p - q) + g^*(q)).$$

17.4 Separation Theorems and Support Functions

Suppose that K is a convex closed subset of a Hilbert space. Then $\forall x \in X$, there exists a unique solution $\pi_K(x) \in K$ of the best-approximation problem

$$\|x - \pi_K(x)\| = \inf_{y \in K} \|x - y\|. \quad (1)$$

This is characterised by the variational inequality

(i) $\qquad\qquad\qquad \pi_K(x) \in K$

(ii) $\qquad\quad \forall y \in K, \ \langle \pi_K(x) - x, \pi_K(x) - y \rangle \leq 0. \quad (4.2)$

The mapping $\pi_K : X \to K$ is continuous and satisfies

$$\|\pi_K(x) - \pi_K(y)\| \leq \|x - y\|$$
$$\|(1 - \pi_K)(x) - (1 - \pi_K)(y)\| \leq \|x - y\|.$$

The mapping π_K is called the 'projector of best approximation' onto K.

The function $\sigma_K : X^* \to \mathbb{R} \cup \{+\infty\}$ defined by

$$\forall p \in X^*, \quad \sigma_K(p) := \sup_{x \in K} \langle p, x \rangle \in \mathbb{R} \cup \{+\infty\} \tag{3}$$

is called the *support function* of K and its domain

$$b(K) := \operatorname{Dom} \sigma_K := \{p \in X^* | \sigma_K(p) < +\infty\}$$

is called the *barrier cone* of K.

If K is a *convex closed* subset of a Hilbert space X and if $x_0 \notin K$, then there exist $p \in X^*$ and $\varepsilon > 0$ such that

$$\sigma_K(p) := \sup_{y \in K} \langle p, y \rangle \leq \langle p, x_0 \rangle - \varepsilon. \tag{4}$$

If K is a convex subset of a finite-dimensional space X and if $x_0 \notin K$, then there exists $p \in X^*$ such that

$$p \neq 0 \quad \text{and} \quad \sigma_K(p) := \sup_{y \in K} \langle p, y \rangle \leq \langle p, x_0 \rangle. \tag{5}$$

$$\begin{aligned} \forall p \in X^*, \quad \sigma_K(p) &= \sigma_{\overline{\mathrm{co}}(K)}(p) \\ \sigma_K &= \psi_K^* \end{aligned} \tag{6}$$

- If $K = B$ is the unit ball, $\sigma_K(p) = \|p\|_*$.

- If K is a cone, $\sigma_K(p) = \psi_{K^-}(p)$ and $b(K) = K^-$.

- $b(K)^- = \cap_{\lambda > 0} \lambda(K - x_0)$ (for all $x_0 \in K$).

Any support function σ_K is convex, positively homogeneous and lower semi-continuous. (4.7)

Conversely, any function σ from X^* to $\mathbb{R} \cup \{+\infty\}$ which is convex, positively homogeneous and lower semi-continuous, is the support function of the set

$$K_\sigma := \{x \in X | \forall p \in X^*, \ \langle p, x \rangle \leq \sigma(p)\}. \tag{8}$$

- If K is a convex closed subset, then

$$K = \{x \in X | \forall p \in X^*, \ \langle p, x \rangle \leq \sigma_K(p)\}. \tag{9}$$

- If K is a convex closed cone, then

$$K = (K^-)^-$$

- If K is a closed vector subspace then $K = (K^\perp)^\perp$.

- If $A \in L(X, Y)$ is a continuous linear operator and $K \subset X$, then

$$A(K)^- = A^{*-1}(K^-).$$

In particular, $(\operatorname{Im} A)^\perp = \operatorname{Ker} A^*$.

- If $K \subset L$, then $b(L) \subset b(K)$ and $\sigma_K \leq \sigma_L$.   (4.10)

- If $K_i \subset X_i$ $(i = 1, \ldots, n)$, $b(\prod_{i=1}^n K_i) = \prod_{i=1}^n b(K_i)$ and

$$\sigma_K(p_1, \ldots, p_n) = \sum_{i=1}^n \sigma_{K_i}(p_i).$$

- $b(\overline{\operatorname{co}} \cup_{i \in L} K_i) \subset \cap_{i \in I} b(K_i)$ and $\sigma_{\overline{\operatorname{co}}(\cup_{i \in I} K_i)}(p) = \sup_{i \in I} \sigma_{K_i}(p)$.

- If $B \in L(X, Y)$, then $b(\overline{B(K)}) = B^{*-1} b(K)$ and $\sigma_{\overline{B(K)}}(p) = \sigma_K(B^* p)$.

- $b(K_1 + K_2) = b(K_1) \cap b(K_2)$ and $\sigma_{K_1 + K_2}(p) = \sigma_{K_1}(p) + \sigma_{K_2}(p)$.

- If P is a convex closed cone, then

$$b(K + P) = b(K) \cap P^-$$

and

$$\sigma_{K+P}(x) = \begin{cases} \sigma_K(p) & \text{if } p \in P^- \\ +\infty & \text{otherwise.} \end{cases}$$

- $b(K + \{x_0\}) = b(K)$ and $\sigma_{K+x_0}(p) = \sigma_K(p) + \langle p, x_0 \rangle$.

- If $A \in L(X, Y)$, if $L \subset X$ and $M \subset Y$ are convex closed subsets and if $0 \in \operatorname{Int}(A(L) - M)$ then

$$b(L \cap A^{-1}(M)) = b(L) + A^* b(M)$$ (11)

and $\forall p \in b(K)$, $\exists \bar{q} \in b(M)$ such that

$$\begin{aligned} \sigma_{L \cap A^{-1}(M)}(p) &= \sigma_L(p - A^* \bar{q}) + \sigma_M(\bar{q}) \\ &= \inf_{q \in Y^*} (\sigma_L(p - A^* q) + \sigma_M(q)). \end{aligned}$$

- If $A \in L(X, Y)$, if $M \subset Y$ is convex and closed and if $0 \in \operatorname{Int}(\operatorname{Im}(A) - M)$, then $b(A^{-1}(M)) = A^* b(M)$ and $\forall p \in b(A^{-1}(M))$, $\exists \bar{q} \in b(M)$ satisfying

$$A^* \bar{q} = p \quad \text{and} \quad \sigma_{A^{-1}(M)}(p) = \sigma_M(\bar{q}) = \inf_{A^* q = p} \sigma_M(q).$$

- If K_1 and K_2 are convex closed subsets of X such that $0 \in \operatorname{Int}(K_1 - K_2)$, then $b(K_1 \cap K_2) = b(K_1) + b(K_2)$ and for all $p \in b(K_1 \cap K_2)$, there exist $\bar{p}_i \in b(K_i)$ $(i = 1, 2)$ such that $p = \bar{p}_1 + \bar{p}_2$ and

$$\sigma_{K_1 \cap K_2}(p) = \sigma_{K_1}(\bar{p}_1) + \sigma_{K_2}(\bar{p}_2) = \inf_{p = p_1 + p_2} (\sigma_{K_1}(p_1) + \sigma_{K_2}(p_2)).$$

17.5 Subdifferentiability

Definition. Suppose that $f : X \to \mathbb{R} \cup \{+\infty\}$ is a nontrivial convex function. Suppose also that $x_0 \in \mathrm{Dom} f$ and $v \in X$. Then the limit

$$Df(x_0)(v) := \lim_{h \to 0+} \frac{f(x_0 + hv) - f(x_0)}{h} \tag{1}$$

exists in $\overline{\mathbb{R}}$ and is called the *right derivative* of f at x_0 in the direction v.

It satisfies the properties

$$f(x_0) - f(x_0 - v) \le Df(x_0)(v) \le f(x_0 + v) - f(x_0) \tag{2}$$

and

$$v \to Df(x_0)(v) \text{ is convex and positively homogeneous.}$$

Definition. f is said to be *Gâteaux differentiable* at x_0 if $v \to Df(x_0)(v) := \langle \nabla f(x_0), v \rangle$ is linear and continuous. Then the subset

$$\partial f(x_0) := \{p \in X^* | \forall v \in X, \langle p, v \rangle \le Df(x_0)(v)\} \tag{3}$$

is called the *subdifferential* of f at x_0. The subdifferential is a *convex closed* set (which may be empty, for example, if there exists v such that $Df(x_0)(v) = -\infty$).

If $v \to Df(x_0)(v)$ is a nontrivial *lower semi-continuous* function from X to $\mathbb{R} \cup \{+\infty\}$ then

$$\sigma(\partial f(x_0), v) = Df(x_0)(v). \tag{4}$$

Suppose that f is a nontrivial convex function which is subdifferentiable at x. Then, the following assertions are equivalent:

(a) $p \in \partial f(x)$;
(b) $\langle p, x \rangle = f^*(p) + f(x)$;
(c) $\forall y \in X, f(x) - \langle p, x \rangle \le f(y) - \langle p, y \rangle$. $\tag{5.5}$

If in addition f is lower semi-continuous, then

$$p \in \partial f(x) \Leftrightarrow x \in \partial f^*(p). \tag{6}$$

If f is a convex function which is continuous on the interior of its domain, then f is right differentiable and subdifferentiable on $\mathrm{Int}\,\mathrm{Dom}\, f$ and satisfies the following properties:

(a) $(x, u) \in \text{Int Dom } f \times X \to Df(x)(u)$ is upper semi-continuous;
(b) $\exists c > 0$ such that $Df(x)(u) = \sigma(\partial f(x), u) \le c\|u\|$;
(c) $\forall x \in \text{Int Dom } f$, $\partial f(x)$ is non-empty and bounded;
(d) the set-valued map $x \in \text{Int Dom } f \to \partial f(x)$ is upper hemi-
continuous.   (5.7)

If f is a nontrivial, convex, lower semi-continuous function, then

(a) f is subdifferentiable on a dense subset of the domain of f;
(b) $\forall \lambda > 0$, the set-valued map $x \to x + \lambda \partial f(x)$ is surjective and
its inverse $J_\lambda := (1 + \lambda \partial f(\cdot))^{-1}$ is a Lipschitz mapping with
constant 1.   (5.8)

If $f : X \to \mathbb{R} \cup \{+\infty\}$ and $g : Y \to \mathbb{R} \cup \{+\infty\}$ are nontrivial, convex and
lower semi-continuous, if $A \in L(X, Y)$ and if $0 \in \text{Int}\,(A\text{Dom } f - \text{Dom } g)$, then

$$\partial(f + g \circ A)(x) = \partial f(x) + A^* \partial g(Ax). \tag{9}$$

In particular, if $0 \in \text{Int}\,(\text{Dom } f - \text{Dom } g)$, then $\partial(f + g)(x) = \partial f(x) + \partial g(x)$.
If $0 \in \text{Int}\,(\text{Im } A - \text{Dom } g)$, then $\partial(g \circ A)(x) = A^* \partial g(Ax)$.
If $K \subset X$ is convex and closed and if $0 \in \text{Int}\,(K - \text{Dom } f)$, then $\partial(f|_K)(x) = \partial f(x) + N_K(x)$.

If $f_1, \ldots, f_n$ are n convex lower semi-continuous functions and if $x_0 \in \cap_{i=1}^n \text{Int Dom } f_i$, then

$$\partial\left(\sup_{i=1,\ldots,n} f_i\right)(x_0) = \overline{\text{co}} \bigcup_{i \in I(x_0)} \partial f_i(x_0) \tag{10}$$

where $I(x_0) := \{i = 1, \ldots, n | f_i(x_0) = \sup_{j=1,\ldots,n} f_j(x_0)\}$.

Suppose that $f : X \times Y \to \mathbb{R} \cup \{+\infty\}$ is a nontrivial convex function and
that $g : Y \to \mathbb{R} \cup \{+\infty\}$ is defined by

$$g(y) := \inf_{x \in X} f(x, y). \tag{11}$$

If $\bar{x} \in X$ satisfies $g(y) = f(\bar{x}, y)$, then $q \in \partial g(y)$ if and only if $(0, q) \in \partial f(\bar{x}, y)$.

17.6 Tangent and Normal Cones

Definition. Suppose that K is a convex subset. If $x \in K$ then:

(i) $T_K(x) := \text{closure}\,\left(\bigcup_{h > 0} \frac{1}{h}(K - x)\right)$ is the *tangent cone to K at x*;
(ii) $N_K(x) = \{p \in X | \langle p, x \rangle = \sigma_K(p)\}$ is the *normal cone to K at x*.   (6.1)

We observe that

$$N_K(x) = \partial \psi_K(x) = T_K(x)^-, \qquad T_K(x) = N_K(x)^-. \qquad (2)$$

If $x \in \text{Int}(K)$, then $N_K(x) = \{0\}$ and $T_K(x) = X$.

If $K = \{x_0\}$ then $N_K(x_0) = X$ and $T_K(x_0) = \{0\}$. $\qquad (6.3)$

If $K = B$ (the unit ball) and if $\|x\| = 1$, then

$$N_K(x) = \{\lambda x\}_{\lambda \geq 0} \quad \text{and} \quad T_K(x) = \{v \in X | \langle v, x \rangle \leq 0\}.$$

If $K = \mathbb{R}^n_+$ and if $x \in \mathbb{R}^n_+$, then

$$\begin{aligned}
N_K(x) &= \{p \in -\mathbb{R}^n_+ | p_i = 0 \text{ when } x_i > 0\} \\
T_K(x) &= \{v \in \mathbb{R}^n | v_i \geq 0 \text{ when } x_i = 0\}.
\end{aligned}$$

If $M^n := \{x \in \mathbb{R}^n_+ | \sum_{i=1}^n x_i = 1\}$, then

$$N_{M^n}(x) = \{p \in \mathbb{R}^n | p_i = \max_{j=1,\dots,n} p_j \text{ when } x_i > 0\}$$

$$T_{M^n}(x) = \{v \in \mathbb{R}^n | \sum_{i=1}^n v_i = 0 \text{ and } v_i \geq 0 \text{ when } x_i = 0\}.$$

Formulae. $\qquad (6.4)$

- If $K \subset L$ and $x \in X$, then $T_K(x) \subset T_L(x)$ and $N_L(x) \subset N_K(x)$.
- If $K_i \subset X_i$ $(i = 1, \dots, n)$, then

$$T_{\prod_{i=1}^n K_i}(x_1, \dots, x_n) = \prod_{i=1}^n T_{K_i}(x_i)$$

and

$$N_{\prod_{i=1}^n K_i}(x_1, \dots, x_n) = \prod_{i=1}^n N_{K_i}(x_i).$$

- If $B \in L(X, Y)$, then

$$T_{B(K)}(Bx) = \text{closure}\,(B T_K(x))$$

and

$$N_{B(K)}(Bx) = B^{*-1} N_K(x).$$

- $$T_{K_1+K_2}(x_1 + x_2) = \text{closure}\,(T_{K_1}(x_1) + T_{K_2}(x_2))$$

and

$$N_{K_1+K_2}(x_1 + x_2) = N_{K_1}(x_1) \cap N_{K_2}(x_2).$$

- If $A \in L(X,Y)$, and if $L \subset X$ and $M \subset Y$ are convex closed sets satisfying $0 \in \mathrm{Int}(A(L) - M)$, then

$$T_{L \cap A^{-1}(M)}(x) = T_L(x) \cap A^{-1}T_M(Ax)$$

and

$$N_{L \cap A^{-1}(M)}(x) = N_L(x) + A^* N_M(Ax).$$

- If $A \in L(X,Y)$ and if $M \subset Y$ is a convex closed subset satisfying $0 \in \mathrm{Int}\,(\mathrm{Im}\,A - M)$, then

$$T_{A^{-1}(M)}(x) = A^{-1}T_M(Ax)$$

and

$$N_{A^{-1}(M)}(x) = A^* N_M(Ax).$$

- If K_1 and K_2 are convex closed subsets of X such that $0 \in \mathrm{Int}\,(K_1 - K_2)$, then

$$T_{K_1 \cap K_2}(x) = T_{K_1}(x) \cap T_{K_2}(x)$$

and

$$N_{K_1 \cap K_2}(x) = N_{K_1}(x) + N_{K_2}(x).$$

17.7 Optimisation

We consider the minimisation problem

$$v := \inf_{x \in X} f(x) = -f^*(0) \qquad (*)$$

where f is a nontrivial function from X to $\mathbb{R} \cup \{+\infty\}$. We denote the set of points at which f attains its minimum by $M := \{\bar{x} \in \mathrm{Dom}\, f \,|\, f(\bar{x}) = v\}$.

If f is both lower semi-continuous and lower semi-compact, there exists at least one solution of $(*)$. \hfill (7.1)

If f is strictly convex, there exists at most one solution of $(*)$. \hfill (7.2)

If f is convex, the solutions of $(*)$ are solutions of the inclusion

$$0 \in \partial f(\bar{x}) \qquad \text{(Fermat's rule)} \qquad (3)$$

If f is convex and lower semi-continuous, then

$$M = \partial f^*(0).$$

If f is convex and lower semi-continuous, and if

$$0 \in \text{Int} \left(\text{Dom} \, f^* \right) \tag{4}$$

then there exists a solution of $(*)$.

If f is nontrivial, convex and lower semi-continuous and if $\lambda > 0$, for every $x > 0$, there is a unique solution $J_\lambda(x)$ of the proximation problem

$$
\begin{aligned}
f_\lambda(x) & := \inf_{y \in X} \left(f(y) + \frac{1}{2\lambda} \|y - x\|^2 \right) \\
& = -\inf_{q \in X^*} \left(f^*(-q) + \frac{\lambda}{2} \|q\|_*^2 - \langle q, x \rangle \right).
\end{aligned}
\tag{5}
$$

The mapping J_λ satisfies

(a) $\|J_\lambda x - J_\lambda y\| \le \|x - y\|$

(b) $\|(1 - J_\lambda)x - (1 - J_\lambda)y\| \le \|x - y\|$

(c) $J_\lambda = (1 + \lambda \partial f(\cdot))^{-1}$.

The functions f_λ are convex and continuously differentiable and we have

$$\nabla f_\lambda(x) = \frac{1}{\lambda}(x - J_\lambda x) \in \partial f(J_\lambda x).$$

In addition, we have the *regularisation* property

$$\forall x \in \text{Dom} \, f, \quad f(x) = \lim_{\lambda \to 0} f_\lambda(x) \quad \text{and} \quad x = \lim_{\lambda \to 0} J_\lambda x$$

and the *penalisation* property

(a) $\inf_{x \in X} f(x) = \lim_{\lambda \to \infty} f_\lambda(x)$

(b) If $\partial f^*(0) \ne \emptyset$, $\lim_{\lambda \to \infty} \nabla f_\lambda(x) = 0$.

Suppose we have:

- two Hilbert spaces X and Y;

- two nontrivial, convex, lower semi-continuous functions $f : X \to \mathbb{R} \cup \{+\infty\}$ and $g : Y \to \mathbb{R} \cup \{+\infty\}$;

- a continuous linear operator A from X to Y. $\qquad\qquad$ (7.6)

We consider the two minimisation problems

$$h(y) := \inf_{x \in X} \left(f(x) - \langle p, x \rangle + g(Ax + y) \right)$$

and

$$e_*(p) := \inf_{q \in Y^*} \left(f^*(p - A^q) + g^*(q) - \langle q, y \rangle \right).$$

(a) If

$$p \in \text{Int} \left(\text{Dom} \, f^* + A^* \text{Dom} \, g^* \right) \tag{7}$$

then there exists a solution $\bar{x}$ of the problem $h(y)$ and

$$h(y) + e_*(p) = 0.$$

(b) If we also suppose that

$$y \in \text{Int} \left(\text{Dom} \, g - A \text{Dom} \, f \right) \tag{8}$$

then the following conditions are equivalent:

(i) $\bar{x}$ is a solution of the problem $h(y)$.

(ii) $\bar{x}$ belongs to the subdifferential $\partial e_*(p)$ of the marginal function e_*.

(iii) $\bar{x}$ is a solution of the inclusion $p \in \partial f(\bar{x}) + A^* \partial g(A\bar{x} + y)$.

(c) Similarly, the assumption (7.8) implies that there exists a solution $\bar{q}$ of the problem $e_*(p)$ and the two assumptions together imply the equivalence of the following conditions

(i) $\bar{q}$ is a solution of the problem $e_*(p)$;

(ii) $\bar{q}$ belongs to the subdifferential $\partial h(y)$ of the marginal function h;

(iii) $\bar{q}$ is a solution of the inclusion $y \in \partial g^*(\bar{q}) - A \partial f^*(p - A^* \bar{q})$.

(d) The two assumptions together imply that the solutions $\bar{x}$ and $\bar{q}$ of the problems $h(y)$ and $e_*(p)$, respectively, are the solutions of the system of inclusions

(i)
$$p \in \partial f(\bar{x}) + A^*(\bar{q})$$
(ii)
$$y \in -A\bar{x} + \partial g^*(\bar{q}).$$

17.8 Two-Person Games

Suppose that $f : E \times F \to \mathbb{R}$ is a function of two variables.
    The following conditions on $(\bar{x}, \bar{y}) \in E \times F$ are equivalent

$$\forall (x, y) \in E \times F, \quad f(\bar{x}, y) \le f(x, \bar{y}) \tag{1}$$

and

(i)
$$\inf_{x \in E} \sup_{y \in F} f(x, y) = \sup_{y \in F} \inf_{x \in E} f(x, y)$$

(ii)
$$\sup_{y \in F} f(\bar{x}, y) = \inf_{x \in E} \sup_{y \in F} f(x, y)$$

(iii)
$$\inf_{x \in E} f(x, \bar{y}) = \sup_{y \in F} \inf_{x \in E} f(x, y).$$

The pair $(\bar{x}, \bar{y})$ is then called a *saddle point* of f.

If we suppose that

(i) E is compact

(ii) $\forall y \in F$, $x \to f(x, y)$ is lower semi-continuous

then, there exists $\bar{x} \in E$ such that

$$\sup_{y \in F} f(\bar{x}, y) = \inf_{x \in E} \sup_{y \in F} f(x, y) = \sup_{K \in \mathcal{K}} \inf_{x \in E} \sup_{y \in K} f(x, y) \qquad (2)$$

where $\mathcal{K}$ is the set of finite subsets K of Y.

If we suppose that

(i) E is compact

(ii) $\forall y \in F$, $x \to f(x, y)$ is convex and lower semi-continuous

(iii) $\forall x \in E$, $y \to f(x, y)$ is concave

then, there exists $\bar{x} \in E$ such that

$$\sup_{y \in F} f(\bar{x}, y) = \inf_{x \in E} \sup_{y \in F} f(x, y) = \sup_{y \in F} \inf_{x \in E} f(x, y). \qquad (3)$$

If we suppose that

(i) E and F are convex and compact

(ii) $\forall y \in F$, $x \to f(x, y)$ is convex and lower semi-continuous

(iii) $\forall x \in E$, $y \to f(x, y)$ is concave and upper semi-continuous

then f has a saddle point. $\qquad (8.4)$

If we suppose that

(i) E is compact

(ii) $\forall y \in F$, $x \to f(x, y)$ is lower semi-continuous

(iii) F is convex

(iv) $\forall x \in E$, $y \to f(x, y)$ is concave

then there exists $\bar{x} \in E$ such that

$$\sup_{y \in F} f(\bar{x}, y) = \sup_{D \in \mathcal{C}(E,F)} \inf_{x \in E} f(x, D(x)) = \inf_{C \in \mathcal{C}(F,E)} \sup_{y \in F} f(C(y), y). \qquad (5)$$

Ky Fan's Inequality. If we suppose that K and $\phi : K \times K \to \mathbb{R}$ satisfy

(i) K is convex and compact

(ii) $\forall y \in K$, $x \to \phi(x, y)$ is lower semi-continuous

(iii) $\forall x \in K$, $y \to \phi(x, y)$ is concave

(iv) $\forall y \in K$, $\phi(y, y) \leq 0$,

then there exists $\bar{x} \in E$ such that $\forall y \in K$, $\phi(\bar{x}, y) \leq 0$.   (8.6)

17.9 Set-valued Maps and the Existence of Zeros and Fixed Points

Definitions. A set-valued map C from K to Y is *upper semi-continuous* at x_0 if

$$\forall \varepsilon > 0, \ \exists \eta > 0 \text{ such that } \forall x \in B_K(x_0, \eta), \ C(x) \subset C(x_0) + \varepsilon B. \quad (1)$$

C is *upper hemi-continuous* at x_0 if

$$\forall p \in Y^*, \ x \to \sigma(C(x), p) \text{ is upper semi-continuous at } x_0. \quad (2)$$

C is *lower semi-continuous* at x_0 if, for any sequence x_n converging to x_0 and for all $y_0 \in C(x_0)$, there exists a sequence of elements $y_n \in C(x_n)$ converging to y_0.

Any upper semi-continuous set-valued map is upper hemi-continuous. (9.3)

If $f : X \to \mathbb{R} \cup \{+\infty\}$ is convex and lower semi-continuous and if $\text{Int Dom} f \neq \emptyset$, then $x \in \text{Int Dom} f \to \partial f(x) \subset X^*$ is upper hemi-continuous.   (9.4)

Banach–Picard Fixed-Point Theorem. If K is a complete metric space and the mapping D is a contraction ($\exists k < 1$ such that $\forall x, y \in K$, $d(D(x), D(y)) \leq kd(x, y)$) from K to itself, then D has a unique fixed point.   (9.5)

Suppose that C is a set-valued map from a complete metric space K to itself. Suppose that there exists a nontrivial positive function f from K to $\mathbb{R} \cup \{+\infty\}$ such that

$$\forall x \in K, \ \exists y \in C(x) \text{ such that } f(y) + d(x, y) \leq f(x). \quad (6)$$

Then one of the two assumptions

(a) f is lower semi-continuous

(b) the graph of C is closed

implies that C has a fixed point.

Three-Poles Lemma. We consider n closed subsets F_i of the simplex $M^n := \{x \in \mathbb{R}_+^n | \sum_{i=1}^n x_i = 1\}$. If $\forall x \in M^n$, $x \in \cup_{\{i|x_i>0\}} F_i$ then $\cap_{i=1}^n F_i \neq \emptyset$.

$$(9.7)$$

Brouwer's Fixed-Point Theorem. If K is a convex compact subset of a Hilbert space and if D is a continuous mapping from K to itself, then D has a fixed point.

$$(9.8)$$

Gale–Nikaïdo–Debreu Theorem. Suppose that C is a set-valued map from $M^n = \{x \in \mathbb{R}_+^n | \sum_{i=1}^n x_i = 1\}$ to $\mathbb{R}^n$ satisfying

(i) C is upper hemi-continuous

(ii) $\forall x \in M^n$, $C(x) = C(x) - \mathbb{R}_+^n$ is convex and closed

(iii) $\forall x \in M^n$, $\sigma(C(x), x) \geq 0$.

Then there exists $\bar{x} \in M^n$ such that $0 \in C(\bar{x})$.

$$(9.9)$$

Brouwer–Ky Fan Theorem. Suppose that K is a convex compact subset of X and that C is an upper hemi-continuous set-valued map from K to X with convex closed values. If we suppose that

$$\forall x \in K, \ \ C(x) \cap T_K(x) \neq \emptyset$$

then

(a) $\exists \bar{x} \in K$ such that $0 \in C(\bar{x})$
(b) $\forall y \in K, \exists \hat{x} \in K$ such that $y \in \hat{x} - C(\hat{x})$.

$$(9.10)$$

Fixed-Point and Surjectivity Theorem. Suppose that $K \subset X$ is a convex compact subset and that $D : K \to X$ is an upper hemi-continuous set-valued map with convex closed values.
(a) If D is *re-entrant* in the sense that

$$\forall x \in K, \ \ D(x) \cap (x + T_K(x)) \neq \emptyset$$

(in particular any $D : K \to K$ is re-entrant) then D has a fixed point $x_* \in K$.
(b) If D is *salient* in the sense that

$$\forall x \in K, \ \ D(x) \cap (x - T_K(x)) \neq \emptyset$$

then

(i) D has a fixed point $x_* \in K$
(ii) $\forall y \in K, \exists \hat{x} \in K$ such that $y \in D(\hat{x})$.

$$(9.11)$$

Suppose we have:

- two convex closed subsets $L \subset X$ and $M \subset Y$;

- a convex compact subset $P \subset Y^*$;

- a continuous mapping $c : L \times P \to Y$,

satisfying the following conditions:

(i) $\qquad\qquad\qquad\qquad \forall x \in L, \; p \to c(x,p)$ is affine

(ii) $\qquad\qquad\qquad\qquad \forall x \in L, \; \forall p \in P, \; c(x,p) \in T_L(x)$

and

(i) $\qquad\qquad\qquad\qquad L \cap A^{-1}(M)$ is compact

(ii) $\qquad\qquad\qquad\qquad 0 \in \mathrm{Int}\,(A(L) - M)$

(iii) $\qquad\qquad\qquad\qquad \forall y \in M, \; N_M(y) \subset \bigcup_{\lambda \geq 0} \lambda P$

together with

$$\forall x \in L, \; \forall p \in P, \; \langle p, Ac(x,p) \rangle \leq 0.$$

Then there exist $\bar{x} \in L$ and $\bar{p} \in P$ satisfying

$$A\bar{x} \in M \text{ and } c(\bar{x}, \bar{p}) = 0. \tag{12}$$

Leray–Schauder Theorem. We consider a convex compact subset $K \subset \mathbb{R}^n$ with a non-empty interior and a set-valued map C from $K \times [0,1]$ to $\mathbb{R}^n$ which is upper hemi-continuous with convex closed values. We suppose that

(i) $\qquad\qquad\qquad\qquad \forall x \in K, \; C(x,0) \cap T_K(x) \neq \emptyset$

(ii) $\qquad\qquad\qquad\qquad \forall \lambda \in [0,1[, \; \forall x \in \partial K, \; 0 \notin C(x,\lambda).$

Then there exists $\bar{x} \in K$ such that $0 \in C(\bar{x}, 1)$. $\qquad\qquad$ (9.13)

Suppose that $K \subset X$ is a convex compact subset and that $C : K \to X^*$ is an upper semi-continuous set-valued map with non-empty, convex, compact values. Then:

$$\exists \bar{x} \in K \text{ such that } 0 \in C(\bar{x}) - N_K(\bar{x}). \tag{14}$$

Suppose that $K \subset X$ is a convex compact subset and that $C : K \to K$ is an upper hemi-continuous set-valued map with non-empty, convex, closed values. We consider a function $\phi : K \times K \to \mathbb{R}$ satisfying

(i) $\qquad\qquad\qquad\qquad \forall y \in K, \; x \to \phi(x,y)$ is lower semi-continuous

(ii) $\qquad\qquad\qquad\qquad \forall x \in K, \; y \to \phi(x,y)$ is concave

(iii) $\qquad\qquad\qquad\qquad \forall y \in K, \; \phi(y,y) \leq 0.$ $\qquad\qquad$ (9.15)

Suppose that the set-valued map C and the function ϕ are linked by the property

$$\{x \in K \mid \sup_{y \in C(x)} \phi(x, y) \leq 0\} \text{ is closed.}$$

Then there exists a solution $\bar{x} \in K$ of

(i) $\bar{x} \in C(\bar{x})$

(ii) $\sup_{y \in C(\bar{x})} \phi(\bar{x}, y) \leq 0.$

We consider a finite covering $\{A_i\}_{i=1,\ldots,n}$ of a metric space E. There exists a continuous partition of unity subordinate to this covering, in other words, there exist n *continuous* functions $a_i : E \to [0, 1]$ such that

(i) $\forall x \in E, \quad \sum_{i=1}^{n} a_i(x) = 1$

(ii) $\forall i = 1, \ldots, n, \quad \text{support}(a_i) \subset A_i.$ (9.16)

where $\text{support}(a_i) := \text{closure}\{x \in E \mid a_i(x) \neq 0\}$.

We consider two mappings F and G from M^n to $\mathbb{R}^m$ satisfying

(i) the components f_i of F are convex and lower semi-continuous;

(ii) the components g_i of G are concave, positive and upper semi-continuous;

(iii) $\exists p \in M^m$ such that $\forall x \in M^n$, $\langle p, F(x) \rangle > 0$;

(iv) $\exists x \in M^n$ such that $\forall i = 1, \ldots, n$, $g_i(x) > 0$.

(a) Then there exist $\delta > 0$, $\bar{x} \in M^n$ and $\bar{p} \in M^m$ such that

(i) $\forall i = 1, \ldots, n, \quad \delta f_i(\bar{x}) \leq g_i(x)$

(ii) $\forall x \in M^n, \quad \langle G(x) - \delta F(x), \bar{p} \rangle \leq 0$

(iii) $\forall i = 1, \ldots, n, \quad \bar{p}_i(\delta f_i(\bar{x}) - g_i(\bar{x})) = 0.$ (9.17)

(b) The number $\delta > 0$ is defined by

$$\frac{1}{\delta} = \sup_{p \in M^m} \inf_{x \in M^n} \frac{\langle p, F(x) \rangle}{\langle p, G(x) \rangle} = \inf_{x \in M^n} \sup_{p \in M^m} \frac{\langle p, F(x) \rangle}{\langle p, G(x) \rangle}. \qquad (*)$$

If $\lambda > 0$ and $x \in M^n$ satisfy the inequalities $\lambda f_i(x) \leq g_i(x)$, $\forall i = 1, \ldots, n$, then $\lambda \leq \delta$.

(c) For all $\mu > \delta$ and for all $y \in \text{Int}(\mathbb{R}^m_+)$, there exist $\beta > 0$ and $\hat{x} \in M^n$ such that

$$\forall i = 1, \ldots, n \quad \mu f_i(\hat{x}) - g_i(\hat{x}) \leq \beta y_i.$$

We consider two matrices F and G from $\mathbb{R}^n$ to $\mathbb{R}^m$ satisfying

(i) the coefficients g_{ij} of G are non-negative;

(ii) $\forall i = 1, \ldots, m,\ \sum_{i=1}^{n} g_{ij} > 0$;

(iii) $\forall j = 1, \ldots, n,\ \sum_{i=1}^{n} f_{ij} > 0$.

Then there exist $\bar{x} \in M^n$, $\bar{p} \in M^m$ and $\delta > 0$ such that

(i) $$\delta F \bar{x} \leq G \bar{x}$$
(ii) $$\delta F^* \bar{p} \geq G^* \bar{p}$$
(iii) $$\delta \langle \bar{p}, F\bar{x} \rangle = \langle \bar{p}, G\bar{x} \rangle.$$

Moreover, for all $\mu > \alpha$ and all $y \in \text{Int}\,\mathbb{R}_+^m$, there exists $\hat{x} \in \mathbb{R}_+^n$ such that

$$\mu F\hat{x} - G\hat{x} \leq y. \tag{18}$$

Suppose that F is a mapping from M^n to $\mathbb{R}^n$ satisfying

(i) the components f_i of F are convex and lower semi-continuous;

(ii) $\exists p \in M^n \cap \text{Int}(\mathbb{R}_+^n)$ such that $\forall x \in M^n$, $\langle p, F(x) \rangle > 0$;

(iii) if $x_i = 0$ then $f_i(x) \leq 0$.

Suppose that G is another mapping from M^n to $\mathbb{R}^n$ satisfying

(i) the components g_i of G are concave and lower semi-continuous;

(ii) $\forall x \in M^n$, $\forall i = 1, \ldots, n$, $g_i(x) > 0$.

We consider the number $\delta > 0$ defined by $(*)$ (above). Then there exist $\bar{x} \in M^n \cap \text{Int}(\mathbb{R}_+^n)$ and $\bar{p} \in M^n \cap \text{Int}(\mathbb{R}_+^n)$ such that

(i) $$\delta F(\bar{x}) = G(\bar{x})$$
(ii) $$\forall x \in M^n,\ \langle \bar{p}, G(x) - \delta F(x) \rangle \leq 0.$$

If $\mu > \delta$ and $y \in \text{Int}(\mathbb{R}_+^n)$ are given, then there exist $\beta > 0$ and $\hat{x} \in M^n$ such that

$$\mu F(\hat{x}) - G(\hat{x}) = \beta y \tag{19}$$

Perron–Frobenius Theorem. Suppose that G is a positive matrix.

(a) G has a strictly positive eigenvalue δ and an associated eigenvector $\bar{x}$ with strictly positive components.

(b) δ is the only eigenvalue associated with an eigenvector of M^n.

(c) δ is greater than or equal to the absolute value of all other eigenvalues of G

(d) The matrix $\mu - G$ is invertible and $(\mu - G)^{-1}$ is positive if and only if $\mu > \delta$.   (9.20)

We consider a mapping H from $\mathbb{R}^n_+$ to $\mathbb{R}^n$ satisfying

(i) the components h_i of H are convex, positively homogeneous and lower semi-continuous;

(ii) $\exists b \in \mathbb{R}$ such that $\forall x \in \mathbb{R}^n_+$, $bx_i > h_i(x)$;

(iii) $\forall x \in M^n$, $\exists q \in M^n$ such that $\langle q, H(x) \rangle > 0$.

Then $\forall y \in \text{Int } \mathbb{R}^n_+$, $\exists x \in \text{Int } \mathbb{R}^n_+$ such that $H(x) = y$.   (9.21)

Theorem (Surjectivity of the M Matrices). Suppose that H is a matrix from $\mathbb{R}^n$ to $\mathbb{R}^n$ satisfying

$$\forall i \neq j, \quad h_{ij} \leq 0. \tag{22}$$

The following conditions are equivalent

(a) $\forall x \in M^n$, $\exists q \in M^n$ such that $\langle q, Hx \rangle > 0$;

(b) H is invertible and H^{-1} is positive;

(c) H^* is invertible and H^{*-1} is positive.

References

Aliprantis, C., Brown, D. and Burkinshaw, O. 1989, Exitence and optimality of competitive equilibria. Springer-Verlag

Arrow, K.J. and Hahn, F.M. 1971, General Competitive Analysis. Holden Day, New York

Aubin, J.-P. 1977, Applied Abstract Analysis. Wiley Interscience, New York

Aubin, J.-P. 1979a, Applied Functional Analysis. Wiley Interscience, New York

Aubin, J.-P. 1979b, Mathematical Methods of Game and Economic Theory. North-Holland, Amsterdam

Aubin, J.-P. 1982, Méthodes Explicites de l'Optimisation. Dunod, Paris

Aubin, J.-P. 1991, Viability Theory. Systems and Control 16, Birkhäuser

Aubin, J.-P. 1994, Initiation à l'Analyse Appliquée. Masson

Aubin, J.-P. 1996, Neural Networks and Qualitative Physics: A Viability Approach. Cambridge University Press, Cambridge

Aubin, J.-P. 1997, Dynamical Economic Theory. Springer-Verlag

Aubin, J.-P. 1998, Mutational and morphological analysis: tools for shape regulation and optimization. Birkhäuser, Boston, Basel

Aubin, J.-P. In preparation, La mort du devin, l'émergence du démiurge.

Aubin, J.-P. and Cellina, A. 1984, Differential Inclusions. Springer, Heidelberg

Aubin, J.-P. and Ekeland, I. 1984, Applied Nonlinear Analysis. Wiley Interscience, New York

Aubin, J.-P. and Frankowska, H. 1990, Set-valued Analysis. Systems and Control, Birkhäuser,

Aumann, R.J. and Shapley, L.S. 1974, Values of Non-atomic Games. Princeton University Press, Princeton

Auslender, A. 1976, Optimisation: Méthodes Numériques. Masson, Paris

Axelrod, R. 1984, The ecolution of cooperation. Cambridge University Press

Balasko, Y. 1988, Foundations of the theory of general euilibrium. Academic Press

Balinski, M.L. and Young, M.P. 1982, Fair Representation. Yale University Press, New Haven

Baumol, W. 1970, Economic dynamics. Macmillan

Beer G. 1993, Topologies on closed and closed convex sets. Kluwer Academic Publisher

Benetti, C. and Cartelier, J. 1980, Marchands, salariat et capitalistes. Maspero

Bensoussan, H., Hurst, G. and Naslund, B. 1974, Management Applications of Modern Control Theory. North-Holland, Amsterdam

Bensoussan, A. 1982, Stochatic control by functional analysis methods. North-Holland

Bensoussan, A. and Lions J.-L. 1978, Applications des inéquations variationnelles en contrôle stochastique. Dunod.

Bensoussan, A. and Lions J.-L. 1982, Contrôle impulsionnel et inéquations quasi-variationnelles. Dunod, Paris

Berge, C. 1959, Espaces Topoloqigues et Fonctions Multivoques. Dunod, Paris

Bernhard, P. 1976, Commande Optimale, Décentralisation et Jeux Dynamiques. Dunod, Paris

Brezis, H. 1973, Opérateurs Maximaux Monotones et Semi-groupes de Contractions dans les Espaces de Hilbert. North-Holland, Amsterdam

Brezis, H. 1983, Analyse Fonctionelle et Applications. Masson, Paris

Brock, W.A. and Malliaris, A.G. 1989, Differential equation, stability and chaos in dynamic economics. North Holland

Browder, F.E. 1976, Nonlinear Operators amd Nonlinear Equations of Evolution in Banach Spaces. American Mathematical Society, Providence

Burger, E. 1963, Introduction to Theory of Games. Prentice Hall, Englewood Cliffs

Canard, N.-F. 1802, Principes d'économie politique.

Cartelier, J. 1996, La monnaie. Flammarion

Cea, J. 1971, Optimisation, Théorie et Algorithmes. Dunod, Paris

Chandler, A.D. 1977, The visible hand: The managerial revolution in American business. Belknap Press

Chandler, A.D. 1990, Scape and scope. Belknap Press

Cherene, L.J. 1978, Set valued dynamical systems and economic flow. Springer-Verlag

Ciarlet, P. 1982, Introduction à l'Analyse Numérique Matricielle et à l'Optimisation. Masson, Paris

Clarke, F.H. 1983, Optimization and Non-smooth Analysis. Wiley Interscience

d'Autume, A. 1985, Monnaie, croissance et déséquilibre. Economica

Day, R.H. 1994, Complex Economic Dynamics, Vol. I, An introduction to dynamical systems and market mechanims. MIT Press

Day, R.H. to appear, Complex Economic Dynamics, Vol. II, An introduction to macroeconomic dynamics. MIT Press

de Montbrial, T. 1972, Économie Théorique. Presses Universitaires de France, Paris

Debreu, G. 1959, Theory of Value. Wiley, New York

Demianov, V.V. and Malosemov, V.N. 1972, 'Introduction to the Minimax'. Nauka, Moscow

Demianov, V.V. and Vassilev, L.V. 1981, 'Nondifferentiable Optimisation'. Nauka, Moscow

Dieudonné, J. 1978, Abrégé d'Histoire des Mathématiques 1700–1900, vols. 1–2. Hermann, Paris

Dieudonné, J. 1982, History of Functional Analysis. North-Holland, Amsterdam

Dixmier, J. 1981, Topologie Générale. Presses Universitaires de France, Paris

Ekeland, I. 1974, La Théorie des Jeux et ses Applications à l'Économie Mathématique. Presses Universitaires de France, Paris

Ekeland, I. 1979, Éléments d'Économie Mathématique. Hermann, Paris

Ekeland, I. and Temam, R. 1974, Analyse Convexe et Problèmes Variationnels. Dunod, Paris

Fisher, F. M. 1983, Disquilibrium foundations of equilibrium economics. Cambridge University Press

Florenzano, M. 1981, L'Équilibre Économique Général. Éditions du C.N.R.S., Paris

Frankowska, H. to appear, Set-Valued Analysis and Control Theory. Systems and Control: Birkhäuser, Boston, Basel, Berlin

Freeman, C. 1982, The economics of industrial innovation. Pinter Publishers

Fudenberg, D. and Tirole, J. 1991, Game theory. MIT Press

Gomez, G. L. 1992, Dynamic probabilistic models and social structures. Reidel

Green, R. and Laffont, J.J. 1979, Incentives in Public Decision Making. North-Holland, Amsterdam

Heal, G. 1973, The Theory of Economic Planning. North-Holland, Amsterdam

Hildenbrand, W. 1974, Core and Equilibria of a Large Economy. Princeton University Press, Princeton

Hildenbrand, W. and Kirman, A. 1975, Introduction to Equilibrium Analysis. North-Holland, Amsterdam

Hiriart-Urruty, J.-B. and Lemarechal, C. 1994, Convex Analysis and Minimization Algorithms. Springer-Verlag (2 volumes)

Hofbauer, J. and Sigmund, K. 1988, The theory of evolution and dynamical systems. Cambridge University Press, London Math. Soc. # 7

Ingrao, B. and Israel, G. 1987, La mano invisibile. Laterza

Intriligator, M.D. 1971, Mathematical Optimization and Economic Theory. Prentice Hall, Englewood Cliffs

Ioffe, A.D. and Tikhomirov, V.M. 1979, Theory of Extremal Problems. North-Holland, Amsterdam

Istratescu, V.I. 1981, Fixed-point Theory. D. Reidel Pub. Co., Dordrecht

Isaacs, R. 1965, Differential games. Wiley

Kalecki, 1971, Selected essays and the dynamics of the capitalistic economy. Cambridge University Press

Keynes, J. M. 1936, The general theory of employment, interest and money. in The Collected Writings of John Maynard Keynes. Vol. VII (MacMillan 1973)

Kuratowski, K. 1958, Topologie, vols. 1 and 2 4th. ed. corrected. Panstowowe Wyd Nauk, Warszawa. (Academic Press, New York, 1966,

Laurent, P.J. 1972, Approximation et Optimisation. Hermann, Paris

Lions, J.L. 1969, Quelques Méthodes de Résolution des Problèmes aux Limites Non Linéaires. Dunod, Paris

Luenberger, D.G. 1969, Optimization by Vector-space Methods. Wiley, New York

Luenberger, D.G. 1973, Introduction to Linear and Nonlinear Programming. Addison-Wesley, Reading

Malinvaud, E. 1969, Leçons de Théorie Micro-économique. Dunod, Paris

Mas-Collel, A. 1985, The theory of general economic equilibrium: a differential approach. Cambridge University Press

Marshall, A. 1920, Principles of economics. Macmillan

Moulin, H. 1980a, Fondation de la Théorie des Jeux. Hermann, Paris

Moulin, H. 1980b, Théorie des Jeux pour l'Économie et la Politique. Hermann, Paris

Moulin, H. 1980c, La Stratégie du Vote. Éditions du C.N.R.S., Paris

Moulin, H. and Fogelman-Soulié, F. 1981, La Convexité dans les Mathématiques de la Décision. Hermann, Paris

Nelson, R. and Winter, S. 1982, An evolutionary theory of economic change. The Belknap Press of Harvard University Press

Nikaïdo, H. 1968, Convex Structures and Economic Theory. Academic Press, New York

Nirenberg, L. 1974, Topics in Nonlinear Functional Analysis. Lecture notes, New York University

Owen, G. 1968, Game Theory. Saunders

Pareto, V. 1909, Manuel d'Économie Politique. Girard et Brière

Pchenitchnyi, B.N. 1980, 'Convex Analysis and Extremal Problems'. Nauka, Moscow

Polanyi, K. 1968, The great transformation. Beacon Press

Rockafellar, R.T. 1970, Convex Analysis. Princeton University Press, Princeton

Rockafellar, R.T. 1983, La Théorie des Sous-gradients et ses Applications à l'Optimisation. Presses de l'Université de Montréal, Montreal

Rockafellar, R.T. and Wets, R.B. 1997, Variational Analysis. Springer-Verlag

Saari, D.G. 1994, Geometry of Voting. Springer-Verlag

Scarf, H.E. and Hansen, P. 1973, The Computation of Economic Equilibria. Yale University Press

Schwartz, J.T. 1961, Lectures on Mathematical Methods in Analytical Economics. Gordon and Breach, New York

Schwartz, J.T. 1969, Nonlinear Functional Analysis. Gordon and Breach, New York

Schwartz, L. 1970, Topologie Générale et Analyse Fonctionelle. Hermann, Paris

Shelling, T.C. 1960, The Strategy of Conflict. Harvard University Press, Cambridge

Shumpeter, J. 1934, Theory of economic development. Harvard University Press

Sigmund, K. 1993, Games of Life. Oxford University Press

Treves, F. 1967, Topological Vector Spaces, Distributions and Kernels. Academic Press, New York

von Neumann, J. and Morgenstern, O. 1944, Theory of Games and Economic Behaviour, Princeton University Press, Princeton

Walras, L. 1874, Éléments d'Économie Politique Pure. Corbaz, Lausanne

Zangwill, W.I. 1969, Nonlinear Programming. Prentice Hall, Englewood Cliffs

Index

accepted
–, multilosses   198, 216
–, multistrategies   197
action game   214
affine   21
aggressive strategy   112
allocation   169
approximate minimisation   15
Arrow–Debreu–Nash Theorem   192
asymptotic
–, centre   313, 362–363
–, cone   51
atomicity axiom   217, 229, 300
average fitness   207
average gain   208

balanced
–, family   166, 198
–, game   199, 225
–, growth   180
balancing   166, 202, 225
Banach   34
Banach–Picard Fixed-point Theorem
    18, 415
barrier cone   51–52, 406
Battle of the Sexes   115
behaviour
–, non-cooperative   189–190
–, profile   211
behavioural quantity   211
Best Approximation Theorem   28
best compromise solution   195
biconjugate   36, 37
biloss mapping   107
bistrategy   102
–, consistent   103
Broad Separation Theorem   31

Brouwer's Fixed-point Theorem   104,
    141, 143, 162, 416

canonical decision rule   108, 192
Carathéodory's Theorem   241
Caristi's Fixed-point Theorem   17
change function   173
Clarke differentiable   88
coalition   196–197, 209
–, equilibrium   214
–, fuzzy   210
–, generalised   211
–, social   212
collective
–, Budgetary rule   174
–, stability   109
commodity   168, 179
complement (of a toll set)   53
completeness   15
concave   21
cone
–, asymptotic   51
–, barrier   51–52, 406
–, contingent   272
–, negative polar   48, 315
–, normal   60, 70–73, 97–98, 409–411
–, tangent   60, 70–73, 97–98, 330,
    378–381, 409–411
conjugate function   37, 43–48, 245–254,
    404–405
conservative
–, strategy   110–111, 126, 130–134, 194
–, vector   111
consistent   189
–, bistrategy   103
–, multistrategies   189
constraints   80–82, 99, 192
consumption   168

contingent
–, cone   272
–, epiderivative   269
continuous partition of unity   135–137, 141
continuously differentiable   88
contraction   18
convex function   21–34, 240–245, 401, 403–404
convexification strategy   105–106
cooperative game   197–231, 297–300
coordination game   115
core   197, 198, 221, 222, 224–231
core (of a toll set)   53
Cournot's Duopoly   116–123
Cournot's equilibrium   119
cyclically monotone set-valued map   258

Debreu–Gale–Nikaïdo Theorem   148–149
decentralisation, price   82–84, 167, 173
decentralised, price   169
decision rule   102–104, 120, 189
–, canonical   108, 192
–, optimal   137–142
–, Stackelberg   121
demand map   170, 344, 345, 398
derivative
–, Clarke directional   87
–, right   62, 88, 408
differentiable
–, Fréchet   88, 92
–, Gâteaux   62, 92
Dirac measures   53
disjunct   33
domain   10, 401
domain (of a toll set)   53
dual problems   36, 77
duality   323, 372
duality interval   126

economic equilibrium   167–179, 285–289
efficiency axiom   217, 229, 300
Eigen-Schuster's hypercycle   207
Ekeland's Theorem   15
epigraph   10–11, 401
equilibrium

–, coalition   214
–, Cournot   119
–, economic   167–179, 285–289
–, Nash   108, 190
–, non-cooperative   108, 119, 190
–, social   192
–, Stackelberg   121–123
–, state   214
–, static   102
–, Walrasian   170, 344, 397
equilibrium of replicator systems   206
Euler–Lagrange inclusion   78
evaluation function   106
evolutionary stable equilibrium   206
exchange economy   168
existence of zeros of set-valued map   150–152, 415–420
extremal point   253

Fenchel's Theorem   39–43, 319, 368
Fermat   57
Fermat's rule   58, 76–80, 99, 327, 375
finite game   112–116
Fisher-Wright-Haldane's model   207
fitness matrix   207
fixed-point theorem   154–155
–, Banach–Picard   18, 415
–, Brouwer's   104, 141, 143, 162, 416
–, Caristi   17
–, Kakutani   154
–, Kakutani–Fan   155
Fréchet differentiable   88, 92
function
–, change   173
–, conjugate   37, 43–48, 245–254, 404–405
–, convex   21–34, 240–245, 401, 403–404
–, evaluation   106
–, gamma-convex   316, 366
–, indicator   10, 402
–, locally Lipschitz   87–99, 268–275, 401
–, loss   106
–, lower semi-compact   13–15, 401
–, lower semi-continuous   11–13, 37–39, 401
–, marginal   238, 303, 350

-, support   36, 48–52, 405–407
-, upper semi-continuous   12
-, utility   106
-, worst-loss   110
fuzzy
-, coalition   210
-, game   209–231, 297–300
-, set   210

Gâteaux differentiable   62, 88, 92, 408
Gale–Nikaïdo–Debreu Theorem   338, 389, 416
game
-, action   214
-, balanced   199, 225
-, Chicken   114
-, cooperative   197–231, 297–300
-, coordination   115
-, finite   112–116
-, fuzzy   209–231, 297–300
-, inessential   299
-, market   222
-, n-person   189–208, 290–297
-, non-cooperative   192–193
-, regular   218
-, simple   231
-, subadditive   220
-, two-person   101–123, 413–415
-, two-person zero-sum   125–142, 275–279
-, weighting   219
gamma-convex function   316, 366
Gaussian toll sets   55
generalised
    gradient   268
-, coalition   211
-, gradient   87–99, 221
gradient   62, 88
-, generalised   87–99, 221, 268
graph   18, 145, 301
growth rate   180

Hamiltonian system   78

inclusion   75, 76, 143–166, 280–285
indicator function   10, 402
indicators   53
individual stability   108
inequality

-, Ky Fan's   140, 275–279, 337, 388–389
-, quasi-variational   160–162
-, variational   157–159, 342
inessential game   299
inf-compact   13
inf-convolution   45, 237
inner semicontinuous   147
inverse of set-valued map   145

Kakutani's Fixed-point Theorem   154
Kakutani–Fan Fixed-point Theorem   155
Knaster–Kuratowski–Mazurkiewicz
    lemma   163, 416
Ky Fan's inequality   140, 275–279, 337, 388–389
Ky Fan's Theorem   275–279

Lagrange multipliers   78, 83–84, 97, 270
Legendre transformation   64
Leray–Schauder Theorem   159–160, 417
locally Lipschitz function   87–99, 268–275, 401
loss function   106
Lotka-Volterra equation   208
lower section   11
lower semi-compact function   13–15, 401
lower semi-continuous function   11–13, 37–39, 401
lower semi-continuous set-valued map   146, 415–420

Maintenon (marquess of)   101
marginal function   238, 303, 350
marginal properties   75–86, 261–268
market game   222
Maynard-Smith' dynamic game   208
measure Maslov   56
membership   210
membership cost functions   53
metagame   192–193
minimax theorem   134, 277, 335–337, 386–388
-, Ky Fan   140, 275–279, 337, 388–389
-, von Neumann   275–279

minimisation problem   9–19, 35–56,
    75–86, 99, 235–300, 411–413
minimising sequence   235
minimum, virtual   110, 194
Minkowski   34, 36
mixed strategy   105, 134
monotone set-valued map   343, 395
Moreau transform   55
multilosses
–, accepted   198, 216
multistrategies
–, accepted   197
–, consistent   189
Méré (Antoine Gombaud, chevalier de)
    101

n-person game   189–208, 290–297
Nash equilibrium   108, 190
Nash's Theorem   191
negative polar cone   48, 315
non-cooperative
–, behaviour   189–190
–, equilibrium   108, 119, 190
–, game   192–193
nonlinear equations   143–166, 280–285
normal cone   60, 70–73, 97–98, 409–411
normal form (of game)   106–108,
    190–191

open covering   135
open image theorem   311, 360–362
optimal decision rule   137–142
optimisation   411–413
orthogonal
–, projector   315, 365–367
–, subspace   48
outer semicontinuous   147

pareto optima   108–110, 193–195
partial order   107
participation   210
partition of unity
–, continuous   135–137
peaceable strategy   112
penalisation   84–86, 412
Perron–Frobenius Theorem   184–186,
    420
population genetics   207
prebiotic evolution   207

price
–, decentralisation   82–84, 167, 173
–, decentralised   169
–, simplex   169
Prisoner's dilemma   112–113
probability simplex   205
projector of best approximation   28,
    405
proper mapping   317, 367–368
Proximation Theorem   27–31, 84
pseudo convex   274

quasi-convex   22
quasi-variational inequality   160–162

re-entrant   154, 416
redundant-players axiom   218, 229, 300
regular game   218
regularisation   84–86, 412
replicator system   205
replicator systems for linear growth
    rates   207
restriction   10
right derivative   62, 88, 408

saddle point   125–130, 134, 413
salient   154, 416
Scarf's Theorem   199
section   11, 401
Separation Theorem   31
–, Broad   31
set-valued map   144, 301–303, 347–350
–, cyclically monotone   258
–, existence of zeros   150–152, 415–420
–, inverse   145
–, lower semi-continuous   146, 415–420
–, monotone   343, 395
–, upper hemi-continuous   145,
    415–420
–, upper semi-continuous   145, 301,
    415–420
Shapley
–, (K-K-M-S)Theorem   162–166
–, value   218, 227, 230
share-out rule   217
side payments   216–224
simple game   231
Slater condition   249
Smith, Adam   167

social
–, coalition   212
–, equilibrium   192
solution (of the game)   221
stability
–, collective   109
–, individual   108
Stackelberg
–, disequilibrium   123
–, decision rule   121
–, equilibrium   121–123
static equilibrium   102
strategic form (of game)   190–191
strategy
–, aggressive   112
–, conservative   110–111, 126, 130–134,
    194
–, mixed   105, 134
–, pair   102
–, peaceable   112
strict   10
strictly convex   22, 401
strictly Fréchet differentiable   88, 92
subadditive games   220
subdifferentiability   408–409
subdifferential   57–73, 254–261
subgradient   58, 62
support   135, 282
–, function   36, 48–52, 405–407
surjectivity of the M matrices   187–
    188, 420
symmetry axiom   217, 229, 300
system under constant organization
    205

tangent cone   60, 70–73, 97–98, 330,
    409–411
tangential condition   149–150
Three-Poles Lemma   162
toll boundary   53
toll set   53
two-person game   101–123, 413–415
two-person zero-sum game   125–142,
    275–279

upper hemi-continuous   144–148
upper hemi-continuous set-valued map
    145, 415–420
upper semi-continuous function   12

upper semi-continuous set-valued map
    145, 301, 415–420
utility function   106

value (of a game)   125–130, 334
value, conservative   110
variational inequality   157–159, 342
viability   152–154
virtual minimum   110, 194
von Neumann model   179–188, 290
von Neumann's Theorem   134, 181,
    275–279

Walras Law   171
–, collective   171
–, tâtonnement   172
Walras, Léon   167
Walrasian
–, equilibrium   170, 344, 397
–, mechanism   169–173
weighting game   219
worst-loss function   110

Graduate Texts in Mathematics

continued from page II

65 WELLS. Differential Analysis on Complex Manifolds. 2nd ed.
66 WATERHOUSE. Introduction to Affine Group Schemes.
67 SERRE. Local Fields.
68 WEIDMANN. Linear Operators in Hilbert Spaces.
69 LANG. Cyclotomic Fields II.
70 MASSEY. Singular Homology Theory.
71 FARKAS/KRA. Riemann Surfaces. 2nd ed.
72 STILLWELL. Classical Topology and Combinatorial Group Theory.
73 HUNGERFORD. Algebra.
74 DAVENPORT. Multiplicative Number Theory. 2nd ed.
75 HOCHSCHILD. Basic Theory of Algebraic Groups and Lie Algebras.
76 IITAKA. Algebraic Geometry.
77 HECKE. Lectures on the Theory of Algebraic Numbers.
78 BURRIS/SANKAPPANAVAR. A Course in Universal Algebra.
79 WALTERS. An Introduction to Ergodic Theory.
80 ROBINSON. A Course in the Theory of Groups.
81 FORSTER. Lectures on Riemann Surfaces.
82 BOTT/TU. Differential Forms in Algebraic Topology.
83 WASHINGTON. Introduction to Cyclotomic Fields.
84 IRELAND/ROSEN. A Classical Introduction to Modern Number Theory. 2nd ed.
85 EDWARDS. Fourier Series. Vol. II. 2nd ed.
86 VAN LINT. Introduction to Coding Theory. 2nd ed.
87 BROWN. Cohomology of Groups.
88 PIERCE. Associative Algebras.
89 LANG. Introduction to Algebraic and Abelian Functions. 2nd ed.
90 BRØNSTED. An Introduction to Convex Polytopes.
91 BEARDON. On the Geometry of Discrete Groups.
92 DIESTEL. Sequences and Series in Banach Spaces.
93 DUBROVIN/FOMENKO/NOVIKOV. Modern Geometry – Methods and Applications. Vol. I. 2nd ed.
94 WARNER. Foundations of Differentiable Manifolds and Lie Groups.
95 SHIRYAYEV. Probability, Statistics, and Random Processes.
96 CONWAY. A Course in Functional Analysis.
97 KOBLITZ. Introduction in Elliptic Curves and Modular Forms.
98 BRÖCKER/TOM DIECK. Representations of Compact Lie Groups.
99 GROVE/BENSON. Finite Reflection Groups. 2nd ed.
100 BERG/CHRISTENSEN/RESSEL. Harmonic Analysis on Semigroups: Theory of Positive Definite and Related Functions.
101 EDWARDS. Galois Theory.
102 VARADARAJAN. Lie Groups, Lie Algebras and Their Representations.
103 LANG. Complex Analysis. 2nd ed.
104 DUBROVIN/FOMENKO/NOVIKOV. Modern Geometry – Methods and Applications. Part II.
105 LANG. $SL_2(R)$.
106 SILVERMAN. The Arithmetic of Elliptic Curves.
107 OLVER. Applications of Lie Groups to Differential Equations.
108 RANGE. Holomorphic Functions and Integral Representations in Several Complex Variables.
109 LEHTO. Univalent Functions and Teichmüller Spaces.
110 LANG. Algebraic Number Theory.
111 HUSEMÖLLER. Elliptic Functions.
112 LANG. Elliptic Functions.
113 KARATZAS/SHREVE. Brownian Motion and Stochastic Calculus. 2nd ed.
114 KOBLITZ. A Course in Number Theory and Cryptography.
115 BERGER/GOSTIAUX. Differential Geometry: Manifolds, Curves, and Surfaces.
116 KELLEY/SRINIVASAN. Measure and Integral. Vol. I.
117 SERRE. Algebraic Groups and Class Fields.

Graduate Texts in Mathematics

118 PEDERSEN. Analysis Now.
119 ROTMAN. An Introduction to Algebraic Topology.
120 ZIEMER. Weakly Differentiable Functions: Sobolev Spaces and Functions of Bounded Variation
121 LANG. Cyclotomic Fields I and II. Combined 2nd ed.
122 REMMERT. Theory of Complex Functions. *Readings in Mathematics.*
123 EBBINGHAUS ET AL. Numbers. *Readings in Mathematics.*
124 DUBROVIN/FOMENKO/NOVIKOV. Modern Geometry – Methods and Applications. Part III.
125 BERENSTEIN/GAY. Complex Variables. An Introduction
126 BOREL. Linear Algebraic Groups.
127 MASSEY. A Basic Course in Algebraic Topology.
128 RAUCH. Partial Differential Equations.
129 FULTON/HARRIS. Representation Theory. A First Course. *Readings in Mathematics.*
130 DODSON/POSTON. Tensor Geometry.
131 LAM. A First Course in Noncommutative Rings.
132 BEARDON. Iteration of Rational Functions.
133 HARRIS. Algebraic Geometry. A First Course.
134 ROMAN. Coding and Information Theory.
135 ROMAN. Advanced Linear Algebra.
136 ADKINS/WEINTRAUB. Algebra: An Approach via Module Theory.
137 AXLER/BOURDON/RAMEY. Harmonic Function Theory.
138 COHEN. A Course in Computational Algebraic Number Theory.
139 BREDON. Topology and Geometry.
140 AUBIN. Optima and Equilibria. An Introduction to Nonlinear Analysis.
141 BECKER/WEISPFENNIG/KREDEL. Gröbner Bases. A Computational Approach to Commutative Algebra.
142 LANG. Real and Functional Analysis. 3rd ed.
143 DOOB. Measure Theory.
144 DENNIS/FARB. Noncommutative Algebra.
145 VICK. Homology Theory. An Introduction to Algebraic Topology. 2nd ed.
146 BRIDGES. Computability: A Mathematical Sketchbook.

147 ROSENBERG. Algebraic K-Theory and Its Applications.
148 ROTMAN. An Introduction to the Theory of Groups. 4th ed.
149 RATCLIFFE. Foundations of Hyperbolic Manifolds.
150 EISENBUD. Commutative Algebra with a View Toward Algebraic Geometry.
151 SILVERMAN. Advanced Topics in the Arithmetic of Elliptic Curves.
152 ZIEGLER. Lectures on Polytopes.
153 FULTON. Algebraic Topology: A First Course.
154 BROWN/PEARCY. An Introduction to Analysis.
155 KASSEL. Quantum Groups.
156 KECHRIS. Classical Descriptive Set Theory.
157 MALLIAVIN. Integration and Probability.
158 ROMAN. Field Theory.
159 CONWAY. Functions of One Complex Variable II.
160 LANG. Differential and Riemannian Manifolds.
161 BORWEIN/ERDÉLYI. Polynomials and Polynomial Inequalities.
162 ALPERIN/BELL. Groups and Representations.
163 DIXON/MORTIMER. Permutation Groups.
164 NATHANSON. Additive Number Theory: The Classical Bases.
165 NATHANSON. Additive Number Theory: Inverse Problems and the Geometry of Sumsets.
166 SHARPE. Differential Geometry: Cartan's.
167 MORANDI. Field and Galois Theory.
168 EWALD. Combinatorial Convexity and Algebraic Geometry.
169 BHATIA. Matrix Analysis.
170 BREDON. Sheaf Theory. 2nd ed.
171 PETERSEN. Riemannian Geometry.
172 REMMERT. Classical Topics in Complex Function Theory.
173 DIESTEL. Graph Theory.
174 BRIDGES. Foundations of Real and Abstract Analysis.
175 LICKORISH. An Introduction to Knot Theory.
176 LEE. Riemannian Manifolds.
177 NEWMAN. Analytic Number Theory.

Graduate Texts in Mathematics

178 CLARKE/LEDYAEV/STERN/WOLENSKI.
     Nonsmooth Analysis and Control Theory.
179 DOUGLAS. Banach Algebra Techniques
     in Operator Theory. 2nd ed.
180 SRIVASTAVA. A Course on Borel Sets.
181 KRESS. Numerical Analysis.
182 WALTER. Ordinary Differential Equations.
183 MEGGINSON. An Introduction to Banach
     Space Theory.

184 BOLLOBAS. Modern Graph Theory.
185 COX/LITTLE/O'SHEA. Using Algebraic
     Geometry.
186 RAMAKRISHNAN/VALENZA. Fourier
     Analysis on Number Fields.
187 HARRIS/MORRISON. Moduli of Curves.
188 GOLDBLATT. Lectures on Hyperreals.
189 LAM. Lectures on Rings and Modules.

Printing: Mercedesdruck, Berlin
Binding: Buchbinderei Lüderitz & Bauer, Berlin